The Aubin Academy Master Series
AutoCAD MEP 2012

Paul F Aubin
Darryl McClelland, LEED AP
Martin Schmid, PE
Gregg Stanley

G3B • Press

The Aubin Academy Master Series: AutoCAD MEP 2012

Paul F. Aubin, Darryl McClelland, LEED AP, Martin Schmid, PE, and Gregg Stanley

ISBN-13: 978-1461141266

ISBN-10: 1461141265

G3B Press

c/o Paul F. Aubin Consulting Services

P.O. Box 223

Oak Lawn, IL 60454

USA

To learn more about titles by G3B Press, the book's authors and other offerings by Paul F Aubin Consulting Services, please visit **www.paulaubin.com**. Updates are posted to the blog section of the site. Please use Contact link to send email.

Notice to the Reader

Publisher does not warrant or guarantee any of the products described herein or perform any independent analysis in connection with any of the product information contained herein. Publisher does not assume, and expressly disclaims, any obligation to obtain and include information other than that provided to it by the manufacturer. The reader is expressly warned to consider and adopt all safety precautions that might be indicated by the activities described herein and to avoid all potential hazards. By following the instructions contained herein, the reader willingly assumes all risks in connection with such instructions. The publisher makes no representations or warranties of any kind, including but not limited to, the warranties of fitness for particular purpose or merchantability, nor are any such representations implied with respect to the material set forth herein, and the publisher takes no responsibility with respect to such material. The publisher shall not be liable for any special, consequential, or exemplary damages resulting, in whole or part, from the readers' use of, or reliance upon, this material.

The views expressed herein are solely those of the authors/presenters and are not those of Autodesk, Inc., its officers, directors, subsidiaries, affiliates, business partners, or customers.

Contents at a Glance

Contents

SECTION II—WORKING WITH MEP OBJECTS

SECTION III—CONTENT AND DISPLAY

Preface

WELCOME

Within the pages of this book you will find a comprehensive introduction to the methods, philosophy, and procedures of the AutoCAD MEP software. AutoCAD MEP is an advanced and powerful engineering design and documentation software package covering the disciplines of HVAC, Plumbing, Piping, and Electrical design. By following the detailed tutorials contained in this book, you will become immersed in its workings and functionality.

There are two basic goals to this book: (1) shorten the AutoCAD MEP learning curve, and (2) help you develop sound methods and procedures. All anyone needs for success is a proper understanding of how the program functions and a clear understanding of what the program can and cannot do. This coupled with good procedure may be the magic key to success in *Mastering* AutoCAD MEP.

AUTOCAD AND AUTOCAD MEP: WHAT IS THE DIFFERENCE?

AutoCAD MEP (AMEP) is built on top of AutoCAD Architecture, which is built on top of AutoCAD. As such, it includes the standard features of AutoCAD including its powerful drafting tools, External References, Paper Space, Layers and Blocks. AutoCAD Architecture includes a collection of objects representing the most common architectural components, such as Walls, Doors, Windows, Stairs, Roofs, Columns, Beams, and much more. AMEP expands further on this by providing Ducts, Pipes, Fittings, Equipment, Devices, Wires, Conduits, Pumps, Sprinkler Heads, and much more. All of those objects are able to take advantage of Display Control (purpose-built display based on object function and common engineering drawing conventions), Connections (physical rule-based linkage between objects related in a system), and Styles (collections of parameters applied to objects as a group). The base

AutoCAD package does not offer such objects or functionality; instead, it relies on generic geometric components such as lines, arcs, and circles, which need to be assembled by the operator to represent the architectural or engineering items being designed and organized manually, often through complex layer and file schemes.

Success in completing most tasks requires a combination of understanding of one's goals, ample time and planning, and access to the right tools. Although knowledge and planning are critically important, having the proper tool for the job can often determine the overall success or failure of a given undertaking. A handsaw and a power saw are both capable of cutting wood. However, the power saw is generally capable of creating a better cut in less time, provided the operator knows how to use it properly. Used improperly, the results can be dire. The situation is the same in creating engineering documents. While both AutoCAD and AutoCAD MEP can accomplish the job, AMEP is designed specifically for engineering design/drafting and will generally do a better job in less time, provided, of course, that the user knows how to use it properly. AutoCAD, while capable of producing engineering documents, is not designed specifically for this task (it is more general purpose). Having purchased this book, you probably already own AMEP or have access to it at work. Read on and begin gaining the knowledge needed to use AMEP properly!

WHAT IS AN INTELLIGENT OBJECT?

An *intelligent* object is an entity within AMEP that is designed to behave as the specific "real-world" object after which it is named. The creation of a floor plan in generic AutoCAD involves a process of drafting a series of lines and curves parallel to one another to represent ducts, pipes, and other elements. This process is often time-consuming and labor-intensive. When design changes occur, the lines must be edited individually to accommodate the change. Furthermore, a plan created this way is two-dimensional only. When sections are needed, they must be created from scratch from additional lines and circles, which maintain no relationship to the lines and circles that make up the original plan. Further, all coordination tasks must be performed manually since disconnected two-dimensional drawings offer little assistance in determining clashes and interference between systems.

'n contrast, AutoCAD MEP includes *true* engineering objects. These objects are referred to in software as AEC Objects (AEC stands for Architecture, Engineering, and Construction). ˙ than drafting lines as in the example above, AMEP includes a true Duct object. This object has all of the parameters of an actual duct built directly into it. You can even calculate air flow! Therefore, one need only assign the values to these parameters to add or modify the Duct within the drawing. In addition, the Duct object can be represented two-dimensionally (in 1-Line or 2-Line) or three-dimensionally, in plan or in section, using a single drawing element. This means that unlike traditional drafting, which requires the Duct to be drawn potentially several times, an AMEP Duct need only be drawn once, and then "represented" differently to achieve each type of drawing (plan, 1-Line, 2-Line and section). You can even generate Schedules directly from the objects in the drawing which report the properties of the objects directly—you do not have to manually type in all the fields of the Schedule. An even greater advantage of the AEC object is that if it is edited, it changes in all views and schedules. This is the advantage of its being a single object, and it provides a tremendous productivity boon. With lines, each view remains a separate drawing; therefore, edits need to be repeated for each drawing type—a definite productivity drain.

Objects also adhere to built-in rules that control their behavior under various circumstances. Ducts and Pipes, for example, automatically add fittings when you turn corners. Spaces (rooms) know to grow and shrink when their controlling edges are reshaped and have dynamically calculated areas and volumes. Wires and Devices attach to circuits and the loads of these circuits report calculated totals. Tags remain attached and continue to report their associated data even across XREFs (separate but linked drawing files). These and many other relationships are programmed into the software. The intelligence of the object extends even further. AEC objects may have graphical and non-graphical data attached to them, which can be linked directly to schedules and reports. All of these features allow us to elevate our ordinary model to a Building Information Model (BIM).

Intelligent objects make the process of creating engineering drawings more efficient and streamlined. Mastery of objects begins with understanding their properties, their styles, and their rules. Mastery of AutoCAD MEP begins with mastery of individual objects but, more importantly, requires mastery of the interrelationship of objects and the procedures and best practices required to take full advantage of them. Through the process of learning AMEP, you will learn to construct a Building Information Model—an interconnected series of objects and rules used to generate all of the required engineering documentation and communication, including loads, calculations, and other critical data, which is greater than the sum of its parts.

WHO SHOULD READ THIS BOOK?

The primary audience of this book is users who have AutoCAD experience and wish to begin getting the most from AutoCAD MEP. This book is well suited to existing AMEP and new users alike. Specifically, this includes anyone who currently uses AutoCAD to produce engineering construction documentation or design drawings. Mechanical, Plumbing, Fire Protection, Electrical, and other building design engineering professionals, as well as building industry CAD professionals stand to benefit from the information contained within.

YOU SHOULD HAVE SOME AUTOCAD BACKGROUND

Although no prior knowledge of AMEP is required to read and use this book, this book assumes a basic level of AutoCAD experience. At the very least, you should be familiar with the basics of drafting, layers, blocks, XREFs, object snaps, and plotting.

WHAT YOU WILL FIND INSIDE

Section I of this book is focused on the necessary prerequisite skills and underlying theory behind AMEP. This section is intended to acquaint you with the software and put you in the proper mindset. Section II introduces each major discipline in its own chapter and discusses the tools, settings, and objects supporting that discipline. Content is a critical part of AutoCAD MEP. Section III is dedicated to understanding and creating various kinds of AMEP content. Section IV explores annotation and other features specific to construction documentation and coordination.

WHAT YOU WON'T FIND INSIDE

This book is not a command reference. This book approaches the subject of learning AMEP by exposing conceptual aspects of the software and providing extensive tutorial coverage. No attempt is made to give a comprehensive explanation of every command or every method available to execute commands. Instead, explanations cover broad topics of how to perform various tasks in AutoCAD MEP, with specific examples coming from engineering practice. The focus of this book is the design development and construction documentation phases of engineering design. AutoCAD commands and features and AutoCAD Architecture commands and features are not covered. If you wish to learn more about AutoCAD Architecture, please pick up a copy of *The Aubin Academy Master Series: AutoCAD Architecture* by Paul F. Aubin.

STYLE CONVENTIONS

Style Conventions used in this text are as follows:

Text	AutoCAD MEP
Step-by-Step Tutorials	1. Perform these steps.
Menu picks	**Application menu > Save As > AutoCAD Drawing**
Dialog box and palette input	For the length, type **10'-0"**.
Keyboard input	Type **DuctAdd** and press ENTER. Type **599** and press ENTER.
File and Directory Names	*C:\MasterMEP 2012\Chapter01\Sample File.dwg*

> **CAD MANAGER NOTE:**
> Especially for CAD Managers—there are many issues of AMEP usage that are important for CAD Managers and adherence overall to office standards. Throughout the text are notes to the CAD Manager titled "CAD Manager Note." If you are the CAD Manager, pay particular attention to these items because they are designed to assist you in performing your CAD Management duties better. If you are not the CAD Manager, these notes can help give you insight into some of the salient CAD management issues your firm may be facing. If your firm does not have a dedicated CAD Manager, pay close attention to these points because these issues will still be present, only there will not be a single individual dedicated to managing these issues and solving relevant related problems as they arise. If CAD management is not within your interests or responsibilities, you can safely skip over these notes.

UNITS

This book is written in Imperial units. Metric datasets and references are not provided.

HOW TO USE THIS BOOK

The order of chapters has been carefully considered with the intention of following a logical flow. If you are relatively new to AMEP, it is recommended that you complete the book from beginning to end. However, if certain chapters do not pertain to the type of work that you or your firm performs, feel free to skip those topics. However, bear in mind that not every procedure will be repeated in every chapter. For example, if you are an Electrical Engineer, you may choose to skip Chapter 5 – Mechanical Systems. However, many procedures are common across disciplines, and even though you may not be a Mechanical Engineer, you may find valuable tips or techniques in the chapters devoted to other disciplines. Therefore, for the best experience, you are encouraged to read the entire book. After you have completed your initial pass of the tutorials in this book, keep the book handy as it will remain a valuable desk resource in the weeks and months to come.

BOOK DATASET FILES

Files used in the tutorials throughout this book are available for download from **www.paulaubin.com**. Most chapters include files required to begin the lesson, and in many cases a completed version is provided as well that you can use to check your work. This means that you will be able to load the files for a given chapter and begin working. When you install the downloaded dataset, the files for *all* chapters are installed automatically. The files will install into a folder on your *C:* drive named *MasterAME 2012*. Files *must* be installed in this folder. The root folder will contain a folder for most chapters and several other folders with files used by multiple chapters. Please note that in some cases, a particular chapter may *not* have its own folder or its own drawing files. This is intentional.

Should updates be required, a notice will be posted to **www.paulaubin.com/blog**.

> NOTE: Please note that the accompanying dataset contains *only* drawing (DWG) and other related resource files necessary to complete the tutorial lessons in this book. The provided dataset does *not* contain the AutoCAD MEP software. Please contact your local reseller if you need to purchase a copy.

To download and install the files, please do the following:

1. In your web browser, visit: www.paulaubin.com.

2. Click on the Books link at the top.

3. Click on the link for the book whose files you wish to access.

   Downloads will be listed in the "Downloads" section of the page.

4. There may be more than one item to download. Click each item and follow the instructions of your browser to download each one.

5. Run the WinZip EXE file and unzip the files to your C Drive.

The default unzip folder is named *C:\MasterAME 2012* on your hard drive. Unzipped files will utilize approximately 135 MB of disk space.

CAUTION:
Please do not move the files from this location; if you do, the Catalogs may not function properly. Moving any of the other files can also cause issues with project files. See the "Repathing Projects" topic below.

PROJECTS

The AutoCAD MEP Drawing Management tools (Projects) are used throughout this text. Please do not open and save files outside the Project Navigator unless directed to do so in the chapter's instructions. Although there is no physical difference between a drawing file created inside a project and one created outside a project, procedurally, there are large differences. Please follow the instructions at the start of each chapter regarding how to install and load the current project files.

Completed versions of the exercises are typically provided alongside the original file with the suffix *Complete* after their name. They are provided for you to compare the complete version with your own to check your progress.

REPATHING PROJECTS

In some cases when you load a project, you will be prompted to repath the project. This occurs when the project has been moved from its original location. If you move the dataset files to a location other than *C:\MasterAME 2012*, a message will appear asking you if you want to repath the project. If you receive this message, click "Repath the Project Now." This is very important because the project files will not function properly if you ignore this message. It is possible to postpone the decision, but some files may not function properly until you repath.

SERVICE PACKS

It is important to keep your software current. Be sure to check **www.autodesk.com** on a regular basis for the latest updates and service packs to the AutoCAD MEP software. Having the latest service packs installed will help ensure that your software runs trouble-free.

AutoCAD MEP also has the Info Center at the top right corner of the application frame. This tool will alert you when updates and information are available.

WE WANT TO HEAR FROM YOU

We welcome your comments and suggestions regarding this and and any of our books. Please visit **www.paulaubin.com** and click the Contact link to send an email using the form provided. You can reach all four authors using this form. Also be sure to visit the blog as updates to the book's content will be posted there as soon as they become available.

ABOUT THE AUTHORS

Paul F. Aubin is the author of many CAD and BIM book titles including the widely acclaimed: *The Aubin Academy Mastering Series: Revit Architecture, AutoCAD Architecture, AutoCAD MEP and Revit MEP* titles. Paul has also authored video training both on his Web site and for lynda.com **www.lynda.com/paulaubin**. Paul is an independent architectural consultant who travels internationally providing Revit® Architecture and AutoCAD® Architecture implementation, training, and support services. Paul's involvement in the architectural profession spans over 20 years, with experience that includes design, production, CAD management, mentoring, coaching, and training. He currently serves as Moderator for Cadalyst magazine's online CAD Questions forum, is an active member of the Autodesk user community, and has been a top-rated speaker at Autodesk University (Autodesk's annual user convention) for many years. This year Paul is speaking at the Revit Technology Conference in both the US and Australia. His diverse experience in architectural firms, as a CAD manager, and as an educator gives his writing and his classroom instruction a fresh and credible focus. Paul is an associate member of the American Institute of Architects. He lives in Chicago with his wife and three children.

Darryl McClelland, LEED AP has 26 years of practical design experience in MEP engineering. Although his primary focus was the design of mechanical systems, he spent 11 of those 26 years designing electrical and plumbing systems as well. He also ran his own engineering business for eight years. His design experience ranges from complex research laboratories and institutional facilities to medical and professional office buildings, and everything in between. He is a graduate of Purdue University and an active member of ASHRAE, ASPE, and a LEED AP.

Martin J. Schmid, P.E. works with customers to implement best practices using AutoCAD MEP and Revit MEP. In his current role, in addition to customer interaction, he works with product design and product management to convey industry needs and trends. Mr. Schmid has also worked in various roles in a variety of architecture and engineering firms, including electrical designer, engineering coordinator, and application developer. In addition to product and industry expertise, Mr. Schmid applies the API's of Autodesk's products to automate processes and solve customer problems. Mr. Schmid has presented internally to coworkers, at Autodesk University, industry conferences, and as a consultant to design firms and 3rd party application developers.

Gregg Stanley has over twenty two years experience in Mechanical Process Design focused on Water Wastewater treatment systems using AutoCAD based solutions since Release 1.1. He has also been in the position as a CAD Manager responsible for developing and instituting company specific customized applications, CAD standards and training. Mr. Stanley has written and presented several training classes on AutoCAD and AutoCAD MEP both internally to coworkers, as an independent consultant and at Autodesk University and has designed and tested software for the engineering industry for the last 5 years.

The views expressed herein are solely those of the authors/presenters and are not those of Autodesk, Inc., its officers, directors, subsidiaries, affiliates, business partners, or customers.

DEDICATION

Paul's dedication: This book is dedicated to my wife Martha. Thank you for your boundless love and support.

Darryl's dedication: This book is dedicated to Anne Marie and Bryan. Thank you for all your support and love all these years. I would not be where I am today without it.

Martin's dedication: Thanks to my wife Carrie for her patience and support all these years. To Jolie and Calista, I hope you too find something you enjoy and are passionate about. Keep smiling!

Gregg's dedication: This book is dedicated to my wife Jennifer and my children, Amanda, Nicholas, and Michael. I just want to thank you for all the happiness, love, precious moments, and support.

ACKNOWLEDGMENTS

The authors would like to thank several people for their assistance and support throughout the writing of this book. There are far too many folks in Autodesk's AEC Industry Group to mention. Thanks to all of them but, in particular, Toby Smith, Armundo Darling, Rebecca Richkus, Eric Grey, Pat Jenakanandhini, Peter Blixt, Hans Granden, Jaydeep Dave, Liang Zhu, Jitender Uppal, Sue Gorte, Steve Butler, Simon Jones, Gary Ross, Bryan Otey, Yishu (Kevin) Xu, Bo Noren, Lynn Poliquin, Danny Hubbard, Lars-ake Johansson, Aaron Gardner, Anna Oscarson, Rick Foster, Jonathan Gilbert, Paul Sweet, Adam Weick, Laura Gutwillig, Christina Persson, Joshua Benoist, Jeremy Smith, Jian (Stewart) Li, Mike Myles, Jason Bishop, Mike Vose, Ann-Marie McKenna, Tony Sinisi, Jason Martin, and all of the folks at Autodesk Tech Support.

Thanks to Stacy Masucci. This edition of the book would not be possible without Stacy's efforts to transfer it to our care. Thank you Stacy.

Thanks also go out to Scott Cote of AECOM and to Natalia Khaldi of ARUP for listening to Gregg's ideas and providing insight on the topics in the book. Gregg would also like to thank David Derocher at East Coast CAD CAM.

Martin also wishes to thank John Birge, Steve and Doug Alvine, Mark Koblos, Brian Uhlrich, and Betty Feldman, for each providing me the opportunity to fill different roles and see various sides of the industry. Thanks also to Steve Moser, my academic advisor, and Mark Frost for opening the world of Autodesk to me.

Finally, Paul would like to thank Martin, Darryl, and Gregg. Having an architectural background, I could not have even contemplated writing such a book without your assistance. It continues to be a pleasure to work with each of you. Thank you.

Cover Design: Michael Brumm

Author Illustrations: Ron Bailey

Cover Artwork: iStockphoto.com

Introduction and Methodology

WHAT'S IN THIS SECTION?

This section introduces the methodology of AutoCAD MEP. Many concepts will be familiar to the seasoned AutoCAD user; many concepts will be new. If you are a current AutoCAD user, skim through this section looking for concepts unique to AutoCAD MEP (AMEP), particularly in Chapter 2.

If you do not have AutoCAD experience, please read this entire section. Many basic AutoCAD skills are assumed (please see the Preface for details); therefore, it may also benefit you to complete some basic AutoCAD tutorials prior to reading this section.

SECTION I IS ORGANIZED AS FOLLOWS:

Chapter 1

The User Interface

INTRODUCTION

This chapter is designed to get you acquainted with the user interface and work environment of AutoCAD MEP. Collectively, all aspects of the user interface and work environment are referred to as the "workspace." In addition to the workspace, this chapter will also explore any necessary AutoCAD skills required for successful usage of AMEP. If you did the Quick Start tutorial prior to this chapter, then you are already familiar with some of the objects and features of AMEP. Read on to begin understanding the logic of the workspace, and what user interface skills are required to be successful with AMEP.

OBJECTIVES

- Understand the AutoCAD MEP environment.
- Gain comfort with the user interface.
- Explore the ribbon.
- Understand unique MEP Interface items such as MEP Snaps
- Assess your existing AutoCAD skills.

THE AUTOCAD MEP WORKSPACE

AutoCAD MEP is an MEP-*flavored* version of AutoCAD. The workspace of AutoCAD MEP (AMEP) offers a clean and streamlined environment designed to put the tools and features that you need to use most often within easy reach, while allowing for endless customization for those whose needs vary. As such, it shares many similarities with core AutoCAD. However, there are some distinct differences. For instance, AMEP has its own collection of highly specialized ribbon tabs and tool palettes. You'll explore the AMEP workspace here and later cover some of the traditional AutoCAD elements. The AutoCAD items' focus is on those things that are critical to typical AMEP usage and success. For more detailed information on the AutoCAD workspace, commands and features, consult the online help or a book specifically on AutoCAD.

THE DRAWING EDITOR

Chapter 1 is not specifically formatted as a tutorial, but you can follow along in AutoCAD MEP as you read its topics. The AMEP drawing editor includes many features and controls. Here is a simple overview of the most important features (see Figure 1.1). For more information on interface features, choose **User Interface Overview** from the Help drop-down menu at the top right corner of the screen.

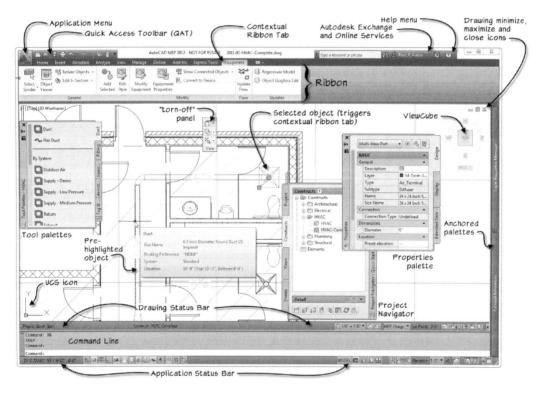

FIGURE 1.1

Major components of the AutoCAD MEP drawing editor

Consistent with most Windows software applications, the AMEP screen is framed with the Application menu, Quick Access Toolbar (QAT), InfoCenter and ribbon along the top edge; the Windows minimize, maximize and close icons in the top right corner and an application status bar along the bottom edge. In addition to these Windows standards, the AMEP screen also includes the command line, typically docked along the bottom edge of the screen just above the application status bar and tool palettes. Above the command line sits the drawing status bar, which is similar in appearance to the application status bar, but differs in function (see Figure 1.2). The ribbon, command line and tool palettes are critically important interface elements in AMEP and will be elaborated on in topics below. If you are a seasoned AutoCAD user, you are already very familiar with the command line. However, as we will see in the topic below, we have a very viable alternative to the command line called "dynamic input." Other notable elements of the AMEP screen include the UCS icon, the Scale and Display Configuration pop-up menus and the main drawing editor window (see Figure 1.1 and Figure 1.2). Several of these key interface items warrant further discussion and are elaborated on in the topics that follow.

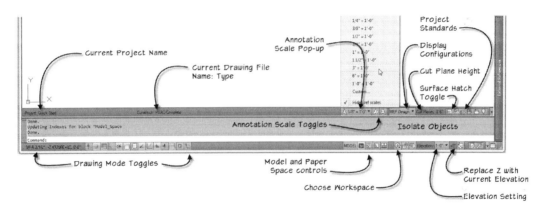

FIGURE 1.2

The drawing and application status bar areas

Application Status Bar

The application status bar runs across the length of the bottom edge of the drawing editor and includes a series of Drawing Mode toggles such as Snap Mode, Grid Display, Polar Tracking and Dynamic Input. Each of these modes helps you to control cursor movements and make drawings more accurate. Many of these topics are covered below and elsewhere in this manual; you can also look them up in the online help.

The next cluster of icons to the right allows you to move between Model and Layouts within the current drawing as well as between all open drawings. To the right of that is the Workspace

Switching pop-up menu and the Toolbar/Window position controls, which enable you to lock certain elements of your workspace and prevent them from accidentally being moved or turned off. If you want to maintain the look of your custom user interface (CUI), this tool can be a big help.

With Elevation control, you can quickly set the current Z Elevation in the drawing and then toggle the automatic substitution of this Z value for all clicked points. This can be very handy when working in 2D to keep things "flat" if 3D objects are present in the drawing. It can also be helpful in 3D to avoid inaccurate Z snapping based on view direction.

Drawing Status Bar

The drawing status bar stays attached to the bottom edge of the drawing window and reveals information about the current drawing. As you can see in Figure 1.2, this includes the project name, drawing type and name, and if the current drawing belongs to a project in AutoCAD MEP. (In Figure 1.2, on the left, the current project is "Quick Start," and in the middle the drawing is a Construct named "HVAC-Complete.") The Drawing Management tools in AMEP were touched upon briefly in the Quick Start tutorial, and are covered in more detail in Chapter 3. On the right side, you will find the Annotation Scaling controls, which will be explored in later chapters. The Current Display Configuration menu allows you to change the currently active Display Configuration within the current drawing window (in model space or viewport in a paper space layout). Display Configurations are covered in greater detail in Chapter 2. The Cut Plane height control displays the current height and allows you to change it without opening the Display Manager. Next to this, an icon tray provides quick access to a number of features, including the Surface Hatch toggle, Layer Key Overrides and Object Isolation. If the current drawing is part of a project in AMEP and drawing standards are enabled, the Drawing Standards icon appears next. Use it to configure standards in an AMEP project and synchronize the current drawing to the standards. If there are external references in the file, the Manage Xrefs icon will appear at the far right of the icon tray. External references (XREFs) are links to other drawing files. Details and techniques on their usage will be covered throughout this book

InfoCenter

The InfoCenter at the top right corner of the screen provides several means to find or receive information. Some—but not all—of these features require a live Internet connection. The Search feature allows you to do a keyword search on a user-customizable list of resources, including Help. You can sign in to your Autodesk account directly on the InfoCenter and then access the Autodesk Exchange. Autodesk Exchange gives access to the subscription center, product videos and updates. You can also open the Help or use the drop-down menu to access specific Help features. Sign in and choose Account Settings to configure your account to suit your needs.

In addition to the resources provided in the InfoCenter, be sure to check out the authors' blog for any updates to this book. You can visit the blog at: **www.paulaubin.com/blog**. We post any tips, text updates or dataset updates as they occur so please check the blog often.

THE AUTOCAD MEP USER INTERFACE

Now that you have explored some of the common elements of the AMEP workspace, it is important to have a look at the most common ways to interface with the product. The Application menu, Quick Access Toolbar and ribbon replace the pull-down menus and toolbars as a means of starting commands. Tool palettes allow you to both start commands and import content and styles. Contextual ribbon tabs and right-click context menus provide easy command access when editing existing objects. In addition, you will also frequently interface with objects directly on screen using dynamic dimensions and grip editing. As you interact with your drawings and models, it will be necessary to move fluidly around your screen and be comfortable viewing the model from all views, zoomed in and out. All of these items will be addressed in this topic.

APPLICATION MENU

File access and management tools are grouped under the Application menu (adorned by the large AutoCAD "A" icon). Click on the big "A" to open the Application menu. At the very top, you will find a command search feature. Type in the name of a command and it will search the Application menu, static ribbon tabs, any current contextual ribbon tab and the Quick Access Toolbar and display the results of any matches, including the location. This can be a great help when first learning where things are found in the ribbon.

If this is your first time launching AutoCAD MEP, the right side of the Application menu will be empty. But as you open and close files, the list of recent files will begin to populate. AMEP remembers the last several files and/or projects you had open and shows them here. You can even click the pushpin icon to permanently "pin" a particular file to the menu, making it easier to load next time (see the left side of Figure 1.3). Right at the top of the Application menu are two icons to switch the list from Recent Documents to currently Open Documents. These icons are pointed out in the figure. If you switch to Open Documents and you have several project files and/or view windows open, you can use the Application menu to switch between open windows (see the middle of Figure 1.3). If you hover over an item on either list, you will get a ToolTip that shows the full path of the file, a thumbnail image and file data.

On the left side of the menu, you will find commands like New, Open, Save and Save As. Submenus on many items, denoted by the arrow at the right, give additional related commands. For example, selecting the Open icon will open the Select File dialog, allowing you to choose a drawing file to open. The sub-menu offers the options to open a project file, a DGN file or an IFC file, in addition to a drawing file (see the right side of Figure 1.3). Hover over each of the icons on the left side and become familiar with the commands available here.

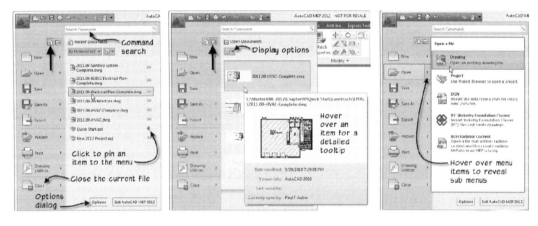

FIGURE 1.3
The Application Menu

> **TIP:** The Drawing Setup dialog, formerly available on the Format pull-down or Open Drawing menu, can be opened from the **Drawing Utilities** sub-menu of the Application menu.

At the bottom of the Application menu two buttons appear: Options and Exit AutoCAD MEP. Exit AutoCAD MEP is self-explanatory. AMEP will prompt you to save your work. Use the Options button to open the Options dialog. This dialog has many program preferences that you can configure. Most of the out-of-the-box settings are suitable for the beginner. There may be some items that you or your CAD Manager will want to adjust. In the coming chapters, several of the MEP-specific options will be discussed. You can also refer to the online help for more information.

QUICK ACCESS TOOLBAR

The Quick Access Toolbar (QAT), as its name implies, is a location for commonly used tools to which you wish to have easy and "quick access." The default QAT includes QNew, Open, Save, Undo, Redo, Plot, Project Browser and Project Navigator. (see the left side of Figure 1.4). You can add buttons to the QAT with the menu on the right end of the QAT itself. The Match Properties command is not part of the default QAT. Simply choose it from the pop-up menu to add it. For other commands, locate them on the ribbon (see the next topic), right-click the tool and choose **Add to Quick Access Toolbar** (see the right side of Figure 1.4). The QAT can be repositioned below the ribbon by choosing **Show Below the Ribbon** from the customize menu at the right end of the QAT, if you are willing to give up the screen space.

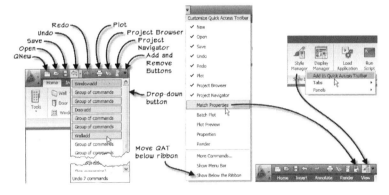

FIGURE 1.4

The Quick Access Toolbar

WORKSPACES

Workspaces are used in AutoCAD MEP to configure various user interface elements to be domain specific. For example, the Home, Annotate, Analyze and Manage ribbons all have different tools and options depending on which workspace is selected. Similarly, the workspace sets a domain specific tool palette group active. To select a workspace, click the workspace menu, and select the desired workspace (see Figure 1.5 in the Quick Start for an example).

FIGURE 1.5

Use the Workspace pop-up icon to change the current Workspace

RIBBONS

One means of issuing commands in AutoCAD MEP is by clicking their tools on the ribbon. The ribbon replaces the traditional pull-down menus and toolbars in the interface. A series of tabs appears just beneath the QAT. Each tab is separated into one or more Panels. Each Panel contains one or more Tools (see Figure 1.6).

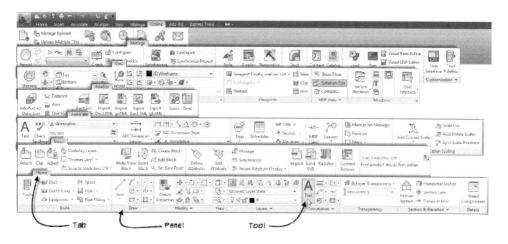

FIGURE 1.6

A look at the AutoCAD MEP ribbon tabs

To navigate the ribbon, click a tab, locate the panel and tool you need and then just click the tool to execute a command. When tutorial instructions are given in this text, you will be directed first to the tab, then the panel and finally the tool. For example, instructions to execute the Duct tool might look something like this:

On the Home tab of the ribbon, on the Build panel, click the *Duct* tool.

In the context of the exercise, when it is obvious which tab or panel will be used, the description might be shortened to something like:

> On the Build panel, click the *Duct* tool.

Or the description might be simply:

> Click the *Duct* tool.

Look to "The Static Ribbon Tabs" topic of the online help for a description of each of the six default static ribbon tabs.

Contextual Ribbon Tabs

In addition to the six default "static" ribbon tabs certain actions you perform in the software will cause other ribbon tabs to appear. These "contextual" ribbon tabs contain tools and commands specific to the item you are creating or editing. For example, if you select a Duct object in the model a Duct contextual ribbon tab will appear to the right of Manage. If you execute the MTEXT tool and begin creating text, a Text Editor tab will appear with the tools and options associated with multi-line text.

When one or more AMEP objects of the same type are selected in the drawing editor, a contextual ribbon tab will be displayed.

6. Launch AutoCAD MEP if it is not already running.

7. On the Home tab of the ribbon, on the Build panel, click the *Duct* tool.

8. Click a point anywhere on the left side of the screen within the drawing area.

9. Move the mouse position to the right side of the screen and click again.

10. Right-click and choose Enter.

Notice that "Enter" is the default option at the top of the menu, but several other options appear as well. Most of the options shown are also available on the Properties palette and the command line.

11. Click directly on the newly created Duct object. It will be highlighted with several grips along its length (see Figure 1.7).

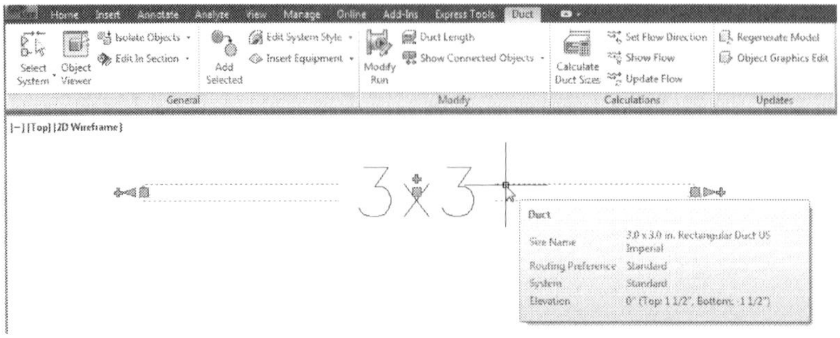

FIGURE 1.7
The Duct contextual ribbon tab

The Duct contextual ribbon tab appears and becomes current. Notice the green shading of the tab and panel titles. The contextual tabs for all AMEP objects will feature this color to distinguish them from the static tabs. Examine the duct-related commands presented on the tab. The static tabs remain available.

12. Right-click and notice that the Duct tab includes most of the Duct-related commands previously available through the right-click context menu.

13. Choose Deselect All from the menu, and notice that the Duct tab disappears.

When more than one object of different types are selected in the drawing editor, a "Multiple Objects" contextual ribbon tab, with basic editing commands not specific to any particular object type, will display.

14. On the Home tab of the ribbon, on the Draw Panel, click on the Line flyout and choose the *Polyline* tool.

15. Click a point anywhere on the left side of the screen within the drawing editor.

16. Move the mouse position to the right side of the screen and click again.

17. Right-click and choose Enter, (or press ENTER).

18. Click somewhere in the upper-right corner of the screen (being careful not to click directly on any object).

19. Move the pointer to the lower left corner of the screen and click again. (Both objects should be highlighted. Look up "Crossing Window Selection" in the online help for more information.)

20. Study the contextual tab that appears (see Figure 1.8).

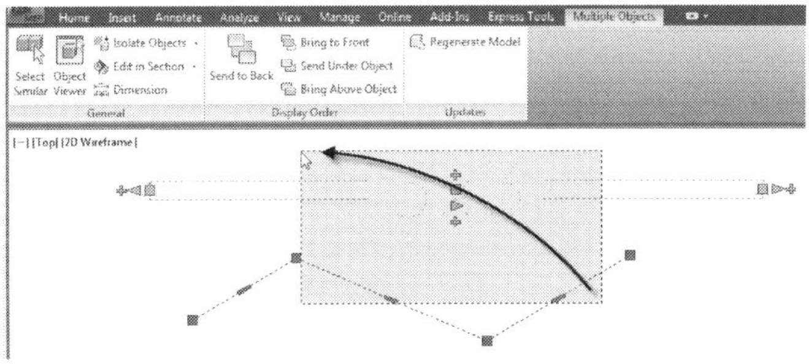

FIGURE 1.8
Multiple Objects contextual ribbon tab

Notice that the tools available are not object-specific.

21. Right-click and note that the context menu also contains only non-object-specific commands.

22. Choose Deselect All from the menu.

Panels

Ribbons are segregated into panels to help further classify and group the various tools. Panels simply group common tools and make it easier to locate the tool you need. If you use a certain tool frequently, you can right-click on it and add it to the QAT as noted above in the "Quick Access Toolbar" topic. If you use all the tools on a particular panel frequently, you can "tear off" the entire panel. This makes the panel into a floating toolbar on your screen. You can drag such a floating panel anywhere you like, even to a secondary monitor if you have one attached to your system. If you "tear off" any panels, AMEP will remember the custom locations of the panels the next time you launch the application (see Figure 1.9). The View panel is not initially docked to the ribbon when first installed; you can return this panel to the far right side of the Home tab, if you want.

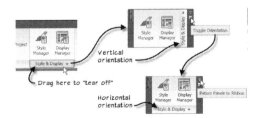

FIGURE 1.9
Tear off ribbon panels and drag them anywhere you like on screen

If you tear off a panel and later wish to restore it, simply move your mouse over the floating panel. This will make gray bars appear on each side. On the left side is a drag bar that you can use to drag the panel around your screen to a new location. On the right side, there are two small icons; the bottom one toggles the orientation of the panel title and the top one restores the panel to its original ribbon tab and location.

> Note: Feel free to customize your interface by tearing off panels if you wish, however all instructions in the tutorials that follow assume that panels are in their default locations on the ribbon tabs and refer to them as such.

You can only tear off panels on the permanent default ribbon tabs. Panels on contextual ribbon tabs cannot be torn off and left floating on screen. However, any of the tools from contextual tabs can be added to the QAT. Refer to the "Quick Access Toolbar" topic above for details.

On the panel title bar (bottom edge of the panel), most panels simply show the name of the panel. In some cases, however, a small icon will appear on the right side of the title. This can be one of two icons. The left side of Figure 1.10 shows a "Dialog Launcher" icon. Clicking an icon such as this will open a dialog. Usually these are settings dialogs that you use to configure several options for a particular type of element.

FIGURE 1.10
Panel with a dialog launcher icon on the left and an expanded panel on the right

On the right side of the figure an expanded Panel is shown. In this case, clicking this icon expands the panel temporarily to reveal additional related tools. Such tools are typically used less frequently than the ones always visible on the panel. Expanded panels are not ideal, but provide a compromise to what would otherwise be overcrowded ribbon panels for those that use them. Use the pushpin icon to pin the expanded panel open if you need to make repeated use of a command in the expanded portion of the panel.

Ribbon View State

The ribbon has three viewing states when docked at the top of the screen. The default state shows the complete ribbon and panels. A portion of the top of the screen is reserved for the ribbon. Click the tabs to switch which tools display, but the same amount of screen space is used regardless of the current tab. This mode makes it easiest to see the tools but uses more precious screen space (see the top of Figure 1.11).

Two alternative states are available that use less screen space. The small icon to the right of the Manage tab is used to toggle to the next state. Click it once to switch to the "Minimize to Panel Titles" state. In this state, ribbon tabs and panel titles are displayed; pass your mouse over a panel title to reveal a pop-up with that panel's tools. Move your mouse (shift focus) away from the panel and it will disappear (see the middle of Figure 1.11).

FIGURE 1.11
Ribbon display states

The final display state shows only the ribbon tabs (see the bottom of Figure 1.11). Click on a ribbon tab to make the tab pop up. Like the panel titles state, if you shift focus away from a tab, it will disappear. It is easy to experiment with each mode and decide on the one you prefer. Simply click the toggle icon once to switch to panel titles, and click it again to switch to tabs. If you wish to return to the full ribbon, click it again. Each time you click, it toggles to the next state.

Tools

Ribbon panels contain tools. The majority of these tools will use one of three types of buttons: Buttons, Drop-down buttons and Split buttons. An example of each of these can be found on the Home tab. An example of a button on the Build panel is the *Plumbing Line* tool (see the top-left of Figure 1.12). Clicking a button simply invokes that tool.

On the Layers panel, the *Layer State* tool is an example of a drop-down button. In this case, if you click the tool, a drop-down list will appear showing the various options for the tool. In the case of the *Layer State* tool, we can choose a previously defined named layer state (if any) from a scrolling list box, or select from the *New Layer State* or *Manage Layer State* tools (see the bottom left of Figure 1.12).

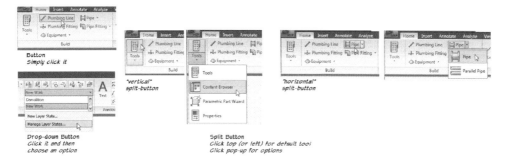

FIGURE 1.12

Examples of the primary button types on the Home tab

Split buttons can be either vertical or horizontal. They appear like the other buttons until you pass your mouse over them, at which point it will be clear that that only part of the button highlights under the mouse. The portion of the button with the small pop-up indicator (small triangle) behaves like a drop-down button. The other side behaves like a normal button. On the Home tab in the Build panel, the *Tools* and *Equipment* are examples of split buttons. *Tools* is oriented vertically; click the top portion for the default tool and the lower portion for a drop-down button. *Equipment* is oriented horizontally. To use the default button, click the right side of the button, or access the drop-down options with the left side of the button (see the right side of Figure 1.12).

Button types you will find include scrolling list boxes, as seen on the *Layer State* drop-down (see the bottom left side of Figure 1.12) or the *Preset View* list box located on the View tab in the Appearance panel. Other button types are text entry boxes, such as the *Seek* command on the Insert tab in the Seek panel, and slider controls, such as that used for the *Locked Layer Fading* control on the Home tab, expanded Layers panel (see Figure 1.13). For the latter, you can either select and drag the bar in the slider control or select the control and key in the desired numeric value.

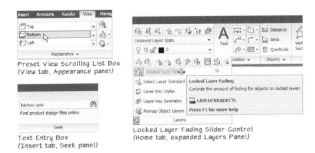

Preset View Scrolling List Box
(View tab, Appearance panel)

Text Entry Box
(Insert tab, Seek panel)

Locked Layer Fading Slider Control
(Home tab, expanded Layers Panel)

FIGURE 1.13

Examples of other button types

Right-Click on the Ribbon

23. Move your mouse over any ribbon tab name and then right-click. (If there is empty space to the right of the ribbon, you can right-click there as well).

Notice the menu that appears (see Figure 1.14).

> Note: If you right click on the ribbon itself instead of the tab name, you will only get the Show Tabs and Show Panels flyouts.

FIGURE 1.14

Ribbon tab right-click menu

The first section is related to Tool Palette Groups (see also Understanding Tool Palette Groups below). If a Tool Palette Group is associated with the ribbon tab on which you right-clicked, the first item will be active and choosing it will open the Tool Palettes, if closed, and set the associated Tool Palette Group current. (If you right-clicked to the right of the ribbon tabs or to the right of the rightmost ribbon panel, the menu will reflect the settings associated with the current ribbon panel.) The Tool Palette Group flyout allows you to associate a Tool Palette Group with a ribbon tab. The checkmark indicates the current association. If you do not want an associated Tool Palette Group, choose None.

> Tip: You can right-click on an inactive tab name and set the Tool Palette Group associated with that tab current. This will not make that ribbon tab current.

The Minimize sub-menu allows you to directly choose one of the three ribbon view states (see Ribbon View State above). The Show Tabs and Show Panels menu items allow you to hide and display the tabs and panels on the ribbon. Items with a checkmark are displayed. Select them from the menu to toggle off their display. Select again to toggle back on.

The Show Panel Titles menu item toggles the display of the panel titles for the full ribbon display. While you can change this setting when in one of the minimize modes, you will only see the effect when the full ribbon display is restored. You can Undock the ribbon from the top of the screen, turning it into a floating palette, which can then be auto-hidden, docked or anchored to the left or right, like any other palette (see Figure 1.22 below). Choosing Close will close the ribbon. You can reopen it with the RIBBON command at the command line.

Customization of the ribbon is beyond the scope of this book; for more information on this topic, refer to the online help.

TOOLTIP ASSISTANCE AND ALT-KEY COMMAND ACCESS

When you pause your mouse over a tool, a ToolTip usually appears. ToolTips give you the name of a tool, a short description and the name of the command. For certain commands, if you continue to hover, an extended tool tip with a more detailed description and, possibly, a descriptive image will appear. You can find settings to control how much ToolTip assistance you want in the Options dialog on the Display tab in the Window Elements area in the upper left. (Access the Options dialog from the Application menu as shown in Figure 1.3 above.) If you uncheck "Show ToolTips," no ToolTip assistance will appear. Figure 1.15 on the top left shows an example of the initial ToolTip you will receive with "Show ToolTips" checked. This will be all you get if you uncheck "Show extended ToolTips." If you enable ToolTips and extended ToolTips, you can specify the time delay between the display of the initial ToolTip and the expanded ToolTip (Figure 1.15, lower left). To see the expanded ToolTip immediately, set the delay to 0. The "Show shortcut keys in ToolTips" toggle allows you to enable or disable the display of shortcut keys in the ToolTips for those commands with a shortcut key assigned.

> **Note:** To obtain more information than the ToolTip displays, press F1 to open the Help directly to the page for that command.

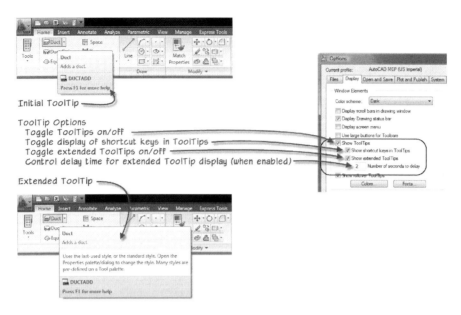

FIGURE 1.15
Configure ToolTip assistance in the Options dialog

Another Windows convention supported by AutoCAD MEP is the ability to invoke ribbon tools with the keyboard using the ALT key and a key letter combination from the desired tool. To try this, press the ALT key. Doing so will place a small label on each tool and tab. Numbers appear on each of the tools on the QAT. Simply press this number to execute that command. Letters appear on each of the Application menu and ribbon tabs. To invoke a tool on a tab, first press the letter for the tab. This will make a new set of letters appear on all the tools. Next press the key or keys shown on the tool. For example, to access the *Display Manager* tool via the ALT key, press the ALT key, and then the letters MA (to select the Manage tab) and then the letters DM (see Figure 1.16). If a drop-down button is involved (like with Style Manager), use the arrows on the keyboard to choose the desired tool and then press ENTER to complete the selection.

> **Note:** Even if the tab you want is current, when using the ALT key, you must still press the keystroke for that tab first.

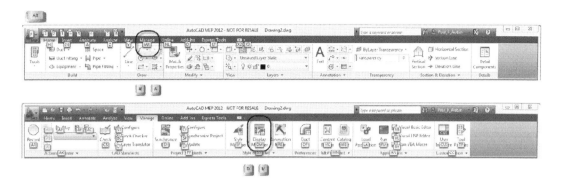

FIGURE 1.16
Press the ALT key to reveal alternate shortcuts

As with all AutoCAD-based programs, you can also customize the *acad.pgp* file and add command aliases for frequently used commands if you like. Use the ToolTip to identify the command name. For more detailed information on command aliases, consult the "Create Command Aliases" topic in the online help.

UNDERSTANDING TOOL PALETTES

Tool palettes provide instant access to a complete collection of AutoCAD MEP tools organized in logical groupings. Tool palettes combine the user-friendly visual icon-based interface of toolbars with the flexibility, power and customization potential of pull-down menus. Simply click on a tool to execute its function (you do not need to drag it). Tools are interactive and many parameters can be manipulated on the Properties palette while the tool is active. Furthermore, properties can be pre-assigned to the tools so that default settings are automatically assigned on tool use. Using the Content Browser, you can add tools and complete palettes to your personal workspace at any time (see more on the Content Browser below).

The default installation of AMEP loads several basic tool palettes populated with a variety of the most commonly used tools. The palettes are organized into tool palette groups (see below). The HVAC tool palette group contains the most basic mechanical object tools. The Duct palette (part of the HVAC tool palette group) contains a variety of duct tools for specific settings. The remaining palettes contain shortcuts to some specific content. Groups are loaded by right-clicking the title bar, and individual palettes are accessed by clicking their tab on the tool palettes.

> Note: If you have installed and are using a content pack other than US Imperial or US Metric, the specific tool palettes and groups you have might vary slightly from the ones noted and pictured in this text.

Using Tool Palettes is intuitive. The following exploratory steps will help you quickly become acquainted with this critical interface item.

1. Launch AutoCAD MEP, if it is not already running.

2. On the Quick Access Toolbar (QAT), click the QNew icon (see Figure 1.17).

FIGURE 1.17
Create a new drawing using QNew

The QNew command will automatically create a new drawing file using your default template. If the Select Template dialog appears when you click QNew, choose the template: *Aecb Model (Imperial Ctb).dwt.*

> CAD MANAGER NOTE: If QNew fails to load a template automatically, open the **Application menu** and click the **Options** button. Click the Files tab. There, expand the Template Settings item and then the Default Template File Name for QNEW item. Finally, select the entry listed there, click the Browse button, and choose your preferred default template. You only need to do these steps once and they will remain in place in the current profile on your machine.

3. If the tool palettes are not loaded, on the Home tab of the ribbon, on the Build panel, click the Tools button (or press CTRL + 3).

Tool palettes can be left floating on screen or can be docked or anchored to the left or right side of the drawing editor. Simply drag the palettes by the title bar to the left or right side of the screen. The title bar will dynamically shift from left to right as you move the tool palette close to either edge of the screen or it will dock to the edge of the screen. (Figure 1.18 shows floating, docked and anchored palettes and Figure 1.19 shows the title bar shifting from the right to left sides.)

4. Right-click the title bar of Tool Palettes and check the setting of "Allow Docking."

A checkmark next to Allow Docking indicates that the palette will dock (attach) when close to the edge of the screen. No checkmark means that it will stay floating even if you move it to the edge of the screen. When Allow Docking is enabled (checked), you will also have the ability to "anchor" the palette. Docked palettes attach to the sides of the screen and reduce the overall width of the drawing area. An anchored palette creates an anchor dock on the left or right side. This small gray strip can contain one or more anchored palettes such as Tool Palettes, Properties or External References. You can even tear off the Command Line from its traditional location at the bottom of the screen and anchor it here. Anchored palettes fly open (such as when Auto-hide is enabled) when you pass the mouse over them (see Figure 1.18).

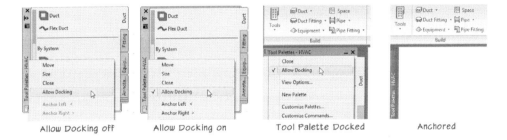

FIGURE 1.18
The Allow Docking feature toggles docking of the tool palettes

5. Test the behavior with Allow Docking on and then with it off.

6. After enabling Allow Docking, try choosing either Anchor Left < or Anchor Right >.

If you dock a tool palette and wish to return it to floating, you can right-click on the title bar and remove the "Allow Docking" checkmark (shown in the third item from the left in Figure 1.18). You can also click on the title bar and drag the palette into the drawing window, or simply double-click on the title bar. The small minus sign icon on the right of the title bar will convert the docked palette to an anchored palette; the "X" icon will close the palette. Right-click the anchored palettes to change the way their labels display.

7. After experimenting, turn off Allow Docking.

8. To see the title bar flip, drag the tool palettes first to the left edge of the screen and then to the right.

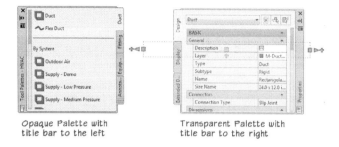

Opaque Palette with
title bar to the left

Transparent Palette with
title bar to the right

FIGURE 1.19

Palettes dynamically justify their title bar to the appropriate edge of the screen and can be made transparent

> Note: Figure 1.19 shows transparency turned on for the palette on the right. To do this, right-click the title bar (or click the small palette menu icon in the top corner of the palette's title bar—shown in Figure 1.22) and then choose **Transparency**. However, this feature can cause a slowdown in performance on some systems, so make sure you test the feature on your system to gauge performance before using it regularly.

Many of the palettes (tool palettes, properties, etc.) have tabs along the edge (or along the top for DesignCenter). Click these tabs to see other tools and options. For the tool palettes, you can customize these tabs and configure their properties; to do so, right-click on a tab (make it current first by clicking on it). When all tabs are not visible, there will be several tabs "bunched up" at the bottom of the tool palette; click there to reveal hidden tabs (see Figure 1.20).

Tool palette tabs can be grouped. A tool palette group includes a small subset of the total available tool palettes. The default installation for US Imperial and US Metric includes six groups: HVAC, Piping, Electrical, Plumbing, Schematic, and Architectural. Groups for other content packs may vary.

9. Click on one or more tabs to switch between different palettes.

10. On the tool palettes, right-click on a tab.

Note the menu options.

11. If all tabs are not shown, click on the bunched-up group of tabs at the bottom to see menu revealing the hidden tabs (see Figure 1.20).

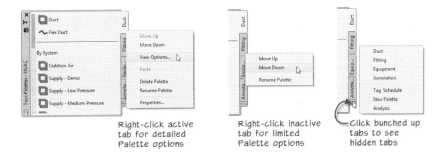

Right-click active tab for detailed Palette options

Right-click inactive tab for limited Palette options

Click bunched up tabs to see hidden tabs

FIGURE 1.20

Accessing palette options and hidden tabs

- **Move Up & Move Down**—Shift the location of the selected tab relative to its neighbors.
- **View Options**—Opens a dialog with options for changing the icon size and configuration displayed on the palette(s) (see Figure 1.21).
- **Paste**—Only available after a tool (from this or another palette) has been copied or cut.
- **Delete & Rename Palette**—Allows you to delete or rename the selected palette.
- **Properties**—Allows you to change the Name and Description of the current palette.

FIGURE 1.21

View options changes icon style and size for this palette or all palettes

Another group of options is available for the entire palette group. In the top corner of the title bar in every palette are three small icons. The first closes the palettes. The second toggles on and off the "Auto-hide" feature of palettes. When this feature is enabled, the palette will automatically collapse to just its title bar whenever the mouse pointer is moved away from the palette. The palette will "pop" back open when the pointer pauses over the title bar again. This same feature can be controlled with the Auto-hide option in the palette Properties menu available by clicking the third icon (in the top corner) or right-clicking on the title bar.

12. On the Tool Palettes, click the small Auto-hide icon (see Figure 1.22).

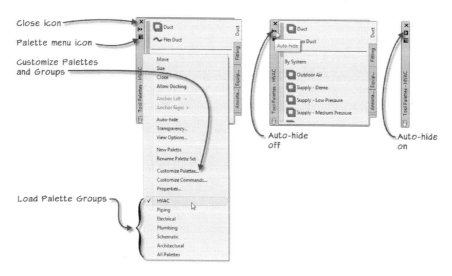

FIGURE 1.22

Access the Properties menu, load Groups and toggle Auto-hide

13. Move your mouse away from the palette.

Notice that the palette collapses to just the title bar (see the far right side of Figure 1.22).

14. Move your mouse back over the collapsed title bar.

Notice that the palette expands again.

15. Click the Auto-hide icon again to turn it off.

Note: For the remainder of this chapter, please turn off the Auto-hide. At the completion of the exercise, you may set it however you want.

16. Click the Properties icon (or right-click the title bar) to display the options menu.

Note the various options.

> **CAD MANAGER NOTE:** Palettes can be made that include any combination of stock and/or user-defined tools. Complete palettes of project-specific tools can be created and subsequently loaded by each member of the project team. Furthermore, these palettes can be linked to a remote catalog location and set to refresh each time AutoCAD MEP is loaded. This will guarantee that project team members always have the latest tools and settings. The customization potential of tool palettes is nearly limitless. For very detailed information on customizing tool palettes (and many other advanced topics), pick up a copy of *Autodesk Architectural Desktop: An Advanced Implementation Guide,* second edition.

UNDERSTANDING TOOL PALETTE GROUPS

As mentioned previously, tool palettes can be organized into groups. Right-click the tool palettes title bar to access other groups. By default, AMEP installs six tool palette groups: HVAC, Piping, Electrical, Plumbing, Schematic, and Architectural. In addition, when an MEP project is loaded, a tool palette group uniquely named for the project may be added (and potentially made current).

17. Right-click on the Tool Palettes title bar and choose **Piping** (to load the Piping tool palette group) from the menu (see the second item in Figure 1.23).

Notice that all the tool palette tabs change to Piping functions.

18. Right-click on the tool palettes title bar again and choose Electrical from the menu to load the Electrical tool palette group (see the third item in Figure 1.23).

19. Right-click on the tool palettes title bar again and choose All Palettes to load palettes from all tool palette groups at once.

Notice that now all the tool palette tabs from all groups appear (not shown in the figure).

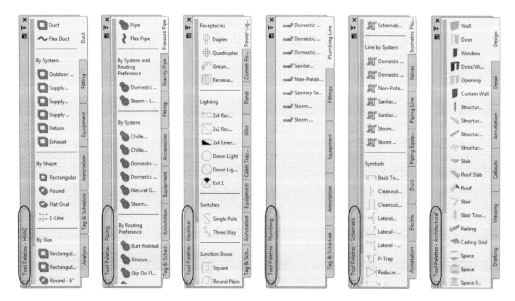

FIGURE 1.23

Six tool palette groups are included out-of-the-box in US Imperial and Metric

You can create your own groups if you wish. To do this, right-click the tool palettes title bar (or click the palette menu icon shown above) and choose **Customize Palettes**. In the Customize dialog, you can create new groups by right-clicking on the right side. Right-click on the left side to create new palettes. Add and remove items from each group by using the drag-and-drop method. The same palette can belong to more than one group.

RIGHT-CLICKING

In AutoCAD MEP, you can right-click on almost anything and receive a context-sensitive menu. In fact, we have just seen several examples in the previous topics on the ribbon and tool palettes. These menus are loaded with functionality.

The next several figures highlight some of the more common right-click menus you will encounter in AMEP. Do take a moment to experiment with right-clicking in each section of the user interface. You will also discover that the typical Windows right-click menus appear in all text fields and other similar contexts (this is used for Cut, Copy, Paste and Select All). Let's explore the right-click.

Right-Click in Drawing Editor (Default Menu)

1. If AutoCAD MEP is not running, launch it now.

   Press the ESC key to clear any commands or object selections.

2. Move the mouse to the center of the screen and right-click. Notice the menu that appears (see Figure 1.24).

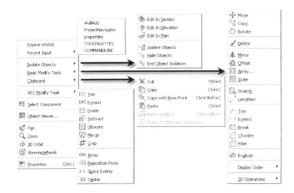

FIGURE 1.24

Default right-click menu

The default right-click menu appears when you right-click in the drawing area with no commands active and no objects selected. It is divided into sections of function. The first item will always show the last command executed and beneath that a flyout list of recent commands. Repeating COMMAND (where COMMAND is the last command run) will give a shortcut to executing the last command. (Figure 1.24 shows the WALLADD command.)

Tip: In addition to these two methods of repeating the last command, you can press the ENTER key or the SPACEBAR to repeat the last command. Also, if dynamic input is on, you can begin typing the first few letters of a command on screen or at the command line and then press the TAB key until the command you need appears. Press ENTER to execute the command.

The next section includes a flyout menu for the Isolate Objects (used to control visibility of selected objects and access the Edit in View functionality) commands. The Basic Modify commands are next, which include all of the common AutoCAD Modify commands, such as Move, Copy and Rotate. Clipboard functions (Cut, Copy and Paste) occupy the next flyout menu. The AEC Modify Tools flyout menu includes a collection of special AMEP editing tools, many of which work on regular AutoCAD entities. The Select Component command allows you to edit the display properties of the components within an AEC object directly on the Properties palette. This will be covered in later chapters. Object Viewer is a separate viewing window for quick study of selected objects. Pan, Zoom and 3D Orbit are the standard AutoCAD navigation commands, and finally, Properties will open the Properties palette if it is not open and make it active if it is already open.

CAD MANAGER NOTE: Many Veteran AutoCAD users continue to lament the loss of the right-click to ENTER and repeat the previous command. Although the behavior of the right-click can be reverted to this style, it is recommended that you do not do this. If you do so, a great deal of necessary AMEP functionality will be lost. Please try the default setting throughout the duration of this book. If after completing the lessons in this manual you are still convinced you will be more productive with the right-click set to ENTER, then at least consider "Time-sensitive right-click" (available on the User Preferences tab of the Options dialog) as an alternative. The Time-sensitive right-click option makes the right-click behave like an ENTER with a "Quick" click of the right button. A "longer click" will display a shortcut menu. This feature will offer a good compromise to many seasoned AutoCAD users. To make this change, choose Options from the Application menu, click the User Preferences tab, and then the right-click Customization button.

Please remember that both the ENTER key and the SPACEBAR on the keyboard function as ENTER within the AutoCAD environment. For veteran AutoCAD users, the old "rule of thumb" still applies. Keep your left thumb on the SPACEBAR for a quick ENTER.

Right-Click in the Command Line

When you right-click in the command line, a small context menu appears (see Figure 1.25). Choosing Recent Commands shows a menu of the last several commands executed. Use this menu as a shortcut to rerun any of these commands. The Copy History command puts a complete list of all command line activity on the clipboard that can then be pasted into any text editing application. You can also access the Options command from this menu. (The Options command is also available on the Application menu.)

FIGURE 1.25

Right-click in the command line (The image shows the command line "torn off" as a palette)

You can also close the command line window. To do this, make the command line a floating window. You can float it by dragging the small double gray bar on its edge, and then releasing when the command line has "undocked" from the edge of the screen. Once the command line is floating, you will see the standard Windows close box (looks like an "X"). Click this box to close the command line. When you do this, a warning dialog will appear (see Figure 1.26).

FIGURE 1.26

You can close (hide) the Command Line Window—use CTRL + 9 to re-display it

---

CAUTION:

It is highly recommended that you use either the command line window or the dynamic input prompts (see below) option. If you disable both of these it will be very difficult to use the software effectively.

RIGHT-CLICK WHILE A COMMAND IS ACTIVE

Most AMEP commands have one or more options. Access them by typing directly in the command line, using dynamic input on-screen prompts, or using the right-click menu.

1. On the Home tab on the Draw Panel, click the Line flyout and choose the *Polyline* tool.

2. Click a point anywhere on the lower left side of the screen within the drawing editor.

3. Move the mouse position to the bottom right side of the screen and click again.

4. Move the mouse to the upper right corner of the screen and click a third time.

5. With the command still active, right-click (see Figure 1.27).

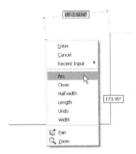

FIGURE 1.27
Right-click within a command (Polyline in this case) to access its options

Compare the menu that appears with the options shown in the command line. You will see many of the same options are available in both places. The same options are also listed in the onscreen prompting if you have dynamic input enabled (see below).

6. From the right-click menu, choose Arc.

7. Move the mouse to the left of the screen and click again.

8. Right-click and choose Close.

Right-Click in the Application Status Bar

The application status bar gives quick access to many of the drafting settings available in AutoCAD MEP. If you wish to customize the default settings of any of these drafting modes, simply right-click the button and choose **Settings** (see Figure 1.28).

Right-clicking the controls on the application status bar to access options

Choose Use Icons to toggle between icons and the "classic" text modes for displaying these status toggles. Note that ORTHO does not offer the Settings choice. The Model and Layout icons replace the Layout tabs that previously appeared along the bottom edge of the drawing window in earlier versions. By right-clicking, you can restore these tabs instead of the icons shown in the application status bar. We will work in model space for most exercises in this book. For the time being, do not click these icons. If you have already clicked the Layout one (named "Work" in the default template with which we started) then you will need to click the Model icon to return to model space (see Figure 1.29).

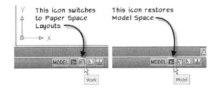

FIGURE 1.29
Click the Model icon to return to model space if necessary

For more information on Model and Paper Space, refer to the online help.

DYNAMIC INPUT

As noted above, the command line is only one way that we can interact with and access command options. Dynamic Input—which places command prompts directly onscreen—gives us many cues and prompts to make the interactive process of creating and manipulating objects more fluid and user-friendly. Dynamic Input has a simple toggle button in the application status bar alongside the other drafting modes like SNAP, GRID and POLAR. If you right-click this toggle, you will find many options to customize the Dynamic Input behavior. Let's explore some of those now.

POINTER INPUT

1. At the bottom of the screen on the application status bar, right-click the Dynamic Input toggle and choose Settings (see Figure 1.30).

Right-click here ⟶

FIGURE 1.30
Right-click Dynamic Input icon to access Dynamic Input Settings

The Drafting Settings dialog will appear with the Dynamic Input tab active.

2. Deselect all checkboxes in this dialog, place a checkmark only in the "Enable Pointer Input" checkbox and then click OK (see Figure 1.31).

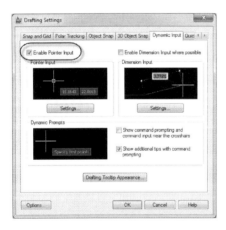

FIGURE 1.31
Enable only the Pointer Input option

This option provides text input fields at the cursor where you can type in coordinates as you draw. All objects in AutoCAD/AMEP exist in a coordinate grid (referred to as the "World Coordinate System" or "WCS"). Coordinate input can be achieved using two different systems to indicate precise locations in the drawing relative to the WCS—Cartesian and Polar. In the

Cartesian system, you input locations using "X" (horizontal) and "Y" (vertical) coordinates. In the Polar system, input is based on a distance (measured in units) and a direction (measured in degrees around the compass). Both systems are valid for input in AMEP, and you can switch on the fly simply by varying your input syntax. The syntax for Cartesian input is: **X,Y**, where *X* and *Y* are input as positive or negative numbers in the current unit system (inches, feet, meters, etc.) and the comma is used to separate them. The syntax for Polar is **D<A**, where *D* equals the distance (nearly always a positive number in the units of the drawing) *and A* is the angle along which this distance is measured in degrees, with the "less than" symbol to separate them. Both systems can optionally add a third coordinate for the Z direction when working in 3D. Much of the input needed in AMEP and this book will use methods simpler than the traditional coordinate input. View the topic: "Use Coordinates and Coordinate Systems" in the online help for more information on coordinate input.

3. If necessary, load the HVAC workspace. On the Home tab of the ribbon, click the Duct tool and then click a point on the screen.

4. Move the mouse around slowly on screen and note the two dynamic prompts that appear (see Figure 1.32).

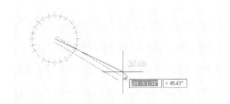

FIGURE 1.32
Pointer Input gives coordinate prompts at the cursor on-screen

5. Type a number such as **10'** on your keyboard—do not press ENTER yet.

Note that the number will automatically appear in the first coordinate field.

By default, AMEP uses Polar coordinates as you can see indicated in the second onscreen prompt. However, you can change this default if you like and you can always input values in either system at any time. After you indicate the first value, type a "<" ("less than" sign) to input the Polar angle next; or, for Cartesian coordinates, type a comma (,) to interpret the first value as an "X" and then input the "Y" value.

6. On your keyboard, type a comma (**,**)

   Notice that the first value "locks" and the second prompt activates.

7. Move the mouse around a bit.

Notice that the first value of 10' X is locked in so that the Duct is constrained in width and your mouse movements only affect the vertical position. Do not click yet.

8. Type in a second value such as **10'** again and then press ENTER (see Figure 1.33).

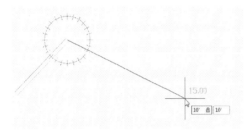

FIGURE 1.33

Pointer movement is currently constrained to the value in the first (locked) field. Use the mouse or type to set the value of the other

> Note: Your screen may look different from the illustrations since the numbers in the illustrations are absolute dimensions. For example, 10' has a unique location in the WCS. Depending on where you click the first point of the Wall drawn in Figure 1.33, this fixed location could be to the right or left of your starting point. Otherwise, everything else should function as indicated.

9. Repeat these steps, but instead of locking the first value with a comma, type the less than symbol (**<**) this time.

10. Move the mouse around on screen.

Notice that this time the length of the Duct has been locked and you can rotate freely around the first point. Your second value will be interpreted as the rotation angle of the Duct this time. Try it!

11. If you have drawn any Ducts, erase them. (Select them, and then press the DELETE key).

DYNAMIC COMMAND PROMPTING

The third option in the Dynamic Input dialog enables command prompting at the crosshairs. You can use this in place of, or in addition to, the command line window. Enable this setting with either or both of the other two settings.

1. Right-click the Dynamic Input toggle once more and choose Settings.

2. Place a checkmark in all three boxes this time, including the "Show command prompting and command input near the crosshairs" box, and then click OK.

3. On the Build ribbon, click the Duct tool.

   Notice the command prompt directly at the cursor (see Figure 1.34). The prompt also appears on the Command Line.

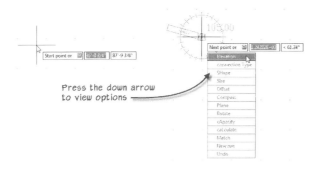

FIGURE 1.34

With Command Prompting enabled, prompts show directly at the cursor

4. Pick a point to start the Duct.

   Notice the Next Point prompt.

5. Pick another point.

6. Press the down arrow on your keyboard.

   A list of command options will appear at the cursor. (This is the same list you would get if you right-clicked at this point.)

7. Press the DOWN ARROW again to begin moving down the list of options. Press ENTER (or click your mouse) to choose the highlighted option.

Continue experimenting as much as you wish. You can return to the Dynamic Input settings at any time and choose different options. For the remainder of this text, we will assume that all three dynamic input options are active. If you want to learn more about Dynamic Input, return to the settings dialog and then click the Help button at the bottom of the dialog.

DYNAMIC DIMENSION INPUT

AutoCAD MEP objects also allow you to specify dimensions directly on screen during layout. This allows you to input distance or angle values into layout AutoCAD MEP components. To use this functionality, follow the previous steps and make sure that "Enable Dimension Input where possible" is checked. Then simply draw objects on screen. A dynamic dimension will appear onscreen. Type the desired dimension value directly into this dimension and then press ENTER to draw the Duct, Conduit or Pipe that length.

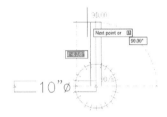

FIGURE 1.35
Specifying Dynamic Dimensions during layout

MEP COMPASS

The MEP Compass provides the ability to route MEP segments at precise angles. The Compass shows tick marks at user definable increments, and snaps cursor pick points at definable angles. The settings for the Compass are available on the View tab on the MEP View panel. Note in Figure 1.36 how the cursor is between 15 and 30 degrees (closer to 30) and the Duct preview is shown snapped to 30 degrees. When using the Compass, it is not necessary to have Ortho on.

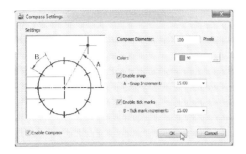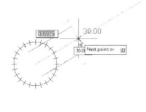

FIGURE 1.36

Compass settings dialog and example of angular constraint

DIRECT MANIPULATION

Direct manipulation refers to the ability to manipulate object geometry directly in the drawing editor without the need to visit a dialog box or even a palette. To do this, you interact with the various grips of the objects. AMEP presents many purpose-specific grip shapes for the various objects. A grip is a visual toggle point on the object that allows you to begin modifying the object in some manner at the grip location. There are several different types of grips in AMEP; Alignment, Elevation, Isometric Plane, Lengthen, Location, Rotate (or Flip). There is also the Edge and Vertex grip that can be used for Spaces. See Chapter 4 for additional information on Edge and Vertex grips. Let's take a look at some of these as well as the basic direct manipulation techniques for them.

1. Using the Duct tool on the Design palette, create a single horizontal Duct on screen.

2. Click on the Duct just created to reveal its grips.

3. Hover your cursor over each grip.

   Do not click the grips yet, simply pass the cursor over each one and wait until the tool tip appears (see Figure 1.37).

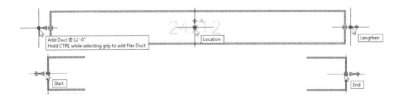

FIGURE 1.37

Hover over a grip to reveal its function and options

Notice that each grip shape serves a different function. Hovering over the grip reveals this function in a tool tip. If there are options to this grip (usually invoked by the CTRL or the ALT key) they will be revealed in the tool tip as well.

4. Click one of the Lengthen grips.

   It is shaped like an isosceles triangle and points away from the end of the Duct.

5. Drag the grip and click to set a new length for the Duct (see Figure 1.38).

FIGURE 1.38

Dragging to a new width with the "Width" grip

6. Try it again, only this time, type a length into the value field (highlighted onscreen) and then press enter to lengthen the Duct by the specified length.

7. This time, hold down ctrl and click a + (Add Duct) grip.

   Notice how this method allows you to add Flex Duct.

8. Try the other grips successively.

Note the difference in behavior between the Start/End, Length and Add Duct grips. With the End (or Start) grip, you have full range of motion and can change the endpoint location as well as the angle of the Duct segment (this is like grip editing an AutoCAD line object). With the Lengthen grip, you can change only the length of the Duct without affecting its orientation (see Figure 1.39). These are both very powerful tools. The Add Duct grip allows you to add rigid or flex Duct, inserting Fittings as necessary.

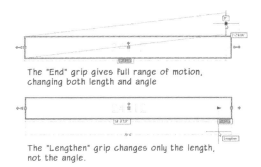

The "End" grip gives full range of motion,
changing both length and angle

The "Lengthen" grip changes only the length,
not the angle.

FIGURE 1.39
Note the difference between the End and Lengthen grips

This quick overview of grips on Duct objects gives you just some idea of the potential of this very powerful interface item. Look for unique grip shapes on every AMEP object. Hover your mouse over them to reveal the tool tip of their function. Fittings and Devices offer some exciting possibilities. We will see several more examples throughout this book.

Edge Grips

Profile-based AMEP objects (such as Spaces) have round grips at their corners and long thin grips at the edges.

9. On the Home ribbon, on the Build panel, click the Space tool.

> **Tip:** If you are not sure which tool is which, pause your mouse cursor over each one and wait for a ToolTip to appear or right-click the palette tab, choose View Options, and then select an icon with text.

10. On the Properties palette, make sure that the "Create Type" is set to Insert, accept all other defaults, and then add a Space anywhere in the drawing.

11. Cancel the Space command, and then click on the Space to reveal its grips.

12. Hover your cursor over each grip.

    Do not click the grips yet; simply pass the cursor over each one and wait until the tool tip appears (see Figure 1.40).

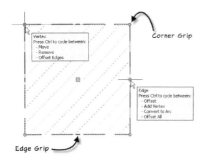

Corner Grip

Vertex
Press Ctrl to cycle between:
- Move
- Remove
- Offset Edges

Edge
Press Ctrl to cycle between:
- Offset
- Add Vertex
- Convert to Arc
- Offset All

Edge Grip

FIGURE 1.40

Explore the Grip Shape Functions of a Profile-based Object

13. Click one of the circular-shaped grips and move it. Click again to complete the move.

14. Click on the same circular-shaped grip and then press the CTRL key once.

   Notice that the shape of the Space changes to reflect the removal of this grip. The operation is not complete, however, until you click the mouse again.

15. Click anywhere to complete the removal of the vertex.

16. Undo and then repeat the process by pressing ctrl twice this time.

   This will offset both adjacent edges instead.

17. Click one of the thin rectangular-shaped grips and move it. Click again to complete the move.

   This operation moves the selected edge while stretching the attached edges.

18. Click one of the thin rectangular-shaped grips again.

19. Press the ctrl key once.

   Notice that a new vertex is formed.

20. Press the CTRL key again.

   Notice that an arc segment is formed.

21. Press the CTRL key once more.

   Notice that all segments will offset around the whole shape.

22. Continue experimenting with any of these functions before moving on.

All of these grip functions work on any AMEP object type that uses closed profile shapes. These include Slabs, AEC Polygons and certain Mass Elements. Feel free to draw other objects and experiment.

MEP Snaps

AutoCAD MEP provides a handful of snaps that are specific to MEP elements. The Schematic snaps apply to Schematic Lines and Symbols. Duct snaps apply to Ducts, Flex Ducts, Duct Fittings and MvParts with duct connectors. Pipe snaps apply to Pipes, Flex Pipes, Pipe Fittings, Plumbing Lines, Plumbing Fittings, and MvParts with pipe connectors. Electrical snaps apply to Panels, Devices and Wires. Wire Ways snaps apply to Conduit, Cable Tray, associated Fittings, and MvParts with conduit and cable tray connectors.

There are two snaps for each type of component, a Curve snap and a Connector snap. The Connector snap is akin to an End snap; it snaps to the end of the segment and at the open end of fittings. The Connector snap also snaps to connectors on MvPart content. These snaps help facilitate Object Snap Tracking as well as auto-layout functionality.

The Curve snap is similar to a Nearest snap. It facilitates connecting to any point along a segment. For two line elements (such as Duct and Pipe), this takes some getting used to because you have to "acquire" the snap by hovering over the edge of the component while the snap shows up along the centerline of the object.

You can temporarily disable an MEP snap by holding down the CTRL key. This is useful if you are working at different planes or in 3D views, you need to route above, behind, or near another element, and don't want the new element to connect to an existing element (see Figure 1.41).

In addition, you can independently toggle AutoCAD and AutoCAD MEP snaps. Use the F3 key for AutoCAD snaps such as Endpoint, Midpoint, Intersection, etc, and the SHIFT and F3 key to toggle the AutoCAD MEP Snaps such as PCON, ECURV, DCON, etc.

FIGURE 1.41

Explore the Grip Shape Functions of a Profile-based Object

THE COMMAND LINE

We have already discussed the Command Line Window above. This text-based interface, typically docked at the bottom of the AutoCAD MEP screen, can be set to display one or more text lines at a time; the default configuration displays three lines. The word "*Command*" will be displayed when the there is no active operation (see Figure 1.42). Most AutoCAD and AutoCAD MEP commands can be typed into this command line area and executed by means of the ENTER key or spacebar. As we have seen in the previous topic, we can enable onscreen dynamic prompting as an alternative or in addition to the command line window prompts. Most common commands can also be executed by choosing them from palettes or ribbon icons. The exact method of command execution is largely a matter of personal preference. As a general rule of thumb, palettes, ribbon icons and onscreen prompting tend to be more user friendly, while the command line tends to provide the most options and allow the fullest automation potential as well as a certain familiarity and comfort for experienced AutoCAD users.

Command Line in its default position docked at the bottom of the screen

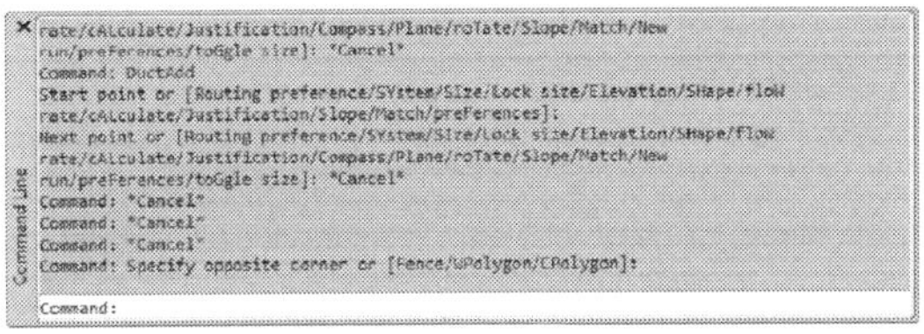

Command Line as a Palette

FIGURE 1.42

The AMEP command line

> **Tip:** As any veteran AutoCAD user can attest, there really is only one golden rule to AutoCAD usage: "Always read your command line."

Despite AutoCAD MEP's heavy reliance on modern interface items such as palettes and direct manipulation, it is still important to understand and pay attention to the command line and/or the onscreen prompts. *Remember this rule, and you will be on your way to success with this software package. Disregard this rule and you will surely struggle and become frustrated.* If you need to see more than three command lines or want to read back through previously executed commands, press F2 to toggle the AutoCAD text window. The AutoCAD text window is a history window allowing you to scroll back through prior commands.

Even though you can turn off the command line and rely solely on the dynamic prompting, many users may find it more comfortable to leave it displayed for those times when command line interaction is preferable. However, the command line does take up a great deal of screen real estate. One solution to this dilemma is the ability to anchor docked palettes. This was discussed briefly above in the "Understanding Tool Palettes" topic. So a good compromise is to tear off the command line from the bottom of the screen, then right-click its title bar and choose **Anchor Right >** or **Anchor Left <**. This will attach it to one or the other side of the screen. When you hover your mouse over the anchor bar at the edge of the screen, the command line (and any other anchored palettes) will fly open for use (see Figure 1.43).

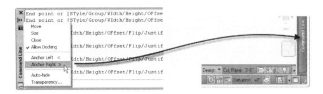

FIGURE 1.43

Anchoring the command line

INTELLIMOUSE

If you have a Windows IntelliMouse, (or any third-party mouse with a middle button wheel and the proper driver) AMEP provides instant zooming and panning using the wheel! If you don't have a wheel mouse this might be a good time to get one. This modest investment in hardware will pay for itself in time saved and increased productivity by the end of the first day of usage. Using the wheel you will have the following benefits:

- **To Zoom**—Roll the wheel.
- **To Pan**—Drag with the wheel held down.
- **To Zoom Extents**—Double-click the wheel. (Do the same type of double-click that you would do with the left button, only on the wheel instead. It takes a little practice at first.)
- **To Orbit in 3D**—Hold down the SHIFT key and simultaneously drag with the wheel held down.

The ZOOMFACTOR command controls the rate of zooming with the wheel. Type ZOOMFACTOR at the command line and then press ENTER. Type the percentage you wish to magnify with each rotation of the wheel on your mouse and then press ENTER again. This setting applies globally, so you only need to do this once. The default setting is 60.

> **Tip:** When typing commands at the command line or on screen dynamic prompts, type the first few letters and then press the TAB key. With each TAB, AutoCAD/AMEP will guess the command you want by completing the rest of the command name. Keep tabbing until it shows the command you want and then press ENTER.

If you have tried to use your wheel to zoom and pan and instead you get a menu with Object Snap settings, you will need to adjust the setting for MBUTTONPAN. This command is a toggle setting that turns on and off the wheel zooming and panning feature. At the command line type MBUTTONPAN and then press ENTER. Be sure the value is set to "1" and press ENTER again. A setting of "1" turns this feature on, while "0" turns it off. If MBUTTONPAN is set properly and the wheel still does not function properly, you may need to adjust the settings in your Mouse applet in the Windows Control Panel. Usually the wheel works best when the wheel button is set to "Autoscroll." You may also need to update or re-install your mouse driver. Check your mouse manufacturer's Web site for complete details on mouse driver installation.

NAVIGATION BAR

The Navigation Bar is a small transparent toolbar that sits near the border of the drawing area. By default it is located on the right side. Contained on this toolbar are the Steering Wheel, Pan flyout, Zoom flyout, Orbit flyout and the ShowMotion control. Let's look at a few of the more common features here. For more detail on any of the items below, refer to the online help.

Zoom, Pan and Orbit

The fastest and easiest way to zoom, pan and orbit is using the wheel on your mouse as discussed in the previous topic. However, the tools on the Navigation Bar offer several alternatives and are worth exploring (see Figure 1.44). If you do not see the Navigation Bar on your screen, you can display it using the User Interface drop-down button on the View tab of the ribbon. Other alternatives also exist to locate the navigation commands. Look on the Home tab for the View panel and the View tab on the Navigate panel. Feel free to experiment with the various zoom, pan and orbit options.

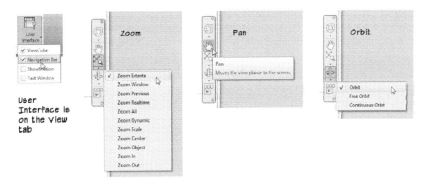

FIGURE 1.44

Display the Navigation Bar on the View tab

Steering Wheels

The Steering Wheels are a series of "tracking menus" that follow your cursor around and give you quick access to selected groups of navigational commands. The simplest, the 2D Navigation wheel (see Figure 1.45), is only available in paper space of a layout, and is the only choice there. In model space, you can choose between three different sets of commands as well as between "full" and "mini" versions of each set.

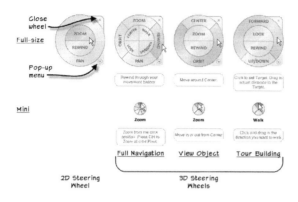

FIGURE 1.45

The Steering Wheels

Activate a Steering Wheel by clicking the Steering Wheel icon on the Navigation Bar or, on the View tab of the ribbon, on the Navigate panel, choose the Steering Wheels drop-down button and select one of the Steering Wheels (see Figure 1.46). Click on the drop-down icon on the full wheels for a pop-up menu of options.

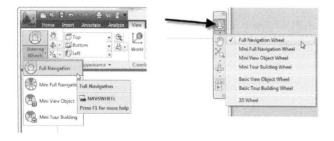

FIGURE 1.46

Activating a Steering Wheel from the ribbon or the Navigation Bar

Try out the Steering Wheels in a drawing with some AMEP objects. Select the desired option on the wheel and hold down the left mouse button while dragging to use the selected feature.

After changing the view several times, try the Rewind feature, which displays a strip of thumbnails of previous views and allows you to return to any one of them (see Figure 1.47).

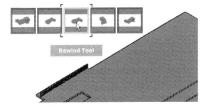

FIGURE 1.47
Using Rewind to return to a previous view

The ViewCube

Another navigational aid is the ViewCube. The ViewCube allows you to quickly move between any of the preset views as well as rotate when in one of the orthogonal views (front, back, left, right, top or bottom). Unlike the Steering Wheels, the ViewCube does not follow the cursor around the screen. The default location is in the upper right corner of the drawing window, but you can specify any of the other three corners by choosing ViewCube Settings from the pop-up menu that displays when you hover over the ViewCube and select the down arrow icon. This menu also allows you to set a particular view as the "Home" view, and then return to it by clicking on the home icon that appears when hovering over the cube. If the ViewCube is not enabled, on the View ribbon tab, on the Navigate panel, expand the panel and choose *ViewCube* (see Figure 1.48).

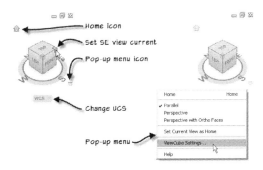

FIGURE 1.48
The ViewCube

> Note: The ViewCube also appears in the AMEP object viewer, the floating viewer available when editing an AMEP object style.

PREREQUISITE SKILLS

The following list of "rules" serves two purposes. For the beginner, it will give you a focused list of topics to research and explore to prepare yourself for the chapters ahead. For the experienced AutoCAD user, treat this list as a self-assessment checklist of prerequisite AutoCAD skills. Use the list to identify areas where you might need a little brushing up. Please note that this book covers all skills necessary to begin using AutoCAD MEP effectively and you are not required to purchase or use any other books or lessons. However, there are several core AutoCAD features and topics that are not covered here in the interest of brevity or because they do not have explicit usefulness for MEP design or production. Keep in mind however that all core AutoCAD functionality is present within AutoCAD MEP. Therefore, any of the dozens of available AutoCAD books, tutorials and videos on the market would provide a nice complement to the lessons covered here.

AUTOCAD SKILLS—"THE RULES"

- **Always read the command line (or dynamic prompts)**—This was covered in the previous section, but repetition always aids in retention. Remember: Read the command line or onscreen prompts (when using dynamic input) and you will be on your way to success with AMEP. Disregard this rule and you will surely struggle and become frustrated.
- **When in doubt, right-click**—This was also covered earlier, but it is no less important than the others, therefore it is also repeated here. This is especially critical for AMEP commands and functions.
- **Draw accurately and cleanly**—The benefits of a cleanly drafted model, where all corners meet, shapes close, and dimensions make sense, and where double lines and broken lines have been avoided, cannot be overstated. In AMEP you will be drafting with Objects instead of lines, but the rule remains the same. Always use tools like Ortho Tracking and Object Snaps. Snap is set to a very small increment and turned on by default in AMEP. This is done to help avoid rounding errors and is recommended. Follow these guidelines and the payoff throughout the process will be tremendous.

- **Draw once; use many**—This rule embodies a major purpose of any computer-aided design software package. This is certainly true for AutoCAD and AMEP. Always look for ways to reuse what you have already created. In MEP this goes beyond the concept of copy and array. In Chapter 2, when we explore the Display System, and Chapter 3 when we look at projects, you will see that "draw once; use many" has numerous applications and interpretations throughout your use of AMEP.

TOOLS AND ENTITIES

Your experience with AMEP will be much more fruitful if you are already familiar with each of the following concepts. While it is beyond the scope of this book to cover each of these topics in detail (as noted above), there are dozens of good AutoCAD books and manuals available. The online help system is also a good resource.

- **Object Snaps**—All AMEP (and AEC including AutoCAD Architecture and Civil 3D) objects take advantage of the standard AutoCAD Object Snaps, and AMEP adds many additional MEP specific snaps.
- **Layers**—Layers are like categories that help organize all of the data within a drawing file. AutoCAD has long used layering to keep drawings organized and manageable. AMEP objects benefit from "automatic" layering. The specific layer used for an object will depend on the layer Standard used in AMEP. For instance, in the United States the default Layer Standard in AMEP is the "AIA Standard" published by the American Institute of Architects. (Actually the Layering Standard installed by default is compliant with recommendations of the most recent version of the AIA Layering as published in the U.S. National CAD Standard.) All layering defaults can be customized to meet your office's layer standard needs. If you use AMEP in a country other than the United States, several other Layer Standards are provided. However, it should also be noted that the Display System controls the display properties of AMEP objects in a much more thorough way than layering alone can. Please see Chapter 2 for an introduction to the Display System.
- **External References**—An external reference (XREF) establishes a file link between two or more drawings in a project set. AMEP is designed to take full

advantage of the XREF functionality of AutoCAD. (With the built-in Drawing Management system, it takes it much further.) This allows you to leverage your existing strategies to their fullest advantage.

- **Layouts (Paper Space)**—Layouts (also called paper space) are most often associated with setting up sheets for plotting. AMEP takes full advantage of this functionality as well. In addition, the advanced display functionality of AMEP can be combined with the functionality of paper space layouts. This allows even more control over output than layouts and viewports alone would allow.

- **Template Files**—Template files provide a starting point for any AutoCAD or AMEP drawing file. By saving desired settings in a template file, you can ensure that new drawings are begun properly and in conformance to company standards. Out-of–the-box templates are used to begin all drawings in this book.

- **Blocks**—Blocks have been a huge part of AutoCAD productivity for years. A block is a collection of objects that have been grouped together and given a name. They then behave as a single object. Blocks continue to play an important role in customizing AMEP objects. Depending on the makeup of your existing block library, you may be able to convert them to useful AMEP parts.

CAD MANAGER NOTE: Your existing block libraries can be converted to AMEP content (making them "draggable") by adding them to tool palettes or by using MEP Content tools found on the Manage ribbon. One must keep in mind, however, the balance between existing 2D blocks, and leveraging 3D block capabilities for physical coordination.

IMPORTANT AUTOCAD TOOLS

This topic includes some important AutoCAD functionality that seasoned AutoCAD users may have overlooked. If you are not familiar with the following tools, you should consult an AutoCAD resource or the online help and become conversant with them.

- **AutoSnap Options**—Parallel and Extension are two Auto Snap features that often go unused by veteran AutoCAD users. Parallel draws a line parallel to an existing line (be it a line of duct, pipe, conduit, cable tray, etc…). This can be set as a running snap, but it works best when used as an override. (To use as an override, hold the SHIFT key down and right-click—this calls the OSNAP cursor menu—and then choose **Parallel**.) If you have truly odd angles in your building, using Parallel may necessitate disabling Snap within the MEP Compass. Extension allows the selection of a point along the extension (trajectory) of an existing line or arc. This provides functionality akin to a *virtual* Extend command.

- **Polar Tracking**—Polar tracking (right-click the POLAR button on the status bar and choose **Settings**) tracks the cursor movement along increments of a set angle, similar to the way ORTHO forces lines to move at multiples of 90°; however, the angle is a user-defined increment such as 45° or 15° (see Figure 1.49). Custom user angles can also be added. Additionally, unlike ORTHO, POLAR does not limit movement to the constraint angles. Instead, it snaps to those angles when the cursor gets close; otherwise it moves freeform. POLAR and ORTHO are not available at the same time. Toggling one on will toggle the other off, and vice versa. Access Drafting Settings from the icons on the Application Status Bar. The default setting for AMEP, is an Increment Angle of 30°, with Additional Angles of 45°, 135°, 225° and 315° added. This gives your cursor all of the positions of the traditional 45 and 30/60/90 triangles used in hand drafting.

FIGURE 1.49
Angle choices available in Polar Tracking

- **Object Snap Tracking**—Uses Object Snap points to set up temporary tracking vectors to align geometry to precise points. On the Object Snap tab of the Drafting Settings dialog (right-click the OSNAP button on the status bar and choose **Settings**), put a checkmark in the "Object Snap Tracking

On" check box to turn it on. A temporary alignment path will project from the various Object Snap points in the drawing. To use this feature, both the OTRACK and OSNAP toggles must be on (buttons pushed in on the status bar). To track, first activate a command, and then hover the mouse over a snap point for a moment. Do *not* click the point. As the cursor moves away from the point, a small plus (+) sign will remain indicating that the point has been "acquired" (see Figure 1.50). Moving horizontally or vertically from this acquired point will enable the tracking feature and keep the cursor lined up with the point. Several points can be acquired, and multiple tracking vectors can be used simultaneously to achieve very precise alignments. Check the online help for more information.

> Note: Object snap tracking potentially affects the elevation of MEP objects. To circumvent this, some controls have lock icons to lock in the specified elevation. Additionally, you can use the "Replace Z value with current elevation" control located on the application frame.

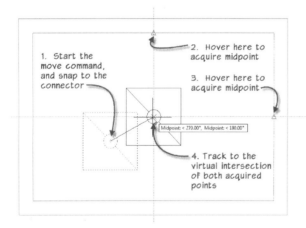

FIGURE 1.50
Acquiring and tracking from temporary track points

AUTOCAD MEP "RULES"

In addition to the summary of basic core AutoCAD skills listed above, the following list summarizes the overall skills you will want to develop and keep in mind in the coming chapters and as you work on projects with AMEP.

- **Edit at the source**—This is really another way of stating the previous rule of draw once and use many. When building projects in AMEP you will always seek the source of a particular piece of data in your Building Information Model (BIM). When editing the source, you are changing it everywhere! This guarantees coordinated models and helps eliminate errors.

- **Never explode objects**—Never explode an AEC object, hatch or dimension. *Never!* If you explode an AEC object, all its "smarts" are gone. This defeats the entire purpose of using AMEP in the first place.

- **Save often**—You only need to lose everything you have done in the last hour once to realize that Auto-save (in the Options dialog) isn't good enough. Here is a simple test you can do any time while you work to see if it is time to save. Any time you can say to yourself: "Boy, I'd hate to draw that again..." it is time to *save*! Use CTRL + S to save quickly. Creating regular backups is always a good idea as well.

- **Work smarter, not harder**—*Just because you can, doesn't mean you should.* Sure you can model all the domestic water plumbing details in a toilet room, or show the wire in safety glass or show all the nuts and bolts on equipment but is this always a *good* idea? AMEP is a wonderful tool, but it can't do *everything* well yet. Learn what it can and can't do well, and learn to discern the often "fine line" between the two. This takes practice, but it is worth the effort. Many examples of this principle will be covered throughout this book.

- **Progressive refinement**—Start with simple, low-detail models and slowly add detail over the course of the project. This process is the basis for all of the tutorials in this book, so you will get plenty of practice.

- **Be consistent**—You will make many small decisions as you work through a project. You will name objects, configure parameters and establish procedures. There is always more than one way to do things; whatever method you choose, implementing it consistently in similar situations will make it much easier to revisit the issue later. It will also make it much easier to work with other team members on the same project. Following the naming conventions established in AMEP can take away some of the guesswork.

- **There are always exceptions to every rule**—Saved the best for last—rules were made to be broken, were they not? These rules can be ignored or broken, but often at a tremendous loss of productivity, and even loss of data in some cases. However, there are times when breaking a rule will make sense; with practice and experience you will learn when it is appropriate. Like Mother always used to say about falling in love: "When the time comes, you'll just know." Until then, however, it's best to stick to the rules.

SUMMARY

✓ The AutoCAD MEP interface is designed to be streamlined, logical and easy to use.

✓ Ribbons provide ready access to most common AMEP tools and functions.

✓ Tool palettes are organized into manageable groups.

✓ AMEP relies heavily on the functionality of the right-click menu for editing commands; when in doubt right-click.

✓ The command line can be hidden and dynamic onscreen prompting enabled instead.

✓ A good foundation in basic AutoCAD skills will enhance your learning of AMEP and help you start off on the right foot.

✓ Several basic concepts like edit at the source, progressive refinement and consistency will provide a solid foundation for all of your efforts with AMEP.

Chapter 2

Conceptual Underpinnings of AutoCAD MEP

INTRODUCTION

AutoCAD MEP (AMEP) is an object-based Computer Aided Design (CAD) software package. It differs from the AutoCAD foundation upon which it is built in a number of ways. The major difference is in its focus on modeling over drafting. The main goal of the software is to facilitate the creation of a virtual building model from which plans, sections, elevations as well as quantities and other data can be readily extracted. The extracted drawings serve as two-dimensional "reports" of the "live" 3D model data. The advantages of this approach are many. From a 2D production point of view, this means less time drafting and coordinating building data because plans and sections are both being generated from the same source data. If the data changes, both plan and section receive the change. Schedules and data reports of quantities, component sizes, materials used, and scores of other property data are also within the realm of possibility and fully accessible. To achieve this level of functionality, it is important to understand a bit about what makes AMEP tick. That is the goal of this chapter. In particular, the focus will be on three major AMEP concepts that are not available in the underlying AutoCAD drafting package. These are: display control, anchors, and object styles. Drawing Management and Property Sets also offer power and flexibility not available within standard AutoCAD; but these topics will be covered extensively in Chapters 3 and 14, respectively.

OBJECTIVES

In this chapter, we will explore the meanings of parametric design, Building Information Modeling (BIM) and object-oriented CAD. Following the steps of a tutorial on the display system, you will learn how to display a single drawing model in many different ways which serve a variety of drawing and documentation needs. By exploring the Style Manager, we will begin to gain comfort with some of the critical conceptual underpinnings of the AMEP software package. Upon completion of this chapter you will be able to:

- Understand objects and their properties.
- Work with the display system.

- Understand object styles.

- Locate and quickly import content from a vast library of provided items.

PARAMETRIC DESIGN

Objects in AMEP are programmed to represent the real-life objects for which they are named. All real-life objects have a series of defining characteristics that determine their shape, size, and behavior. Parametric Design allows us to design while manipulating those real-world parameters directly on the objects.

To get a better sense of what these parameters might be, think of the characteristic to which you would refer if describing the object verbally to a colleague without the benefit of a drawing. Consider, for instance, a duct. If discussing a particular duct needed in a project with the contractor over the telephone, we would rely on descriptive adjectives and verbal dimensions such as "the duct is rectangular, has a particular width and a particular height, and it has insulation." Once we had settled on the duct required and hung up the phone, we would then need to convey graphically in our drawing documents the decisions we had just made regarding that duct (and any others like it). In traditional 2D Design, this would mean translating dimensions and materials into corresponding lines, arcs, circles and/or blocks that represent the required dimensions and materials in the drawing. In AMEP, dimensions and other specifications are simply input into a series of fields and stored with the data for the object. This data remains accessible throughout the life of the object and the drawing via the object's properties. Therefore, the next time we phone the contractor and realize that circumstances on the site have forced us to specify a different duct size, rather than redraw the duct and manually adjust the adjacent fittings as we would in traditional 2D Design, we now simply re-access the properties for that duct and input the new values. Not only does the duct itself update because of this change, but the adjacent fittings connected to the duct update as well. Other linked views such as Elevations and Schedules will also receive the change. This is just one example of parametric design. Object parameters are always available for editing; data never needs to be recreated, only manipulated. Some principles of parametric design are as follows:

- **Draw Once**—In traditional 2D Design, each object needs to be drawn for each required view; therefore the same duct or lighting fixture may need to be drawn two, three or more times. With AMEP, objects need only be drawn once. They are then "represented" in each of the required views of Plan, Section and Elevation.

- **Progressive Refinement**—Complete or final design information is rarely known at the early stages of a design project. Changes occur frequently and often several times. In traditional 2D Design, it is easy to add new information. However, when major design changes occur, drawings must often undergo time-consuming redrafting. With AMEP, designs can be progressively refined over the life of the project. As new data is learned or design changes occur, object parameters may be adjusted appropriately without the need to erase and recreate the drawing. The objects are drawn once, and then modified and refined as required.

- **Style-Based versus Object-Based**—Most AMEP objects make use of styles. A style is a collection of object parameters saved in a named group. When styles are assigned to objects, all properties of the style are transferred to the object in one step. If the style parameters change later, all of the objects using the style will change as well. This is similar to the behavior exhibited by text and dimension styles in traditional AutoCAD. AMEP simply utilizes many more styles, and manages them and their relationships to objects much more completely than the corresponding AutoCAD counterparts. In some cases however, object parameters are assigned directly to the individual objects and *not* controlled by the style. Consider again the duct object as an example. The Duct System style would be used to designate the duct design parameters, such as the surface roughness, and air density. However, ducts can come in a variety of sizes; the roughness is a style-based parameter, while the size is an object-based parameter.

- **Live versus Linked**—Some drawing types in AMEP are edited directly on the "live" model data. This is the case with floor plans or live sections. The display system, discussed later in this Chapter, controls what displays on the screen as we are working in AMEP. Plan views are live. If changes are made to the objects within the plan, those changes will be seen simultaneously on the live model in all other live views. However, some drawing types, namely 2D sections, 2D elevations, and schedules, are "linked." Rather than being a live view of the model, these separate drawings function as "reports" of the model that maintain a link to the live model data. These views must be periodically refreshed to capture changes made to the model. Luckily, there are tools and settings to automate these tasks.

What is AMEP? Simply stated, AMEP is the version of AutoCAD you should be using if you are tasked with the creation of construction documents for the Mechanical, Electrical, and Plumbing (MEP) disciplines in a building design project. AMEP embodies decades of AutoCAD related drafting standard capabilities, provides functionality specifically suited to MEP documentation, and provides flexibility for the editing drawing elements more quickly and easily than manual drafting techniques. Moreover, expediting 2D drafting tasks is not the only benefit; AMEP elements are also 3D objects which make them suitable for interference detection, quantification, and visualization that is not possible in traditional 2D workflows (see Table 2-1).

TABLE 2-1—*TYPES OF CONTENT IN AUTOCAD, AUTOCAD ARCHITECTURE, AND AUTOCAD MEP*

AutoCAD	AutoCAD Architecture	AutoCAD MEP
Templates	Door Styles	Ducts
Dimension Styles	Wall Styles	Duct Fittings
Text Styles	Curtain Wall Styles	Flex Duct
Layer Settings	Window Styles	Duct System Styles
Palettes	Slab Styles	Pipes
Palette Tools	Stair Styles	Pipe Fittings
Blocks,	Structural Member Styles	Flex Pipes
Blocks	Wall Cleanup Styles	Pipe System Styles
and more Blocks	Space Styles	Conduit
	Zone Styles	Conduit Fittings
	Property Set Definitions	Cable Tray
	Property Data Formats	Cable Tray Fittings
	Schedules	Electrical System Styles
	Multi-View Blocks (Equipment and furnishings)	Multi-View Parts (MvParts)
	Multi-View Blocks (Tags)	Wire Styles
		Device Styles
		Panel Styles
		Label Styles
		and more...

AutoCAD provides the familiar user interface, command line, plotting capabilities, snaps and general foundation for AMEP. Everything that you have and are used to using in AutoCAD is in AMEP. AMEP is built on top of AutoCAD Architecture, which provides the Object

Modeling Framework (OMF) – this is a fancy way of saying that in AMEP, you are modeling objects that represent their real world counterparts (Ducts, Electrical Panels, etc.). The components you interact with in AMEP have specific common functionality, namely how their display is controlled. In addition to the familiar layering functionality of AutoCAD, the OMF provides the ability to define multiple representations of an object, and the Display Manager provides the ability to display a particular representation of that object depending on the configuration of the display. For example, in a 3D view, an Electrical Device such as a receptacle may be represented to look like a junction box with a faceplate, whereas in Plan it will be the familiar symbol comprised of a circle with two lines. Similarly, in a 3D view, a Duct will appear as an extruded solid, whereas in Plan all AMEP needs to draw are the edges of the duct as lines.

There are some fundamental differences between AutoCAD and AMEP objects. The most fundamental is that AMEP provides a Display System that allows MEP (and Architectural) objects to be represented differently under various viewing conditions. The most common conditions are 2D, 3D and Section views. Additionally, AMEP objects utilize automatic layering based on industry or office standards. Furthermore, a large catalog of pre-built content with components featuring specialized behaviors is also included. All of these items contribute to what is perhaps the biggest difference between traditional AutoCAD and AMEP; using AMEP effectively requires a change in work process. When you use AutoCAD, you draft individual drawings from individual lines, arcs and circles. There is nothing about two parallel lines drawn in the shape of a duct run that makes them behave like a duct or somehow imparts duct-like awareness upon them. In other words, AutoCAD cannot distinguish between two lines representing a duct from two lines representing a wall to two lines representing a parking space. All are drawn similarly, and to AutoCAD, they are just two lines on a layer.

In AMEP, you do not draft a duct, but rather you "model" a duct. By selecting an appropriate duct, and specifying the characteristics of the run, you build a representation (or model) simulating what will ultimately be built. In this approach you can work in a traditional 2D view or even work three-dimensionally if preferred. There are many advantages to this approach. In the act of laying out your ductwork, you choose type, size, and other characteristics. As you draw the run, it automatically appears on an appropriate layer. Further, you are able to view the resultant model in plan to see a traditional two-dimensional (i.e., two lines) representation of the duct and also view it three-dimensionally. Such three-dimensional views can be used to assist in understanding the design, making design decisions, and identifying clashes using the Interference Detection tool (Analyze ribbon, Inquiry panel). Since a duct or pipe or fixture object simulates the real-world object it is meant to represent, we are able to take advantage of not only the many characteristics of the object itself but also the relationships of elements to one another in the systems as a whole. Traditional AutoCAD offers none of these benefits. Its only strength by comparison is its familiarity and emulation of traditional workflows and procedures. However, once you have immersed yourself in the lessons of this book, you will find yourself as familiar with the AMEP workflow as you previously were at AutoCAD.

To begin a more thorough understanding of the whole of AMEP, the main drawing components available in the software for design modeling and documentation are listed in Table 2-2.

TABLE 2-2—*TYPES OF OBJECTS IN AUTOCAD MEP*

Object Type	Purpose	Key Differentiator from AutoCAD
Duct	Used to represent duct. Available in Rectangular, Round, and Flat Oval	3D components that automatically adjust interconnected components when grip editing.
Pipe	Used to represent pipe.	3D components that automatically adjust interconnected components when grip editing.
Plumbing Line	2D linework to represent plumbing systems.	Typical fittings insert automatically, and interconnected components adjust automatically when editing.
Conduit	Used to represent conduit wireways.	3D components that automatically adjust interconnected components when editing.
Cable Tray	Used to represent cable tray wireways.	3D components that automatically adjust interconnected components when editing.
Hanger	Used to represent structural elements for supporting duct, pipe, conduit, and cable tray systems.	3D components that can be laid out along a run of routed components.
Device	Used to represent building elements such as receptacles, lighting fixtures, switches, security devices, fire alarm devices, and	Devices provide the ability to define power requirements, and may be circuited to panels.

Object Type	Purpose	Key Differentiator from AutoCAD
	communications devices.	
Panel	Used to represent electrical distribution components that provide an origination point for circuits.	Provides a mechanism to circuit devices to panels to maintain electrical connectivity and computed load information.
Wire	Used to annotate schematic wiring between devices.	Tick marks are an embedded part of the wire, and wires may be used to facilitate the circuiting process. Automatically "break" when crossing other wires. (Note: Wires are *not* required for circuiting.)
MvParts	Used to represent general components.	Components with 2D and 3D representations.
MvBlocks (Tags) LA-2	Used to annotate any information about an element (i.e., elevation, circuit, type, etc.)	Automatically update if the underlying data changes.
Labels 12x3	Used to annotate information (typically size) along the length of Duct, Pipe, Conduit, and Cable Tray	Automatically update if the size or other property changes.
Panel Schedules (AutoCAD Tables)	Used to report connected and demand load information for Panel.	Data is extracted directly from the connected components, and may be updated without manually entering data into the schedule.
Schedules	Used to report type and other information about a group of objects.	Data is extracted Property Set Data in Property Set Definitions, and is reported in the schedule.
Schematic Symbol	2D schematic diagramming components.	Schematic components adjust when interconnected linework is modified.

THE DISPLAY SYSTEM

Early in our design careers, we learned the traditional rules of drafting. Those rules governed such things as what a plan or elevation drawing represents, how to create one, and more importantly, what to include and what not to include in making the drawing "read." Although there are accepted universal rules in place, part of the process involves personal style. Therefore, the rules need to be consistent enough to allow them to convey information reliably, and flexible enough to allow for stylistic variation. Amazingly enough, although CAD software such as AutoCAD has revolutionized the way design drawings are created, prior to AMEP, the software offered no specialized tools to assist the designer in achieving the unique graphical look required by architectural documents. Rather, lines were still painstakingly laid out one at a time as they had been in hand drafting, following the internalized prescriptions learned on the job. If a plan, section, and elevation were required to convey design intent, three completely separate drawings needed to be created and, more importantly, coordinated. The display system in AMEP addresses this situation by incorporating the *rules* of drafting directly into the software. Plans, sections, and elevations can now be generated *directly* from a single building model. This reduces rework and redundancy by requiring one set of objects, with three different modes of display (see Figure 2.1). The tools are flexible and fully customizable, so we can fully benefit from this powerful tool and still introduce the nuances of our own personal style into the process.

> CAD MANAGER NOTE: The best way to quickly understand the display system is to start with the drawing template files (DWT) provided with the software. Much of the display system has already been configured for typical design situations in these templates. For most firms, these templates can be used "as-is" or with minor customization. If you are new to AutoCAD MEP, begin with the out-of-the-box templates and display system configurations and then slowly begin customizing them to suit your firm's particular needs as required. This approach guarantees a complete understanding of the tools as you learn by example using suggested settings. The display system is complex, but it is also extremely powerful. In order to fully master the use of AMEP, it is important to become very comfortable with the display system.

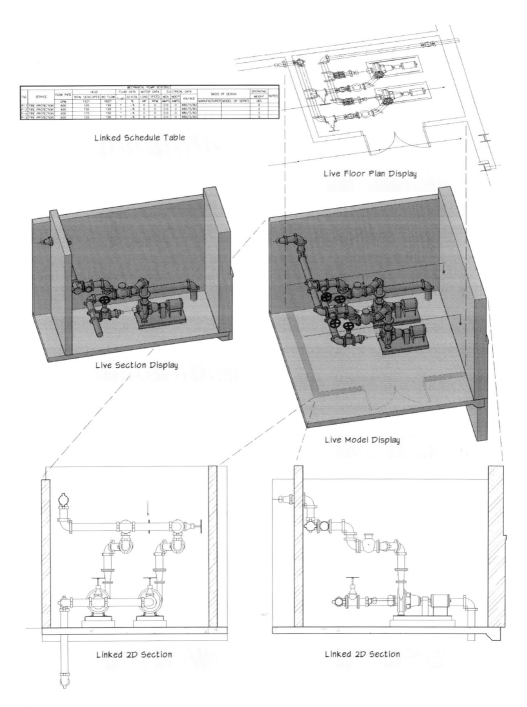

FIGURE 2.1

Generating plan, section, elevation and schedule views from a single model

THE DISPLAY SYSTEM'S RELATIONSHIP TO LAYERS

Objects unique to AMEP are referred to as "AEC Objects" (including objects from AutoCAD Architecture and Civil 3D). They include objects such as Ducts, Devices, Pipes and Schedules. Display control determines how AEC objects are displayed under different viewing conditions and circumstances. (Display control has no effect on AutoCAD entities such as lines, arcs and circles.) Layers are used as a global organizational tool for the management of drawing data, much like a drawing-wide categorization system. All objects (both AutoCAD and AEC) are placed on a layer when they are created. Display control supplements layers in helping you control what is seen and how it is displayed on the screen and in print. Each AEC object contains a series of components. The display control tools determine the display characteristics of each of these components. In some cases, the display properties of individual AEC object subcomponents are in fact handled by layers, although this is certainly not required (and in many cases not desired either). To summarize, Layers know nothing of Display; however, Display settings can optionally include Layers. Layers work on all entities, AutoCAD and AEC alike, but Display works only with AEC objects. Both can be used to control what is seen on screen and in print. Layers do this globally in an absolute way—they cannot respond to the condition of the drawing. Display settings can change if the condition of the drawing meets certain criteria. The Display system has been designed specifically with MEP and Architectural drawing needs in mind; Layers have not.

OVERVIEW AND KEY DISPLAY SYSTEM FEATURES

The display system offers many features and benefits:

- AEC objects display differently under various viewing conditions— Display control settings can dynamically change the display of a building model from plan to elevation to section or 3D model, with a simple change in the viewing direction on screen.
- Fully customizable—Configuration of the display system components and their individual object properties can be customized to suit specialized needs. Customization can be as simple as modifying a setting or two in the configurations provided in the default templates, or as complex as a completely custom-built solution tailored to a project-specific or office-wide need.

- **Understands the nuances of MEP and Architectural drafting**—Object components such as Lining (Ducts), Insulation (Ducts and Pipes), provide very specific control over how elements are displayed. Display by Elevation functionality provides the further ability to automatically hide or change the linework of elements at different elevations.

The Display System Tool Set

The display system tool set consists of a collection of interconnected components. It is important to understand some concepts and terminology related to each of these components before you begin to work with the display system.

- **Object/Subcomponents**—All AEC objects contain one or more subcomponents. These are simply the individual pieces of the object. A duct, for example, contains (among others) the following plan subcomponents: Contour, Insulation, Lining, and Center Line. Just like traditionally drafted AutoCAD entities (lines, arcs and circles), the entire object (the Duct in this case) will be assigned AutoCAD properties such as Layer and Lineweight, while the individual subcomponents may also receive their own individual properties through their object display settings.
- **Display Properties**—Display properties are the collection of display settings for a particular object. These include Visibility mode (On or Off), Layer, Color, Linetype, Lineweight, LTScale, and Plot Style. They also include many object-specific settings like Rise Drop. These are applied as "Drawing Default," "System Style Override", "Style Override" or "Object Override" level.
- **Drawing Default**—In the hierarchy of display settings, the Drawing Default settings come first and establish the baseline for a particular type of object: all walls, all stairs, etc., within a particular drawing.
- **System Style Override**—System Style Override affects a subset of drawing objects that are assigned to a particular System. Establishing the display settings at the object's system level provides a means to distinguish elements that serve a particular purpose (such as Chilled Water Piping vs. Condenser Water Piping or Existing Piping vs. New Piping). System Style-level settings

override the Drawing Default settings. Generally, however, the graphical differences are limited to settings on the layer associated with the system.

- **Style Override**—Style Override affects a subset of drawing objects that belong to the particular style in question only. Style-level settings override the Drawing Default and, if present, the System Style settings. In general, application of style-level overrides should typically be avoided.

- **Object Override**—Object Override affects only a single object selected in the drawing. Object-level settings override those of both the Drawing Default and, if present, the System Style and Style Override. In general, frequent application of object-level overrides should typically be avoided.

- **Display Representation (Display Rep)**—As dictated by the conventional rules of drafting, each type of object has one or more ways in which it may be drawn. Display representations control the behavior of objects under various drawing situations such as Plan and Elevation. Representations also control the specific display characteristics of an object's individual subcomponents from a particular viewpoint (see Figure 2.2).

- **Set**—Set describes which objects will display in a particular drawing situation, and in which mode—Display Representation—they will be displayed. A Set closely approximates a particular type of drawing, such as "Floor Plan" or "Building Section" (see Figure 2.2).

- **Display Configuration**—Configuration controls which Display Representation Set will appear on the screen, as determined by a particular viewing direction. Configurations can be "Fixed View" (same set appears regardless of current view direction) and "View Direction Dependent" (loads a different set based upon view direction). See Figure 2.2.

Putting It All Together

Summarizing how all of these components fit together, objects have one or more Display Representations (display modes). These Representations control the On/Off state, Layer, Color, Linetype, Lineweight, and Plot Style of each of the object's internal subcomponents. These can be assigned in one of four ways: by Drawing Default, System Style Override, Style Override, or Object Override. Most of these settings are available on the Display tab of the Properties palette. Individual objects and their appropriate Representations are grouped together in a Set. Sets are loaded into the viewport based on the conditions outlined in the

Display Configuration. Changing the viewport view direction will trigger the display of appropriate Sets based on these conditions. Figure 2.2 shows two examples of these relationships (a Device in elevation and in plan) in a flow chart style to illustrate the concept. The far bottom left shows the Views toolbar. Next to it is the Display Configuration menu on the drawing status bar (first shown in Figure 1.3). These two items together trigger the rest of the items illustrated. The screen captures in the middle of the figure are from the Display Manager, available on the Manage tab, Style and Display panel. The Display tab of the Properties palette is depicted in two scenarios on the right.

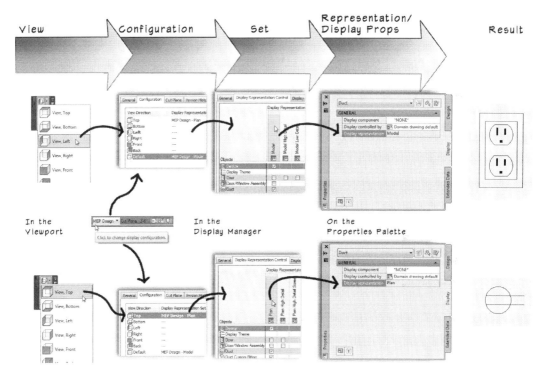

FIGURE 2.2

The relationship of components in the display system

The display system is dynamic. Changes made to its settings are immediately evident in the drawing. Many aspects of its configuration can be set up once and used from one drawing to the next, but certain settings will be in a constant state of flux as project needs dictate. The goal of the display system is to provide a unified tool set for controlling the myriad of display needs in architectural and MEP drawings. The goal here is twofold: reduce some of the tedium associated with drafting, and create a single building model that is capable of representing itself in all the ways necessary to display the drawings required in a document set. With the display system, it is no longer necessary to draw the same data two or three times to accommodate plans, sections, elevations and schedules. Features such as those offered by the display system are central to the concept of Building Information Modeling.

WORKING WITH THE DISPLAY SYSTEM

We begin our exploration of the display system by exploring the settings contained in the default template files. The display system is configured in much the same way in each of the out-of-the-box templates. The dataset used here was created from the *Aecb Model (US Imperial Ctb).dwt* template file.

INSTALL THE CD FILES AND LAUNCH PROJECT BROWSER

The lessons that follow require the dataset files provided for download with this book.

1. Launch AutoCAD MEP 2012 and then on the Quick Access Toolbar (QAT), click the Open icon.

    Refer to "Book Dataset Files" in the Preface for information on installing the sample files included with this book.

2. From the *C:\MasterAME 2012\Chapter02* folder, open the file *Chapter 02.dwg*.

    The current Display Configuration is MEP Design. This is the "standard" Display Configuration that may be used for most MEP design work.

We are going to load several different Display Configurations. Be sure to Zoom and Pan around the drawing a bit after each change to see the complete effect.

LOADING A DISPLAY CONFIGURATION

3. On the drawing status bar, on the right side, open the Display Configuration pop-up menu (it currently reads "MEP Design").

    A menu list of all Display Configurations appears (see Figure 2.3).

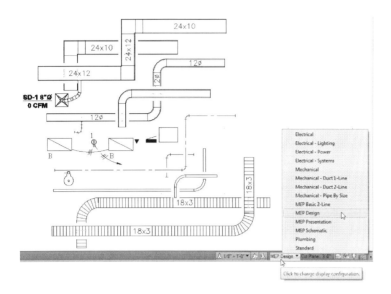

FIGURE 2.3

The Display Configuration pop-up menu on the Drawing status bar

4. Choose **Mechanical** (see Figure 2.4).

   Zoom and pan around the drawing to see the change. Notice that all the electrical objects are screened.

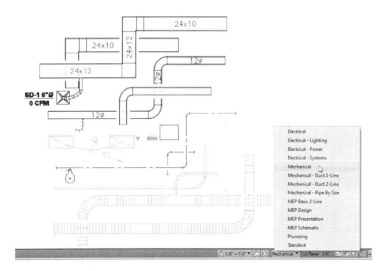

FIGURE 2.4

Load Mechanical from the Display Configurations list

Note: In the following images, tags hide with their associated objects; however, this won't happen in your drawing; you will configure this later.

5. From the Display Configuration pop-up menu on the Drawing status bar, choose **Electrical - Lighting** (see Figure 2.5).

   Zoom and pan around the drawing to see the change. Notice that power devices are hidden, and mechanical components (except for Multi-View Parts) are screened.

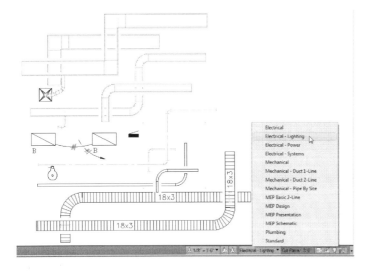

FIGURE 2.5

Load Electrical - Lighting from the Display Configurations list

6. From the Display Configuration pop-up menu on the drawing status bar, choose **Electrical - Power** (see Figure 2.6).

Note: In each case, you might notice a slight delay while the Display Configuration loads and makes the change to the drawing.

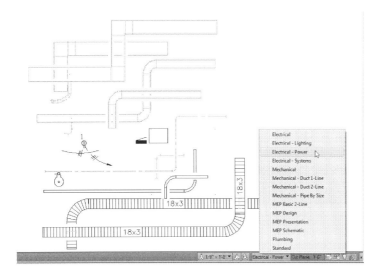

FIGURE 2.6

Load Electrical - Power from the Display Configurations list

Notice the change to the drawing: Lighting Fixtures and Air Terminals are hidden, and Duct and Pipe are screened.

7. From the Display Configuration pop-up menu on the drawing status bar, choose **MEP Basic 2-Line** (see Figure 2.7).

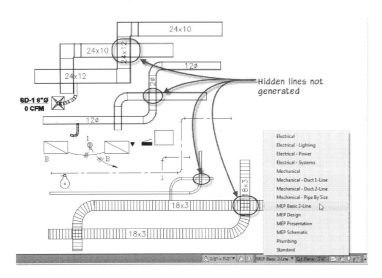

FIGURE 2.7

Load MEP Basic 2-Line from the Display Configurations list

In this configuration, the display is very similar to MEP Design where all components are shown non-screened. However, hidden lines are not computed or drawn, resulting in quicker drawing regeneration time.

As you can see, loading Display Configurations has a broad effect upon the entire drawing. The goal of a Display Configuration is to adjust the display of all AEC objects within the drawing to suit a particular design or presentation situation. Those situations range from floor plans to presentation plans. For most plan drawings, there may be no need to change the Display Configuration; however, you may periodically change the current Display Configuration to help you see things more clearly as you work. You may also find yourself fine-tuning the Display Configurations to suit your own particular workflows, and find yourself using layer control less frequently.

8. Reload the **MEP Design** Display Configuration.

    Notice the drawing's change back to the default Display Configuration.

By loading these Configurations you can begin to see some of the power of the display system; particularly when it comes to displaying the same information in varying graphical formats or in more or less detail.

SCALE DEPENDENT OBJECTS

In addition to the Display Configuration's dynamically modifying the representation of objects, the drawing scale can also have a significant impact on how the drawing appears. As with typical line types in AutoCAD, MEP elements may have subcomponents that are assigned a particular linetype or linetype scale. Changing the scale updates the objects accordingly. Similarly, certain elements are considered annotative, such as Labels, Tags, Wire components, and even the plan representation of some electrical Devices. Thus, the display of such objects dynamically updates to reflect the current drawing scale.

9. On the drawing status bar, on the right side, open the Annotation Scale pop-up menu (it currently reads "1/8" = 1'-0"").

10. A menu list of Annotation Scales appears (see Figure 2.8).

11. Select 1/4" = 1'-0"

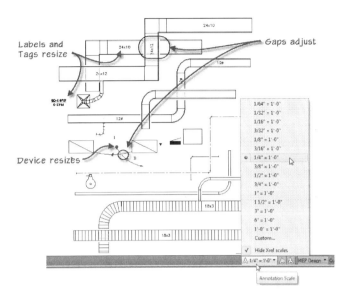

FIGURE 2.8
Select 1/4" = 1'-0" from the Annotation Scale list

Notice how the Labels, Tags, and Receptacles were resized according to the new scale. Note also how the hidden lines and gaps at Ducts, Pipes, Conduit, Cable Tray, and Wires adjusted accordingly.

12. Set the drawing scale back to 1/8" = 1'-0"

UNDERSTANDING HIDDEN LINE

A lot of what AMEP is all about is automating traditional rote drafting tasks; (many of which involve breaking lines to make them appear hidden when under other objects). Most MEP firms have standards documenting on what layer hidden lines should appear, what line type they should be, what color, how much gap should "halo" around crossing objects. Most firms have LISP routines and elaborate button and menu systems to automate the breaking and relayering process to help with consistent drafting output. However, most standards and LISP automation starts to break down when things need to be edited. Unless the Architect you work with gets things right the first time, there is plenty of editing involved in the creation of any set of MEP documents, and as such much grinding of teeth when the beautifully constructed collection of linework you spent the last few days working on needs some simple stretching to reconfigure a duct and pipe system layout in a corridor that the Architect just reconfigured. This is where AMEP really excels. Hidden lines are created when and where they are needed;

they update automatically when things move; settings ensure consistency including colors, layers, gaps, and linetypes; and, for the most part, you can make edits without entering a single command, selecting a single button, or picking a single menu item.

As we already introduced in this chapter, the display characteristics of hidden lines, such as layer, color, and lineweight, are all defined within the respective duct, pipe, conduit, and cable tray system styles. However, certain settings are controlled globally within a drawing.

13. From the Application Menu, choose Options (or right-click in the Command line).

14. Click on the MEP Display Control tab (see Figure 2.9).

FIGURE 2.9
MEP Display Control settings in AutoCAD MEP Options

The Crossing Objects Display section controls a variety of settings related to hidden lines. The Apply Gap inside/outside setting controls if "halo" gaps are created where two objects cross. When these boxes are toggled, the preview image in the upper right corner of this section updates to reflect the effect of the change. The gap indicated by the "B" dimension is defined by default as model length; however, this can be specified as a "plot" length and this is enabled by checking the "Apply Annotation Scale to Gap" option. For example, historically, if you needed a gap of 1/16" at a 1/8" = 1'-0" scale, you would need to refer to a chart hanging by your desk to realize that you need to plug in a value of 6". However, by checking the "Apply Annotation Scale" box and specifying 1/16", the computer can figure out how long in "model units" to make the gap, even if the scale of the drawing changes.

The "Save Hidden Lines in Drawing" option helps increase drawing open performance. This is done by caching the drawn hidden lines in the drawing, instead of re-computing how to draw them when the file opens. Of course, this results in larger files; this is one case where having a larger file actually improves performance!

The last option, "Disable Hidden Lines for Multi-view Parts" does just that (see Figure 2.10).

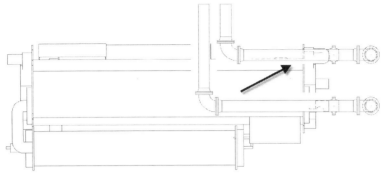

Pipes above and MvPart Chiller with "Disable Hidden Lines for Multi-view Parts"

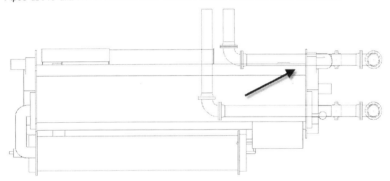

Pipes above and MvPart Chiller with "Disable Hidden Lines for Multi-view Parts" unchecked

FIGURE 2.10

Multi-view Parts can display with or without hidden lines

VIEW DIRECTION DEPENDENT CONFIGURATIONS

So far we have looked only at the changes that occur in a Plan (Top) view. Display control affects the display of objects in all views. A "View Direction Dependent" configuration is tied into the viewing direction in the drawing or the active viewport. This means that as the view direction is changed, the Display Configuration will automatically adjust what is displayed. View Direction Dependent Configurations contain a default setting and at least one other condition tied to any of the six orthographic views: Top, Bottom, Left, Right, Back, and Front. A maximum of six special conditions can be specified; one for each orthographic view. There is always a setting configured for "Default" which will be displayed either when an orthographic view does not have its own override setting, or if a viewing angle other than the six orthographic views (a 3D view) is chosen. The Display Configurations that are in the drawings templates that come with AMEP all have a Top and a Default Display Representation set. The Top view directions are all configured for Plan display, and the Default display configured such that viewing the drawing from any other direction results in a Model (3D) display. The net effect can be most easily demonstrated by orbiting from a Plan view to a 3D view:

15. If you do not already have it open, open the *Chapter 02.dwg* file.

16. The current orientation of the drawing is from a Top view.

17. Hold down the SHIFT key and the middle mouse button. Move the mouse to adjust the view into another orientation.

Note that as you orbit in the drawing (before you release the mouse), all the elements display in a 2D representation. This is how AMEP optimizes its drawing. All MEP elements, when viewed from the Top, typically have a simplified 2D representation. Once you release the SHIFT key, or the middle mouse button, the drawing regenerates as 3D objects. Once in a 3D view, as you orbit, the objects will display as 3D.

FIXED VIEW DIRECTION DISPLAY CONFIGURATIONS

A Display Configuration need not be view direction dependent. It can be assigned a single fixed configuration. In this case, changing viewing direction has no effect on the Display Configuration. Periodically, it is desirable to display a drawing in a 3D/Model view, even when displayed from the Top. There are no Display Configurations that do this by default, but creating it is easy. We will first change the Visual Style to Conceptual where the effect is even more pronounced:

18. On the View panel, from the View pop-up, choose View, Top.

19. From the Visual Styles pop-up, choose the Conceptual Visual Style (see Figure 2.11):

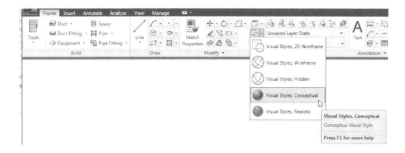

FIGURE 2.11

Load the Conceptual Visual Style

20. Note that even though we have specified the Conceptual Visual Style, in the Plan view, the elements still show as 2D elements with hidden lines.

21. On the Drawing Status Bar, make sure the MEP Design Display Configuration is current.

22. On the Manage tab, on the Style & Display panel, click Display Manager.

23. Expand Configurations, and select MEP Design .

24. Click Copy and then Paste (see Figure 2.12).

This will create a new Display Configuration named MEP Design (2).

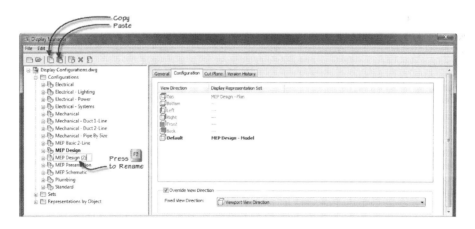

FIGURE 2.12

Copy the MEP Design Display Configuration and rename it

25. Select MEP Design (2), and then press the F2 key to rename to **MEP Design Model Only**.

26. On the Configuration tab, next to Top, choose *None* from the pop-up menu.

27. At the bottom, check the "Override View Direction" checkbox, and make sure "Viewport View Direction" is selected.

    Since the Default is set to MEP Design - Model, when viewing from any direction, even Top, the Model Display Set will be used.

28. Click OK.

29. On the Drawing Status Bar, select the MEP Design Model Only Display Configuration.

Notice that the shading now appears in plan. Try orbiting the model again (step 3 above) and notice that this time the view shows 3D geometry while orbiting. As an exercise for yourself, see if you can create a Display Configuration that shows only the "MEP Design – Plan" Display Representation Set, regardless of the view orientation. Name it **MEP Design Plan Only**. Please note that when this Display Configuration is active, you may still see some Multi-View Parts in a 3D representation. It depends how they were created.

USE THE DISPLAY MANAGER

The Display Manager is the primary interface for managing the display system. The Display Manager can be a little overwhelming, but fortunately you do not need to visit it often. In this lesson, you will learn the basics of the Display Manager. Our aim in the steps that follow is to give a glimpse of the potential this powerful tool offers. What is important here is attaining a general understanding of the concepts presented. In day-to-day production, you are unlikely to need to visit the Display Manager very often, if at all. The exercises in this topic and the next can safely be considered optional if you wish. More likely you will use the tools and techniques covered below starting in the "Edit Display Properties using the Properties Palette" topic.

UNDERSTANDING CONFIGURATIONS

Continue in the *Chapter02.dwg* file.

1. On the Manage tab, on the Style & Display panel, click the *Display Manager* button.

2. Expand (click the plus sign (+) next to) Configurations.

   Study the icons next to the various configurations in the list (see Figure 2.13).

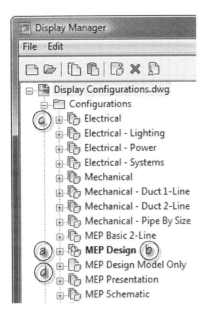

FIGURE 2.13

Understanding the icons used for configurations

a. **AutoCAD Drawing Icon superimposed on icon**—Indicates the default display configuration for the drawing. (All the icons have one or more turned corner pages, this one has the AutoCAD icon on the page). The default is used automatically when creating new layouts. To set the default for the drawing, right-click a configuration in the Display Manager and choose **Set as Drawing Default**. In the default templates, MEP Design is the default.

b. **Bold text**—Indicates that this configuration is active in the current viewport. In Figure 2.13, MEP Design is current in the active viewport.

c. **A single sheet of paper icon**—Indicates that it is a *Fixed View* Display Configuration. (See the "Fixed View Direction Display Configurations" topic above.)

d. **A stack of sheets icon**—Indicates that it is a *View Direction Dependent* Display Configuration. In Figure 2.13, all Configurations except MEP Design Model Only are View Direction Dependent. (See the "View Direction Dependent Configurations" topic above.)

THE CONFIGURATION TAB

3. Click on the MEP Design Display Configuration in the tree view at left to select it.

4. On the right side of the Display Manager, click the Configuration tab (see Figure 2.14).

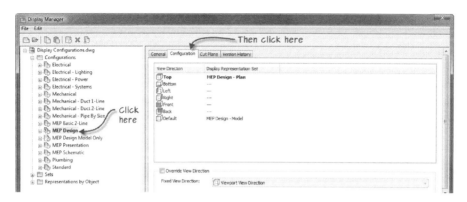

FIGURE 2.14

A View Direction Dependent Display Configuration has multiple entries

The Configuration tab lists each of the six orthographic views and also contains an entry for Default under the heading View Direction. Next to each view direction, a Display Representation Set can be specified. Display Representation Sets are described in detail below. When "Override View Direction" is selected, the Display Configuration is fixed. In this case, we can see that MEP Design is a View Direction Dependent Display Configuration, because Override View Direction is not checked.

There are a few other tabs for Display Configurations: General, Cut Plane and Version History. The General tab contains simply the names and description of the configuration. You can use this tab to edit those values if you wish. Many objects (Ducts, Pipes, Conduit, Cable Tray, etc.) use a cut plane when determining what to draw in plan displays. The Cut Plane tab establishes a single cut plane height that is used by such objects within the drawing. This "Global Cut Plane" is used to help synchronize all of these objects relative to a baseline height in the drawing. If you click the Cut Plane tab, you will see a Cut Plane of 3'-6". For now, we will leave the default settings as is. Version History is used in conjunction with the Project Standards feature.

In this particular configuration, the Top view direction is configured to use the **MEP Design - Plan** Display Set. (Display Sets determine which AEC objects display and how; see the

"Understanding Sets" heading below.) This indicates that whenever the drawing is viewed from the Top, the MEP Design - Plan Display Representation Set will be loaded in the viewport. Next to each of the Bottom, Left, Right, Front and Back view directions no Display Representation Set is specified. When configured to such view orientations, the Set specified for Default is used instead. Additionally, in non-orthogonal views, such as a SW Isometric View, the Set specified for Default will be loaded; in this case, **MEP Design - Model** (see Figure 2.15).

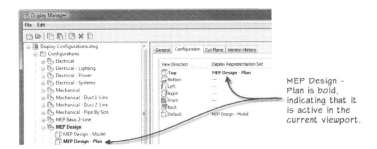

FIGURE 2.15
Examine each of the assignments in the MEP Design Configuration

Nothing in these settings, however, tells a Duct how to look different than a Device. What then, determines why the objects change their graphics when we switch? Let's have a look at Sets to begin to learn the answer.

UNDERSTANDING SETS

A Set determines which objects are displayed on screen and how they are displayed (meaning what Display Representation(s) each should use).

5. Continuing in the Display Manager, expand the *Sets* folder (see Figure 2.16).

   This will reveal all of the Display Representation Sets (or just "Sets") available in this drawing.

FIGURE 2.16
Expand the Sets folder

As noted above, the Plan set is the active display set in the drawing. This is evident both on the Configuration tab of the MEP Design Display Configuration and by the fact that Plan in the Sets list above is bold. The Model tab (model space) is currently set to Top view in the drawing.

Notice the icons next to each Set name:

a) **Bold Text**—indicates that this is the current Set. In Figure 2.16, MEP Design - Plan is current.

b) **A green checkmark in an icon**—indicates that the Set is "in use" by one or more Configurations. The lack of a green square indicates that these have been duplicated from default Sets. Several examples appear in Figure 2.16 such as MEP Schematic – Plan.

c) **A green checkmark in an icon with green square in the corner**—indicates that the Set is "in use" by one or more Configurations. In Figure 2.16, several appear this way. The green square indicates that these are default Sets that are auto-created by the software and cannot be purged.

d) **A small box without a checkmark and no green square**—means that the Set is not currently being used by a Configuration and can be purged. In Figure 2.16, Plan Diagnostic is not in use and can be purged.

e) **A green square in the corner**—Indicates that these are default Sets that are auto-created by the software and cannot be purged. In Figure 2.16, Reflected is not in use and can be purged.

CAD MANAGER NOTE: Like most named objects in AutoCAD, Configurations and Sets can be purged from a drawing only if they are unused. Highlight the item you wish to delete, and then click the Purge icon at the top of the Display Manager dialog box (it looks like a little broom). Be cautious, as no dialog will appear to confirm deletion. Purged items can be restored with Undo. Please note: The Sets with the small green square in the corner cannot be purged even if they are unused. These are default Sets that are auto-created by the software.

---

6. Click on the MEP Design - Model Set in the tree view at left to select it.

Just like the *Configurations* folder above, each Set has four tabs. The General tab serves the same purpose as it did for Configurations. You can use it to change Name and Description. (Default Sets, the ones with the small green square, cannot be renamed.) The Version History tab is used in conjunction with the Project Standards feature. There is also a Display Representation Control tab and a Display Options tab. Let's take a look at the Display Representation Control tab.

7. On the right side of the dialog box, click the Display Representation Control tab (see Figure 2.17).

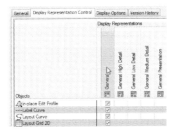

FIGURE 2.17
Access the display control settings of Sets

8. Scroll both horizontally and vertically on the right side of the dialog box.

If necessary, resize the Display Manager window to make the scroll bars appear.

Tip: It is usually a good idea to stretch out the size of this dialog as large as your monitor will allow.

On the left of the Display Representation Control tab is an alphabetical list of all AEC objects. At the top are all of the Display Representations (consolidated from all objects) within the current drawing. Notice that a Set is comprised of a collection of checkboxes. Objects that have one or more of their Display Representations checked in the selected Set are visible on screen or in the current viewport. Objects with no boxes checked are invisible.

Note: Not all Display Representations will be available for all object types. If there is not a box for a particular Display Representation, it means that Representation is not available for that object type.

For example, perhaps you have decided that Labels, while very valuable for plan view, are not desirable in 3D/model views. Using the power of Sets, you can simply turn off Label objects in the MEP Design - Model Display Set while leaving them displayed in the other Sets such as MEP Design - Plan.

9. Locate the Label Curve entry, and click on it.

    This will highlight the complete row in blue.

10. Clear the checkbox in the General column (see Figure 2.18).

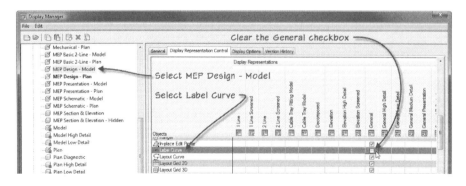

FIGURE 2.18

Turning off the Labels in the MEP Design - Model Set

11. Click OK to accept changes and return to the drawing.

There might be a slight delay as the drawing regenerates the current display. Notice that the Labels have disappeared (see Figure 2.19).

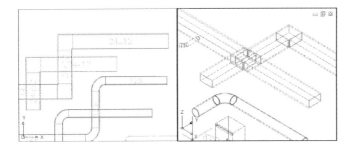

FIGURE 2.19
The Display Configurations file with Labels turned off in the MEP Design - Model Set

It is important to note that this technique is very different from simply freezing the object's layer. Had we frozen the layer, Labels would be invisible in *all* Display Configurations. Note that we only affected the MEP Design - Model Set which is used by the MEP Design configuration. Other Configurations—such as MEP Basic 2-Line—which use other Sets—such as MEP Basic 2-Line - Model—are unaffected. See for yourself; try setting other Display Configurations current:

12. Set the **MEP Basic 2-Line** Display Configuration current using the technique covered above in the "Setting a Display Configuration Current" heading.

Notice that the Labels appear here. Had you frozen the Layer, they would have been invisible here as well.

13. Return to the Display Manager (Manage tab).

14. Expand Sets again and select MEP Basic 2-Line - Model.

15. Highlight Label Curve on the right side again.

Scroll over and take note of the checkmark that appears in the General column.

Although we cleared the General Display Representation for Labels in MEP Design - Plan, there is still a checkmark in General the MEP Basic 2-Line Set. This is why Labels still display in 2-Line display.

16.With Label Curve still selected, clear General and instead put a checkmark in the General Screened column (see Figure 2.20).

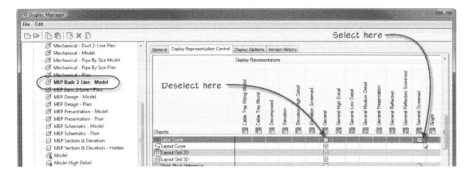

FIGURE 2.20
Making Labels display using the General Screened Display Representation

17.Click OK to accept changes and return to the drawing.

Labels now display using the General Screened Display Representation when the MEP Basic 2-Line Display Configuration is set Current, and the view orientation is 3D (see Figure 2.21).

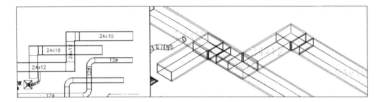

FIGURE 2.21
Labels display screened

This simple exercise shows the versatility of the AMEP object display. With the simple selection of a different display mode in the current Set (or none at all), the drawing can take on a dramatically different look. Of the two techniques shown here, turning off the display of Labels in Model views may be a better option than turning off the associated layer.

The Display Options tab provides the ability to filter the display of certain classifications of objects. In particular, Devices and Multi-View Parts comprise a variety of different building components. Multi-View Parts provide a wide variety of mechanical, electrical, and plumbing content. Devices provide a variety of content for lighting, power, fire alarm and other low-voltage systems. As we saw previously, selecting the Electrical - Power Display Configuration will hide lighting devices, and selecting the Electrical - Lighting Display Configuration will hide power devices. This functionality is controlled by the Classification Filter section of a Display Set's Display Options. Each Display Set can have its own unique set of object classifications toggled off (see Figure 2.22).

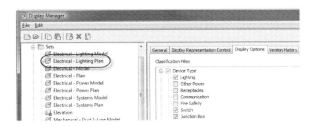

FIGURE 2.22
The Electrical - Lighting Plan Set displays only Lighting, Switch and Junction Box Devices

You may have noticed earlier that when selecting the Electrical - Lighting or Electrical - Power Display Representations the associated tags did not hide when their objects were hidden. This is easy to configure, but first we need to create a Classification Definition for Multi-View Block References:

18. On the Manage tab, on the Style & Display panel, click the Style Manager button.

19. Expand Multi-Purpose Objects.

20. Right-click on Classification Definitions, and choose **New**.

   This will create a new Classification Definition named New Style.

21. Select the New Style.

22. On the right side of Style Manager, click the General tab. Rename the Classification Definition to **Tag Type**.

23. Click the Applies To tab.

24. Check the box for Multi-View Block Reference.

25. Click the Classifications tab.

26. Click the Add button three times.

This will create three new Classifications named New Classification, New Classification (2), and New Classification (3) (see Figure 2.23).

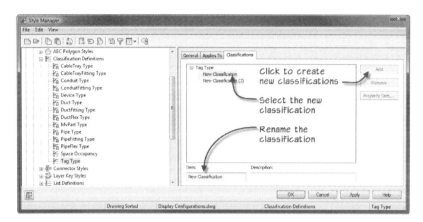

FIGURE 2.23

Use the Add button at add new Tag Type classifications

27. Select New Classification and, in the Item box, rename the classification to **Lighting**.

28. Rename New Classification (2) to **Power**.

29. Rename New Classification (3) to **Air Terminal**.

Next, we need to assign these classifications to the respective tags.

30. Under *Multi-Purpose Objects*, expand Multi-View Block Definitions.

31. Select Aecb_Device_Circuit_Tag.

32. On the Classifications tab, click the browse button next to Tag Type.

33. Select Power, and then click OK.

34. Select Aecb_Light_ID_Tag and, repeating the process, assign it to the Lighting Classification.

35. Select MyAirTerminalTag and, repeating the process, assign it to the Air Terminal Classification.

36. Click OK to close Style Manager.

Now, we can modify the Electrical - Lighting Plan and Electrical - Power Plan Display Sets to hide the unnecessary tags.

37. On the Manage tab, on the Style & Display panel, click the Display Manager button.

38. Expand Sets, and select Electrical - Power Plan.

39. On the Display Options tab, scroll the Classification Filter box to find Tag Types, and uncheck Lighting and Air Terminal.

40. For the Electrical - Lighting Plan Set, uncheck the Power and Air Terminal Tag Types.

41. Click OK to Close Display Manager.

42. Change the Display Configuration to Electrical - Power.

    Notice that the lighting fixture and associated tag are now hidden.

43. Change the Display Configuration to Electrical - Lighting, and note the receptacle and associated tag are hidden.

    As an additional exercise, configure the Mechanical - Plan Set to hide the Lighting and Power Tag Types.

Notice that the wires are not yet classified as lighting vs. power. However, if you were to classify the wires using the above methods, the wires would not quite work as expected. Even if you were to hide the "Power" wire, the "Lighting" wire would still show the gap where the power wire crosses it. Typically, however, lighting and power plans are in separate drawings, and as such, the wires don't *interact* to cause gaps in one another when lighting and power plans are XREFed together, so the method outlined above may be used to avoid using layers to hide such elements.

Certain elements are configured to be not visible by default. For example, if you change the view orientation to a 3D view, certain annotative elements, such as tags and wires are no longer visible.

44. Open Display Manager, and in the tree view, expand Sets, and select the MEP Design - Model Set.

45. On the Display Representation Control tab, scroll down to Wire.

46. Scroll across, and note that none of the checkboxes are checked in the Wires row. Thus, no Wires show up in a Model view.

47. Scroll up and find Multi-View Block Reference.

48. Scroll across, and note the box in the Model column is checked, but the other boxes are not checked.

A Tag is a Multi-View Block, but doesn't show up in a model view. This is because Multi-View Blocks (and other objects) can have multiple representations, and the display system is configured to only show certain Representations in certain Sets. The next topic discusses these Display Representations in more detail.

> As an additional exercise, create a Classification that applies to Label Curves with types for Duct, Pipe, Conduit, Cable Tray, and Plumbing Line. Then, assign the Classifications to the Label Styles in the drawing, and configure the Display Sets accordingly.

UNDERSTANDING DISPLAY REPRESENTATIONS

Objects in AMEP have many different display modes. These are called "Display Representations" (Reps). Each Representation corresponds to a particular drawing type such as Plan or 3D, or a level of detail such as 1-Line or 2-Line.

49. Return to the Display Manager.

50. Expand (click the plus sign (+) next to) Representations by Object (see Figure 2.24).

> This will reveal the complete list of AMEP objects (including ACA objects).

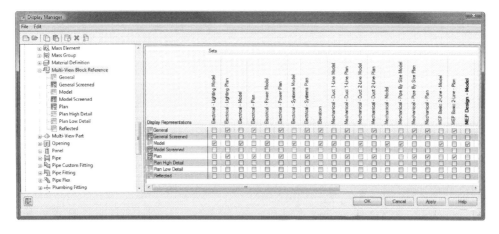

FIGURE 2.24

Expand the Representations by Object list

51. From the list of objects in the tree view at left, select Multi-View Block Reference.

This will reveal a list of all the Display Representations available for this object type in a column on the right side with all the available Sets listed across the top. Checkboxes again appear, indicating specifically which object Representations are utilized in each Set.

For the Multi-View Block Reference, note that the General and Plan Display Representations are used in Plan, but not Model Sets. The Model Display Representation is used in Model Sets, but not Plan sets. If we were to dig into the definition of the tags, we would find that they only have blocks associated with the General Representations; thus, they are not visible in 3D/Model views because they have no blocks associated with the Model Representations. If desired, however, one could associate a block with the Model Representation of the tag to make it visible in 3D/Model views.

An icon appears next to each Display Representation. AMEP includes several "built-in" Display Representations. These are part of the core program and can be configured but not renamed or deleted. Additional user-defined Representations can be created to meet the needs that are not addressed by the default Reps. You can create user-defined Representations by duplicating any of the default Reps. User-defined Representations can later be renamed and, provided they are not being used, deleted. The General Screened Rep for Label Curves (and several other object types in this drawing) are in fact user-defined Reps (see Figure 2.25).

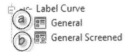

FIGURE 2.25

Icons differ for default permanent Display Representations and user-created ones

a) A "properties" icon—Indicates that it is one of the default (permanent) representations.

b) A "properties" icon with a portrait icon in the corner—Indicates that it is a user-defined representation.

52. Click Cancel to return to the drawing without making changes.

CAD MANAGER NOTE: To create a user-defined Display Representation, you must duplicate an existing one that is similar to the one you wish to create. Select the Rep that you wish to duplicate, right-click it and choose **Duplicate**. Give the new Rep a unique name. As with all named objects, choose your name carefully. Once you have created the custom Display Rep, you can configure its parameters and assign it to a Set. Usually, you will also create a custom Set and Configuration along with your new Rep, but this is not required. See the steps that follow for information on building custom Sets and Configurations.

When you select a particular Display Representation on the left side tree view, the right side of Display Manager shows the display components for the object's selected Display Representation. The Label Curve is a very simple object, and contains only a single display component, called the Label. Other object types, such as a Duct, are much more complex, and may contain dozens of display components. The next section will explore display components in detail.

EDIT DISPLAY PROPERTIES

The parameters of a Display Representation control all of the individual subcomponents of each AEC object. For instance, Ducts have contour, insulation, and lining components, among others. When the display by elevation functionality is enabled, each display component has multiple instances for Below, Low, High and Above conditions (see Figure 2.26).

Note: The display by elevation functionality is described in Chapter 13.

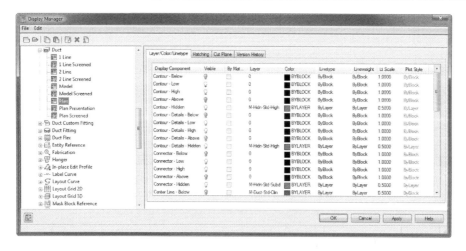

FIGURE 2.26

The Duct Plan Representation contains additional Display Components when Display By Elevation is enabled.

The specific subcomponents that an object type contains may vary for each of its Representations. For instance the Duct objects have Hidden components which are visible when Ducts cross one another at different elevations. However, these components are accessible only within Display Representations intended for Plan views. In the model views and basic 2-line views, it is not necessary to generate hidden lines, and as such, the associated Display Representations do not have these Components. This means that, although the Configurations and Sets determine which objects are visible and invisible on screen, it is the Display Representations that determine exactly how those objects will be drawn graphically. If we wish to edit the way a particular object or objects in the drawing behaves, we often find it easier to edit the object directly. In this case we will work with Duct objects.

1. Make the MEP Design Configuration current and then select the one of the 24x12 duct segments.

2. Open the Properties palette (press CTRL + 1) and then click the Display tab.

3. At the bottom left corner of the palette, click the Select Component icon.

4. Click on the outside edge of one of the 24x12 duct segments (see Figure 2.27) (you may want to zoom in close).

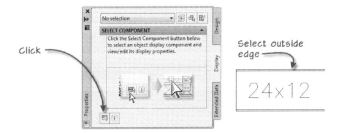

FIGURE 2.27
Use the Select Component icon to select the Duct Insulation hatching

Notice that the palette will reflect the selection of the duct insulation by showing that the Insulation component is selected at the top in the Display component list. Beneath this, all the parameters of the Insulation component such as its visibility, color, material, etc., are listed.

5. Change the color to Color 8.

A confirmation dialog will appear indicating that you are modifying the domain drawing default display level—click OK (See Figure 2.28).

6. Change the Linetype to HIDDEN2.

Notice that the insulation on all the Ducts has been updated to reflect the new color and linetype specifications. Notice, also, that this has happened for all sizes of the supply and return ducts, as well as the fittings. This was what the confirmation dialog was indicating. When you edit display properties at the drawing default level, all objects of that domain are affected, in this case affecting Ducts, Fittings, as well as Custom Fittings and Flex Duct.

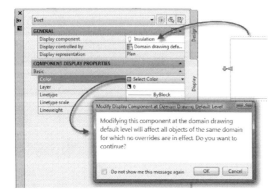

FIGURE 2.28
Change the color of the Display Component on the Properties palette.

7. Press ESC to deselect the Duct or right-click and choose **Deselect All**.

APPLY A STYLE OVERRIDE

Edits to Representations can be applied at four levels: the drawing default, system style, style or object level override. Sets (discussed above) on the other hand can only be edited in the Display Manager and their effects are global. Let's explore the next level of display manipulation: the system style override. Suppose that we wanted to make the insulation on Supply Duct a different color.

8. Select the 24x12 supply duct connected to the air terminal.

9. On the Duct ribbon tab, on the General panel, click Edit System Style button.

10. In the Duct System Definitions dialog, on the Display Properties tab, click the checkbox in the Style Override column for the Plan Display Representation (see Figure 2.29).

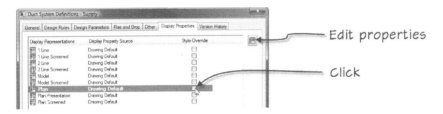

FIGURE 2.29
Add a Style Override to the Supply Duct System Definition

      11. Click the color swatch in the Insulation Display Component row (see Figure 2.30).

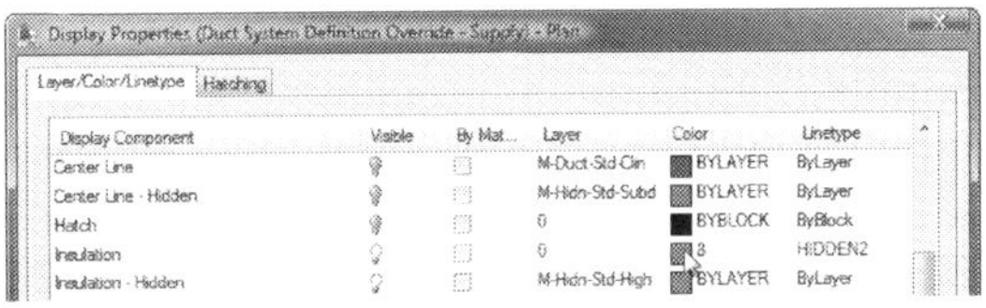

FIGURE 2.30
Modify the color of the Insulation Display Component

      12. Select Red, and then click OK three times to return to the drawing.

Note that the insulation for all the supply ducts is now red, while the return duct system still shows insulation in color 8. Now we will apply a Style override.

      13. Select one of the 24x12 Ducts (doesn't matter if it's supply or return).

      14. On the Duct tab, on the General panel, click Edit Duct Style button (see Figure 2.31).

FIGURE 2.31
Click the Edit Duct Style button

Note that the Duct Style window indicates 24x12 in Rectangular duct. All ducts of this shape and size (which defines the style) will be modified (see Figure 2.32).

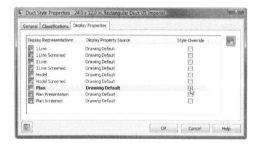

FIGURE 2.32
Add a Style Override to a Duct segment style

15. On the Display Properties tab, Plan row, click the Style Override box.

16. Click the color swatch in the Contour row.

17. Select red, then click OK three times to return to the drawing.

Note that all 24x12 ducts, regardless of system, are now Red. This is what happens when you apply a Style Override to any type of segment (Duct, Pipe, Conduit, Cable Tray). As you will see later in this chapter, a style is a collection of parameters that you apply to an object or objects in your drawing. If you edit the style, the edit will apply to all objects associated with that style. This is why several Duct segments changed even though you selected only one to make the edit.

18. Press ESC to deselect the Duct, or right-click and choose **Deselect All**.

APPLY AN OBJECT OVERRIDE

So far we have seen three of the four levels available to Display Representations. When we changed the insulation to color 8 and HIDDEN2 linetype, it applied to all duct domain objects in the drawing regardless of their system or style. Next we applied a System Style Override to the supply Ducts to color the insulation red. The return ducts did not receive the change, because the change was only applied to the supply system. We then applied a Style Override to 24x12 Ducts. This applied to all ducts of this Style regardless of the system. We will now apply an Object Override to a Duct. When you apply an Object Override, it makes that object completely unique—no longer using the parameters of the drawing default or any attached style overrides.

19. Select any of the 24x12 supply Ducts.

If you closed the Properties palette, right-click, choose Properties and then click the Display tab.

20. Click on Display controlled by setting and then choose **This object** from the pop-up list.

21. In the confirmation dialog, click OK.

22. Toggle on the lightbulb for the Hatch component from the Display component drop list (see Figure 2.33).

At this point you have seen several of these confirmation dialogs. If you wish, you can select the "Don't show me this message again" checkbox to avoid seeing these messages in the future. However, leaving them enabled can serve as a helpful reminder of the level at which the display properties are applied. The choice is up to you.

FIGURE 2.33
Toggle on the Hatch Display Component on the Properties palette

Notice that this time the change applies only to the Duct you selected. At this point, this Duct will no longer respond to display edits made to the system or the style. Furthermore, none of the supply Ducts will respond to edits made to the drawing default of Ducts.

There is one additional point which is very important to note. The four levels of display that we are witnessing here occur for *each* Display Representation. This means if you change the Display Configuration of the drawing to something other than its current MEP Design, everything potentially will change. This means that you can use Drawing Default for some Reps, System Style Override for others, Style Override for others and Object Override for still others—all on the same object and without the need to change layers!

23. From the Drawing status bar, choose **MEP Basic 2-Line** from the Display Configuration pop-up.

Study the results

Notice that all Ducts, including the ones we modified, now display the same again. This is because in the other configurations—which use different Display Sets (see above), which in turn use alternate Representations for Duct objects—have no style or object overrides. If you change the Display Configuration back to MEP Design, the edits you made previously will reappear (see Figure 2.34).

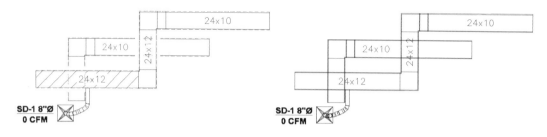

FIGURE 2.34

Overrides in the MEP Design Configuration (left), and unmodified in the MEP Basic 2-Line Configuration (right)

CAD MANAGER NOTE: Much of the configuration of the display system is best left in the realm of office standards. Poll your team and try to determine which Configurations and associated Sets they need. Try to use the out-of-the-box offerings as much as possible. Set up and save any custom displays in your office standard templates. (Using the default templates as a starting point is an excellent way to begin.) You should seek to create the majority of custom user-defined Display Representations required by your team as well. However, individual configuration of the properties of each Display Representation will most likely be tweaked on a regular basis by the users as project needs dictate. Procedures and reasons for doing so are covered throughout the remainder of this book as circumstances demand. If you decide that you need to customize your templates and include custom Display Configurations and Sets, the topic is covered in greater detail in *Autodesk Architectural Desktop: An Advanced Implementation Guide* by Paul F. Aubin and Matt Dillon.

You can experiment further if you wish. Repeat any of these procedures to get a better sense of how all the parts fit together. Generally, it will not be necessary to apply overrides to Systems, Styles, or Objects. However, there are instances when this functionality is just the right tool for the job... such as hatching a duct to indicate a sound attenuator, instead of manually drafting a hatch.

OBJECT STYLES

Virtually all AEC objects use styles to define global object parameters. These can include both physical and display parameters. (We saw examples of style-based display parameters in the previous exercise.) Much like text and dimension styles in core AutoCAD, object styles control all the formatting and configuration of the object. Using styles is a powerful way to control the behavior of objects and quickly make global changes when the design changes. For instance, at any time during project design, a user could simply go back to a style describing a Device and make a modification, such as changing the load from 42va to 32va, or, as we saw above, turn on or off certain components. The change would be reflected throughout the drawing on all wall objects that were associated with that Device style. In most cases, it is best to think of styles as "types" in the same way we commonly distinguish fixture types and diffuser types in a construction document set. Each type needed in the CD set would therefore have a corresponding object style.

WORKING WITH STYLES AND CONTENT BROWSER

Following is an overview of key object style features:

- Editing parameters in the style globally updates all objects within the drawing referencing that style.
- Styles can control physical parameters, display parameters and data property set information used for schedules.
- Styles can be shared between drawings using the tools available in the Style Manager, Tool Catalogs and Tool palettes.

Collections of similar styles are saved in individual drawing files. These drawings can be part of a particular project or a central library accessible to all people in the office. If you create a style in one drawing and wish to use it in another, you can do so easily by saving the style to a tool catalog. This catalog can be accessed by other users and its tools used in any drawing file, making your style easy to use across multiple project files or throughout the entire office. You can also access a large collection of out-of-the-box tool catalogs provided with AMEP. Tool catalogs are accessed from the Content Browser. Content Browser is a Web browser-like tool that is designed specifically to browse, store and access AMEP Styles and Content items. Before going to Content Browser, it is useful to determine what Styles, if any, are contained within the current drawing file.

STYLES IN THE CURRENT DRAWING

All MEP and Architectural elements are based on Styles. Some types of objects, such as Devices and Schematic Symbols can be easily changed from one style to another simply by modifying the Style on the Properties palette. It is easy to determine on what style a particular object within the drawing is based. To do so, we simply click to select the object in the drawing editor and then view the listing for Style on the Properties palette.

1. Continue in the *Chapter 02.dwg* file.

If you did not complete the previous tutorial, follow the steps in the "Install the CD Files and Load Sample File" heading above to install and load the dataset.

Before continuing, load the MEP Design Display Configuration. If you viewed the model in 3D, be sure to return to **View, Top** and **Visual Styles, 2D Wireframe** from the View panel to reset to a simple 2D view.

2. Select one of the 2x4 lighting fixtures.

3. Right-click and choose **Properties**.

4. On the Properties palette, click the Design tab.

   Toward the top, take note of the Style name and the preview image.

5. With the Device still selected, click on the Style preview image on the Properties palette.

   This will activate window to select a different style (see Figure 2.35).

FIGURE 2.35

View the list of Device styles contained in the current drawing

All of the styles on this list are currently available within the current drawing; this is indicated by the Drawing file <Current Drawing> at the top. With Devices, Panels, Schematic Symbols and Plumbing Fittings, you can select a Style from any of the content files included on this list.

CAD MANAGER NOTE: The list of Drawing files for Devices, Panels, Schematic Symbols and Plumbing Fittings is generated based on the files in the paths specified in the "Options" dialog, on the MEP Catalogs tab. There you can view or edit the "Style-Based Content Paths."

6. From the Drawing file list, select Lighting - Fluorescent (US Imperial)

7. From the Category list, select Emergency.

8. Click the 24x48 Emergency Red (the name may be truncated, but if you hover the cursor over a style, the full name will show up in a tool tip.

Note the change to the Device within the drawing.

9. Undo the change by repeating the same steps, and then choosing the original name, or press CTRL + Z to undo.

   The same technique could be used for any style-based object in the drawing. Try it on the Schematic Symbol if you like. Be sure to undo any changes when you are done experimenting.

AMEP relies heavily on other Styles that aren't necessarily specific objects. For example, Systems are based on Styles, as are Schematic Lines and Plumbing Lines. To change the System or Line style of an element, the Style must exist within the drawing.

10. Select one of the 12" diameter pipes.

11. On the Properties palette, Design tab, select the System drop list (see Figure 2.36)

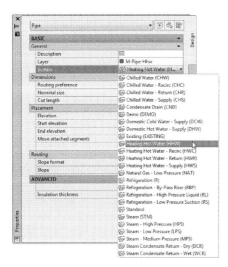

FIGURE 2.36
View the list of Pipe System Styles contained in the current drawing

If you needed to change the Pipe's System to Fire Protection - Wet Pipe, you would first need to import the System Style or create it. (See the "Work with the Content Browser" heading next for more information on importing styles, and refer to Chapter 9 for more information on creating and editing styles.) You can choose a different System Style from the list to change the System of multiple segments and fittings.

CAD MANAGER NOTE: AMEP provides many systems that aren't in the default template files. Assuming the default installation paths, these can be found in:
C:\ProgramData\Autodesk\MEP 2012\enu\Styles\Imperial\System Definitions (US Imperial).dwg

WORK WITH THE CONTENT BROWSER

12. On the Home tab, on the Build panel, click the drop-down button on the Tools button.

13. Choose the Content Browser tool (or press CTRL + 4).

The Content Browser window will open. Content Browser is very similar to a Web browser. It is organized in two panes. Navigation is on the left and the content is displayed on the right. Standard Web browser navigation buttons (Back, Forward, Refresh, etc.) are arrayed across the top of the left pane. Action buttons for creating new catalogs and such are placed at the bottom of the left pane. In the Library home (the main page) there are two such buttons: One creates a new catalog and the other modifies the Content Browser view options (see Figure 2.37).

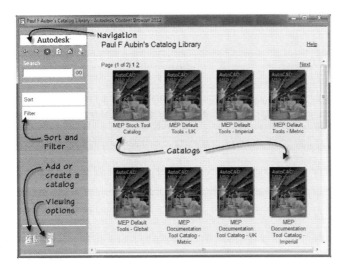

FIGURE 2.37
The Content Browser main home page (specific catalogs vary depending on installed options)

AMEP ships with a vast library of pre-made content. Content items include object styles, symbols and annotation routines. All of these items can be accessed from the Content Browser. The following is a list of each of the catalogs provided with a brief description.

> Note: The exact list of Catalogs available in your Content Browser varies depending on the options chosen during installation of AutoCAD MEP. If you performed a Full Install of the US English version, the following Tool Catalogs should be available:

- **MEP Stock Tool Catalog**—Contains all the standard tools that come with AMEP. There is a tool for each MEP object type and several other core commands. These tools do not reference any particular unit (imperial/metric) system. In most cases, however, you can access the functionality provided by these tools directly on the Ribbon.
- **MEP Default Tools – Imperial/Metric/Global**—Each contains six categories corresponding to the six basic tool palette groups installed in the standard out-of-the-box installation. Each of these categories contains a

backup copy of the installed palettes belonging to the associated units of US Content (imperial/metric), as well as international metric content in the Global library. The palettes contain a mix of stock tools, styles and documentation content.

- **MEP Documentation Tool Catalog – Imperial/Metric/Global**—Each contains a variety of Schedule and Tag styles for a variety of associated model elements.
- **My Tool Catalog**—This catalog is empty and ready to customize for your own use. Use it to store custom tools or your favorites from the standard tools.

AMEP also provides all the content and tools that come with AutoCAD Architecture. A description of each library is provided below:

- **Stock Tool Catalog**—contains all the standard tools that come with AutoCAD Architecture. There is a tool for each architectural object type and several other core commands. These tools do not reference any particular unit system.
- **Sample Palette Catalog – Imperial**—contains four categories corresponding to the four basic tool palette groups installed in the standard out-of-the-box installation. Each of these categories contains a backup copy of the installed palettes belonging to the Imperial units Installation. The palettes contain a mix of stock tools, styles and documentation content.
- **Sample Palette Catalog – Metric**—contains four categories corresponding to the four basic tool palette groups installed in the standard out-of-the-box installation. Each of these categories contains a backup copy of the installed palettes belonging to the Metric units Installation. The palettes contain a mix of stock tools, styles and documentation content
- **Design Tool Catalog – Imperial**—contains tools that refer to all the architectural object styles and AEC design content in Imperial units.
- **Design Tool Catalog – Metric**—contains tools that refer to all the architectural object styles and AEC design content in metric units.
- **Documentation Tool Catalog – Imperial**—contains tools that refer to all the documentation object styles, such as schedule tables and area calculation objects, as well as AEC documentation content in Imperial units.

- **Documentation Tool Catalog – Metric**—contains tools that refer to all the documentation object styles, such as schedule tables and area calculation objects, as well as AEC documentation content in metric units.
- **Visualization Catalog**—contains a large collection of material definitions, lights and cameras for use in rendering.
- **Content and Plug-ins Catalog**—contains links to Web sites containing styles and other utilities and plug-ins.

To open a Catalog, simply click on it. You can then use the links at the left (or right) to navigate through the categories and palettes to the tools. Once you click a Catalog, you can return to the Catalog Library list by clicking on the Home icon on the Navigation bar.

ACCESS A TOOL CATALOG

The following procedure describes how to access content within one of the catalogs.

14. With Content Browser open, click on the MEP Documentation Tool Catalog – Imperial (see Figure 2.38).

An introduction page will appear on the right. Read through it and then proceed to the next step.

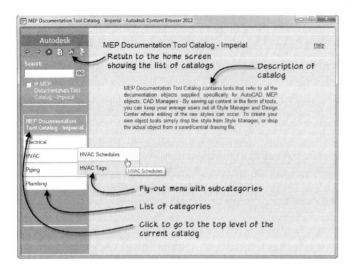

FIGURE 2.38

Navigating through a Catalog

15. Hover your mouse over HVAC.

    A submenu will appear.

16. Click on the HVAC Schedules item in the submenu.

Here you will find several pages of HVAC-related schedule styles ready to drag and drop into your drawings or onto tool palettes. You can scroll through each page if you wish. We are going to drag and drop one of these Schedule style tools into the drawing; it will import the schedule style and associated property sets, and then execute the Schedule command.

17. Locate the Schedule style named Air Terminal Devices.

Each of the tools located in the Content Browser uses an Autodesk technology called "iDrop." iDrop is a Web-based technology that allows content to be dragged into drawings from Web pages, complete with all required data, associated parameters and files. In the case of AEC Styles, all style, material and schedule data properties will be included.

18. Place your mouse over the small eyedropper icon; click and hold down the mouse button and then drag the Air Terminal Devices style tool into the drawing window (see Figure 2.39).

    Wait for the eyedropper to appear to "fill up" before dragging.

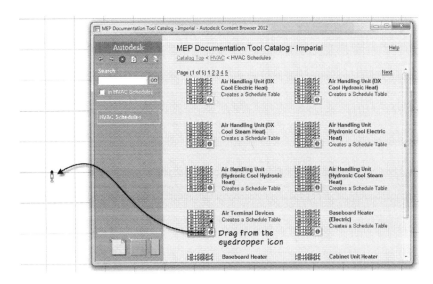

FIGURE 2.39

Using iDrop to drag and drop styles into the drawing

The Schedule command will begin, and all the parameters on the Properties palette will be set to match the tool just dropped.

19. At the "Select objects or Enter to schedule external drawing" prompt type ALL, and then press ENTER twice.

20. Pick a point to place the schedule, and then right-click (or press ENTER) to accept the default scale for the schedule.

Even though the schedule is for Air Terminals, the style is misconfigured to apply to all MvParts, which is why there are three rows in the schedule, instead of one. The following steps will correct this.

21. Select the schedule, and on the Schedule Table ribbon tab, on the General panel, click the Edit Style button.

22. On the Applies To tab, in the Classifications box, expand MvPart Type.

Since no Classifications are checked, the schedule effectively applies to *all* Multi-View Parts.

23. Check Air Terminal, and then click OK.

24. With the Schedule still selected, on the Schedule Table ribbon tab, on the Modify panel, click the Update button.

The Schedule updates, and correctly filters on Air Terminals only.

25. With the Schedule still selected, on the Schedule Table ribbon tab, on the Modify panel, click the Add All Property Sets button.

The Schedule updates again, populating with default values based on the Property Set Definitions attached to the Air Terminal objects. Refer to Chapter 16 for more information on the details of how Property Sets and Schedules work in AMEP.

CONTENT LIBRARY

Once you have gained an understanding of the concepts covered in this chapter, you will no doubt begin to amass a collection of Styles and Display Configurations that will be useful in future projects. In addition, there are thousands of prebuilt styles, and systems included with AMEP to get you started. These resources are accessible via the Content Browser and Style Manager, as we have seen. You can also use the Autodesk Seek Web site accessible directly from MEP on the Insert tab of the ribbon. Just input a keyword in the search field to open Seek and search for content.

We have discussed styles and the display system here. You should already be familiar with AutoCAD blocks. Style-based content and Multi-View Parts are objects that use one or more AutoCAD blocks in different viewing conditions. This allows custom objects to be created that need to look different from various viewing angles. For instance, many MEP elements are drawn differently in plan view than in elevation, such as a plumbing fixtures, air terminals, and electrical devices. Create content to represent these elements intelligently from these different viewing angles. In the display system described earlier, MEP elements respond automatically to these different viewing needs.

We have four chapters devoted to creating custom content later in this book: Chapter 9—Content Creation – Styles, Chapter 10—Content Creation – Equipment, Chapter 11—Content Creation – Parametric Fittings and Chapter 12—Content Creation - Parametric Equipment. Please read those chapters to learn more about building your own custom content.

By default all Styles and Content items are stored in a single root folder on the hard drive. This folder is referred to as the Content Library. Your CAD Manager will set its exact location either on your local machine or on your office network. By default, this path is:

C:\ProgramData\Autodesk\MEP 2012\enu

This path can be edited in the "Options" dialog (Application menu) on the AEC Content tab. The "Tool Catalog Content Root Path" controls the default location for all tools referenced in the Content Browser.

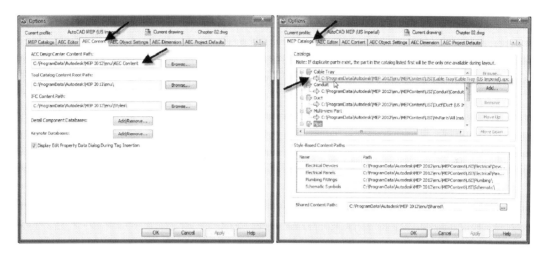

FIGURE 2.40
Configure Content paths in the "Options" dialog

Additionally, on the MEP Catalogs tab, you can specify the MEP Catalog content locations, and as pointed out earlier this tab also provides the Style-Based Content Paths. As new styles and content are accumulated, it is wise to develop a process for adding them to the appropriate content location. The topics of content creation will be covered in more detail in the chapters ahead.

SUMMARY

✓ Usage of AMEP is predicated on a process of parametric design and progressive refinement – instead of draw, erase, redraw, we can draw once and modify many times.

✓ MEP provides a variety of domain specific object types for documenting MEP systems.

✓ The display system allows building models to automatically represent a variety of display modes with a click of the mouse.

✓ Display components, object and system styles, and display settings provide consistency in documentation output.

✓ The Content Browser provides access to additional tools used for documenting your designs.

Chapter 3

Project Navigator for MEP

INTRODUCTION

In this chapter, we will explore the Drawing Management features of AutoCAD MEP (AMEP). Drawing Management is the official formal name for a series of tools that facilitate the collection and management of the various drawing files required in a typical project. Broken into two tools, Project Browser and Project Navigator, the toolset provides many benefits including easy XREF creation and management, centralization of project files, automation of repetitive tasks and more. Project Browser is used to create and load projects and Project Navigator is used for everything else. It is not required that you use Project Navigator (Drawing Management) to use AMEP, but doing so can help encourage the use of standards and procedures in your project teams.

OBJECTIVES

In this chapter, we will explore the Drawing Management features of AMEP. We have already seen a simple project configured using Project Navigator in the Quick Start chapter. In the remaining chapters, we will explore many of the AMEP tools within the context of a small commercial office building project. If you are familiar with Mastering AutoCAD Architecture, you will have seen this building project before. This project represents but one way that Project Navigator can be used to help enhance your AMEP experience. We will explore many of the critical concepts in this chapter, including the following:

- Building comfort with Drawing Management.
- Exploring and understanding provided drawing files.
- Learning to set up callouts, elevations and sections.
- Working with Sheets and Sheet Sets.

BUILDING A DIGITAL CARTOON SET

When the time comes in a project cycle to begin thinking about how many sheets of drawings will be required and what those sheets will contain, it is time to build a digital cartoon set. Just like the traditional paper-based cartoon set, the digital version will help you make good decisions about project documentation requirements and the impact on budget and personnel considerations. One extra advantage of the digital cartoon set is that it is the actual set of electronic files for the project and will evolve as the project develops. This means that the layout of a cartoon set is actually the layout of the real building model and eventual document set! Don't be concerned with the finality that this seems to imply. The documents remain completely flexible and editable, making this approach consistent with the goal of progressive refinement as defined in Chapter 2.

> **CAUTION:**
> Please do not skip this step when setting up your own projects. Establishing the digital cartoon set (or the Building Information Model structure) at the beginning of a project is critical to maximizing the potential of the AMEP tools and methods, and will go a long way toward ensuring success.

Typical projects usually involve a heavy dose of external references (XREFs). Setting these up can be time consuming. However, using the Drawing Management system in AMEP, the use of XREFs is made easy and nearly transparent. You will find that most of the work associated with the attachment and maintenance of the required XREFs is handled automatically and intelligently by the software.

In many ways, the task of creating a digital cartoon set is no more complicated than simply setting up Project Navigator. If you set up all your files early, you can print the digital cartoon set at the start of the project and use it to help staff the project team properly. Furthermore, you will benefit later in the project when your team is not burdened with having to pause productive work to stop and build a required file. If the setup is performed carefully, the team will have most of the files they require at the earliest stages of the project. Naturally, there will always be file needs discovered later in the project, but simply attempting to build everything early will go a long way toward minimizing the quantity of files required later.

DRAWING MANAGEMENT FEATURES AND BENEFITS

As you probably know, AMEP is built on top of the AutoCAD Architecture (ACA) platform. Project Navigator is a tool provided in the core ACA package and, as such, shared by all disciplines that use ACA or AMEP. We note this here simply to make the point that some of the features of the Drawing Management system are more beneficial to the architectural workflow than engineering. This is not to say that the tools are not useful to engineers; rather that we will focus our coverage in this chapter on those tools and features that offer the most benefit to engineers.

To help us understand the Drawing Management system and Project Navigator functionality, a project has been provided with the book CD ROM. The project is a small (30,000 SF) commercial office building. The project is mostly core and shell with some build-out occurring on one of the tenant floors.

The Drawing Management system offers many features and benefits. Using Drawing Management gives you a logical and easy-to-use interface for all your AMEP projects, both large and small. Drawing Management formalizes the relationship between several disparate files used to create the typical project. While Drawing Management is not required to use the other features of AMEP, doing so allows you to gain the fullest benefits from the software. Most of the tools that would otherwise be separate and disparate functions come together into a unified process through the Drawing Management system. Drawing Management offers the following key features:

Level Management—You establish each floor level in the Project Navigator, the software will manage inserting all XREFs at the correct floor heights in all project files.

Clear Delineation of Project Components—Using Constructs, Views and Sheets, AMEP presents a clear standard for the location of model components, adding annotation and plotting.

Ease of Use—Project Navigator introduces drag and drop ease to XREF Management. Simply drag a file from Project Navigator and the software takes care of the rest.

Maintains a Project Database—All project files are tracked and maintained in Project Navigator. In addition, a comprehensive list of project data are maintained and can be fed to project schedules, tags, title blocks and field codes.

Automated View and Sheet creation—You first determine the structure of your building (create its Constructs). Creating plans and sections, adding annotation and building Sheets is easy with a simple wizard interface and drag and drop.

Integration of the AutoCAD Sheet Set functionality—Sheet Set functionality is fully integrated into Project Navigator making the organization, publishing, eTransmitting, archiving and plotting of Sheets easy and powerful.

Automatic Repathing and View Regeneration—Whenever a change to a file location or a file name occurs, Project Navigator can automatically repath all XREFs in the project. In addition, whenever a structural change to the building model is made, such as a change in level or the addition of a new construct, the software will automatically regenerate associated views to reflect the change.

PROJECT BROWSER

The Project Browser is a file browser mechanism that allows you to locate project files (APJ files) anywhere on your computer (local or network). Using the Project Browser, you can locate projects and make them current. You can also move, copy and create new projects in this tool. Whether you need to load an existing project or create a new one, both tasks are accomplished with the Project Browser.

By default, when you install AMEP, a folder named *Autodesk* that contains a subfolder named *My Projects* will be created in the current user's *My Documents* folder. When you open the Project Browser for the first time, it will be set to this location. However, if you completed the Quick Start chapter, it will remember the location of the last project that you had loaded and show that location instead.

It is not necessary that you create projects in the *My Documents* location and, in fact, in a team environment it is preferable to work from a network server location instead. Files installed from this book's dataset are installed in the *C:\MasterAME 2012* folder (see the "Book Dataset Files" topic of the Preface for more information). Even though you will typically work from a server location on your real projects, for the tutorials in this book, it is highly recommended that you work in this location to avoid pathing errors.

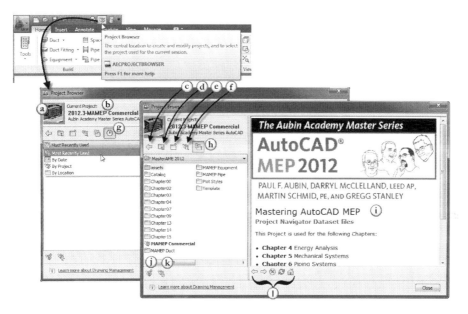

FIGURE 3.1

The Project Browser window

Descriptions of each item in the dialog are as follows:

a. **Project Image**—A custom-defined BMP image can be assigned to the project and displayed here. For example, load the client's logo here.

b. **Current Project Info**—The Name, Number and Description of the current project will display here.

c. **Back**—Click to go back to the previous folder.

d. **Up One Level**—Click to go to parent folder.

e. **Create a New Folder**—Create a new folder in the current location.

f. **Browse Project**—Click to open a standard Browse Window to locate and load project files (AMEP projects have an APJ extension).

g. **Project History**—Click to browse for projects that were previously active (use the drop-down list for additional history view options). While in the History view, you can right-click and remove items from the history list as well as reset the list.

    h. **Project Folder**—Click to browse for projects within the folder tree. This option gives access to My Computer, My Documents and Network locations such as mapped drives and any additional locations that you add to the AEC Project Location Search Path.

    i. **Project Bulletin Board**—The user-defined Project Bulletin Board Web Page, a fully customizable project-specific HTML Web page. AMEP starts with a simple generic page; you can load your own custom one in the Project properties.

    j. **New Project**—Creates a new project within the current folder.

    k. **Refresh Project**—Refreshes the current folder.

    l. **Project Bulletin Board Navigation Tools**—Typical browser functions for the bulletin board page.

> CAD MANAGER NOTE: You can add to the default search path locations used by the Project Browser in the "Options" dialog box. From the Application Menu choose Options, and then click the AEC Project Defaults tab. There you can add paths to the AEC Project Location Search Path. You can also edit the default Project Templates, Project Bulletin Boards and Project Images on this screen. It is recommended at the time of installation that you reset these paths for users to a location on the server where project templates and files are typically stored.

Once you have loaded or created your project, you do not need to open Project Browser again until you need to switch projects. From then on, you will do all your Drawing Management tasks using Project Navigator.

PROJECT NAVIGATOR

The Project Navigator palette (open from the Quick Access Toolbar) provides the complete interface to all files used in a project. Use Project Navigator to set up Levels and Divisions and to create, open and XREF Constructs, Elements, Views and Sheets (see below for terminology). Project Navigator behaves like other AutoCAD palettes and may be docked, floating, transparent and set to auto-hide (see Figure 3.2).

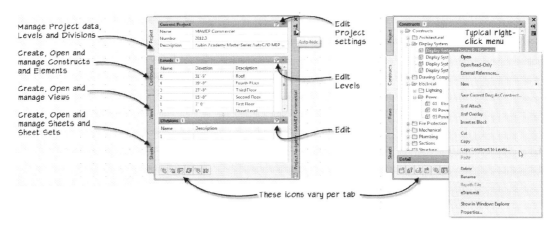

Manage Project data, Levels and Divisions

Create, Open and manage Constructs and Elements

Create, Open and manage Views

Create, Open and manage Sheets and Sheet Sets

Edit Project settings

Typical right-click menu

Edit Levels

Edit

These icons vary per tab

FIGURE 3.2

The Project Navigator palette

Projects are organized into several kinds of files named: Constructs, Elements, Views and Sheets. The Project Navigator palette contains tabs to manage each of these (Constructs and Elements are both found on the Constructs tab) and the Project tab for managing the project's overall parameters, such as Levels and Divisions. Icons appear along the bottom of each tab for common functions. More tools and commands are available in the right-click menu of the file list (shown on the right of Figure 3.2). Right-click menus vary on each tab like the bank of icons along the bottom. Take some time to pause over each tool to see a tooltip explaining the function and right-click in several locations to become familiar with the menus.

PROJECT NAVIGATOR TERMINOLOGY

Projects in AMEP consist of a collection of drawing files and project information files saved together in a common location. Taken together, this collection of graphical and non-graphical data is used to assemble a complete Building Information Model (BIM). The graphical drawing data associated with the project fall into four types of AMEP drawing (DWG) files: *Constructs, Elements, Views* and *Sheets*. Each is defined below. The non-graphical project information files include a single Autodesk Project Information file (APJ) that contains the basic framework of a project, a Sheet Set file (DST) that determines the organization and configuration of the list of printed Sheets within the project, and several individual project data (XML) files (one per drawing) describing how each individual drawing fits into the overall project structure.

When you create a project, you enter the basic descriptive information, such as Name, Description and Project Number. The next task is to determine how the building will be subdivided into Levels and Divisions. Levels are the floor Levels and divide the building horizontally. You can also subdivide the building laterally into Divisions. Although you can edit Levels and Divisions at any time, it is typical to establish the Levels and Divisions that will describe the basic framework of your project at the onset. These tasks are performed in the Project Navigator.

> **Tip:** If Project Navigator did not appear automatically onscreen when you created the project, click the **Project Navigator** icon on the QAT.

LEVELS AND DIVISIONS (PROJECT FRAMEWORK)

- **Level**—A horizontal separation of the building model data. A Level is typically an actual floor level in a building. Levels can be established for actual building stories, and also for mezzanines, basements and other partial levels. You also use Levels to establish Grade level, Roofs and Datum levels (see Figure 3.3).

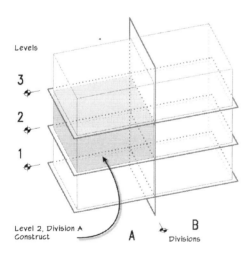

FIGURE 3.3

Divide your building into Levels and Divisions to create your project structure

■ **Division**—A vertical separation of the building model data. Divisions are typically used to articulate a physical separation such as a wing, an annex, or an addition to a building. Divisions can be used to subdivide large floor plates into pieces that are more manageable in size (See Figure 3.3). Projects frequently have only a single division.

> Note: See next topic for a definition of Construct.

MODEL AND SHEET FILES

Standard industry practice and the US National CAD Standard (NCS) recommend the creation and maintenance of two types of files: Model files and Sheet files. This practice is widely used in the industry and offers many benefits.

■ **Model File**—A file containing actual building data drawn at full size (1 to 1 scale). This is a file in which all of the *day-to-day* work is typically performed. In the AMEP Drawing Management system, Constructs, Elements and Views can all be considered Models in the traditional sense (see definitions later).

■ **Sheet File**—A file that is used *exclusively* for printing drawings. No data is saved in this file. It typically contains only a title block and external references to the project's various Model files.

Model files are referenced to Sheet files for printing (or when appropriate, other Model files). Most daily work is performed in Model files. In contrast, Sheet files exist solely for printing final documentation sets for distribution. One or more Model files are "gathered" by the Sheet file, composed on a title block sheet, scaled with the proper Display Configuration (see Chapter 2) active, layers and objects visible, and then printed. The Sheet is saved in this state so that documents can be printed again any time, at a moment's notice. To perform physical edits and design changes, return to Model files and perform the changes there. Those changes will appear in the Sheet file the next time the XREFs in that Sheet are reloaded, which happens automatically when the Sheet is opened.

> **CAUTION:**
> The Sheet file's "ready-to-print" status is maintained only if all project team members agree to work only in Model files and not in the Sheet files.

CONSTRUCTS, VIEWS AND SHEETS (PROJECT DRAWING FILES)

AMEP formalizes the creation of project files based on the Model/Sheet concept in the Drawing Management system and accompanying procedures. This system incorporates the industry-standard use of Model and Sheet files and introduces an additional layer of granularity (Elements, Constructs and Views) to help formalize the process.

- **Element**—(Mostly used in Architecture) A discrete piece of a design *without* explicit physical location within the building. Often an Element represents components that are repeated more than once in the design (typicals). Elements are also ideal for storage of project resources that you wish to have readily accessible. Elements are drawing files that are XREFed to other files (Constructs, Views, Sheets, or even other Elements) as project needs dictate.

- **Construct**—A discrete piece of a design *with* explicit physical location within the building. More specifically, it is a unique piece of the building occurring within a particular zone (Division) on a specific floor (Level) of the building. Constructs are not specific to any particular type of drawing. They are not "Plans" or "Sections" but rather Models that can be used to generate plans, sections, or any other type of drawing. A Construct is distinguished from an Element by its uniquely identifiable physical location within the building. Constructs are drawing files that are XREFed to other files (other Constructs, Views and Sheets) as project needs dictate.

- **View**—A slice of the building model configured to match a standard engineering drawing type. A View gathers all of the Constructs (and their nested Elements) required to correctly represent a specific portion (or *slice*) of the building. Views are akin to a particular type of drawing such as a "plan" or "section," and will contain those project annotations like notes, dimensions and tags appropriate to the drawing type and scale in question. Views are drawing files containing XREFs of Constructs that are, in turn, XREFed into Sheets as project needs dictate. Views provide a bridge between general model data in Constructs and the Sheets upon which they will be printed.

For example, imagine a three-story commercial building. You would have at least one Construct for each floor and discipline, although there could be, and often are, several. For instance, you might have a separate Lighting and Power Construct. A similar structure would be established for each of the other floors. When you were ready to begin creating construction documents to communicate your design, you would start creating drawing-specific Views. A View allows you to create a unique snapshot of a portion of the building. For instance, if you wished to work on the Third Floor East Wing in section, you would create and work in a View that would gather and correctly represent all of the Constructs required by that *physical* portion of the building. Views can be made to accommodate the creation of Plans, Sections, Elevation, Schedules, Details, and even full 3D Models. Be careful, however, to distinguish the "working" nature of Views from the "output" or plotting nature of Sheets. In other words, we still perform edits in Views, but Sheets are for plotting only.

TO VIEW OR NOT VIEW...

AMEP offers many automated routines that make creating section Views or detail Views quick and easy. Plans, on the other hand, offer some unique challenges with respect to Views. Since many AMEP objects offer automatic annotation as an integral part of the object, such as Duct and Pipe labels and Device tags, many engineering firms opt to annotate directly in Constructs for plan drawings. Doing so offers advantages and disadvantages. With annotation so closely related to the model elements in plan drawings, working with both in the Construct offers a significant advantage. The disadvantage of placing annotations in the Construct is that this may require some layer manipulation when referencing other discipline Constructs in your own. Not using views will force you to set up Sheets manually as the Project Navigator drag-and-drop procedure only works when dragging Views to Sheets. Also, cross-referenced Callouts will not link up properly when bypassing Views. Ultimately you will have to decide which approach offers the greater benefit to your firm and projects and build an office procedure to support it. It is possible that you may use views simply as an intermediate step to setting up your Sheets.

- **Sheet**—A "just for printing" view of the building model. While Elements, Constructs, and Views can all be considered *Models* as defined by NCS/AIA, the AMEP Sheet exactly emulates the purpose and intent of the NCS/AIA recommended Sheet file as noted previously. A Sheet file will gather all required building model components (Views and/or Constructs) and compose them on a title block sheet at a particular scale, ready to print.

There are those who argue that annotation and dimensions ought to be placed in the Sheets. Some go further to propose that these items be placed in the Layout (Paper Space) on top of viewport images of the project files. While both these approaches are certainly possible, the AMEP Drawing Management toolset instead supports the approach championed by this text as indicated in the earlier definition of "Sheet." It is the position of this text that Sheets should be set up once and maintained from then on as "for plotting only" files. The goal is to provide a set of files (one for each physical paper sheet in a document set) that is always ready to be opened and printed with no advance notice or *tweaking* required.

When project team members are allowed (or encouraged) to *work* in sheet files, it is possible or even likely that they will leave the drawings in a state that is less than ideal for "ready printing." For instance, one might close the drawing with model space active, or change the LtScale, or accidentally forget to freeze or thaw the correct layers. These are just some examples of the types of mundane settings that, if incorrectly set at the time of plotting, can result in the necessity of reprinting a drawing. Not only is this frustrating to the person making the plots, but it needlessly wastes time, paper and money. It is therefore *strongly* recommended that Sheets be used for plotting only, and all daily work be done in Constructs and/or Views. To help you achieve this goal, the "Match Sheet View Layers to View" feature when turned on (in the "Project Properties" dialog) will synchronize all changes in the View files to the corresponding Sheet viewports automatically.

FILE NAMING STRATEGY

File naming strategies and organization vary from company to company, (and region to region). However, as different as the naming scheme in one firm might be from that of another down the block, there are many similarities from firm to firm. In most cases, you will be able to adapt easily your existing scheme into the use of Project Navigator without too much difficulty. While AMEP imposes no "required" method for naming files, there are certain guidelines that can assist you in the naming your files. The recommendations made by the NCS are the most prolific (in the United States). Therefore, these guidelines will be suggested and utilized throughout this chapter and the rest of the book. More specific information can be found in that publication. Every attempt has been made to follow the NCS recommendations wherever possible and appropriate. We will see, however, that certain modifications to the

NCS naming recommendations are necessary to accommodate the specific needs of building information modeling when working with the AMEP drawing management system. It is hoped that the intent is still discernable if not directly applied.

While the NCS naming is utilized throughout the tutorials in this book, specific best practice guidelines and recommendations will also be made as appropriate. If your firm has specific naming guidelines in place that differ from the NCS, feel free to utilize your firm's naming scheme for files created in this text instead. The exact file name that you choose should not negatively impact the intent or function of the file, assuming that the naming scheme used is logical and understandable and that naming guidelines mentioned herein have been considered.

> Note: References made throughout this book to the NCS summarize the overall intent. However, for the complete explanation of these recommendations and all supporting materials, you are encouraged to purchase and refer to the NCS documents directly. For complete information or to purchase a copy of the NCS, visit: *http://www.nationalcadstandard.org*.

Below is a summary of potential file names based on the recommendations of the NCS as adapted to the Project Navigator and suggestions made herein. An NCS Model file name is composed of a Discipline, plus a drawing type Code, followed by an enumeration (which typically corresponds to the floor number, though not always). Elevations, for instance, would just be numbered sequentially in any logical order.

For example:

M-FP01 = Mechanical First Floor Plan

M-SC01 = Mechanical Sections (first group of section. There could be others named 02, 03, etc.)

The hyphen is used between discipline and code, but not between code and number.

NCS recommends the following codes for "Model" files Use these designations for View file names:

FP = Floor Plan	XP = eXisting Plan	SH = ScHedules
SP = Site Plan	EL = ELevation	3D = isometrics/3D
DP = Demolition Plan	SC = SeCtion	DG = DiaGrams
QP = eQuipment Plan	DT = DeTail	

Technically, "FP" would be used for any type of plan and they would simply be enumerated. Therefore, the first floor plan might be M-FP01, while the first floor ceiling plan would be M-FP02. However, M-FP02 for a ceiling plan on the first floor is confusing to most people, and it is highly likely you would run out of codes with only two digits. Therefore, the following abbreviations are used almost universally by most firms instead:

CP = reflected Ceiling Plan
RP = fuRniture Plan
EP = Enlarged Plan

While "A-3D01" would technically be correct for the name of a composite model, the following code is recommended instead:

CM = Composite Model

So the file would be named A-CM01 (You could also use A-CM00 for the first composite model in a set. It is not that critical with which number you start).

The NCS knows nothing about Constructs. Since their names indicate drawing type and function, they are best suited to use as View file names. Consider descriptive names for Constructs such as:

01 HVAC
01 Power
01 Lighting
01 Fire Protection

For Sheets, the NCS system is fine. They recommend the number of the Sheet; which typically includes the discipline code as well. The number is in two parts, the first digit is a code indicating drawing type, and the remaining two are an enumeration. 1 = Plans (Horizontal Views)

1 = Plans (Horizontal Views)

2 = Elevations (Vertical Views)

3 = Sections (Sectional Views)

4 = Large Scale Views (Plans, Sections & Elevations that are not Details)

5 = Details

6 = Schedules and Diagrams

7 = Used Defined

8 = User Defined

9 = 3D Representations (Isometrics, Perspectives & Photographs)

Examples:

M-101 – (First Floor Mechanical Plan)

M-102 – (Second Floor Mechanical Plan)

P-101 – (First Floor Plumbing Plan)

E-101 – (First Floor Electrical Plan)

The codes are very broad, so it is certainly possible to run out of numbers on a big job. Many firms develop similar systems of naming, but conceptually they are usually very consistent with these. One factor to consider, while there is neither functional nor philosophical reason why existing sheet file naming in your firm cannot simply be adopted, the integration of the Sheet Set Manager within Project Navigator introduces some hard coded assumptions about Sheet file naming which may be different than your current system. When you name a new Sheet on Project Navigator's Sheets tab, you input both the Sheet number and the Sheet Title. These two values, which will ultimately feed two different fields on the title block, are by default concatenated together to form the Sheet File name. It is likely that this varies from how you name sheets outside of Project Navigator. Whether it will prove problematic is another matter. In most cases, this change in naming will have little to no impact on your workflow or process, considering the level of automation that Sheet Set Manager brings to the table. Do give this new approach some serious consideration before adopting a contrary policy that may actually prove harder to manage.

PROJECT SETUP SCENARIOS

Overall, there are a few basic common project start-up scenarios. In most cases, you will be using background files from some other source. That source may be an internal or external Architect, client provided files or a combination of these. In this topic, we will address the overall scenarios in broad brush with the aim of discussing how these scenarios impact your strategy with regard to usage of Project Navigator.

Scenario 1—Backgrounds are provided in Project Navigator. This scenario has two sub-scenarios:

- AE firm: Project Navigator is being used in-house and files are shared on the same network.
- Outside firm: Project Navigator is used by the Architect and the project files have been provided.

Scenario 2—Backgrounds are provided as AutoCAD files (not in Project Navigator). This scenario also has two sub-scenarios. Despite the Architect's lack of Project Navigator, you can:

- Set up a Project Navigator project and import the AutoCAD backgrounds.
- Decide not to work in Project Navigator and use traditional XREF and project management techniques.

Again, these are broad overall scenarios and there are likely many variations that you will encounter. Let's dig a little deeper into each scenario and discuss some of the advantages and disadvantages of each.

AE FIRM USING PROJECT NAVIGATOR IN-HOUSE

If your engineering department is part of an AE firm, and your architectural department is using Project Navigator for your project, it will be easy for you to work directly with them in the same project. The workflow will mirror traditional workflows with which you are already familiar. Each discipline in the project will have its own folders and files. You will XREF backgrounds created by the Architect and use them to create your work. Since you are both working in the same Project Navigator environment, you will always have the latest backgrounds, and XREF paths will remain synchronized. The only setup required at the start of the project is to create a folder for each new discipline as it comes on board, and then create the required Constructs, Views and Sheets.

There is one issue worth mentioning with respect to discipline folders. It is common for many firms to create a folder for each discipline at the root of the project folder, and then within each discipline folder there are folders for each major drawing type, such as plans, sections and schedules. When using Project Navigator, there are four required folders at the top level of the project: *Constructs*, *Elements*, *Views* and *Sheets*. You cannot rename or delete these folders. Also, you cannot remove the requirement for these files to exist. Therefore, you will have to place your discipline folders within the *Constructs*, *Views* and *Sheets* folders.

In summary, the basic procedure when working in a multi-discipline firm is as follows:

- The project is created and configured by the CAD Manager, IT department or another designate.
- A folder(s) is created within each of the main root folders for each discipline (see Figure 3.4).

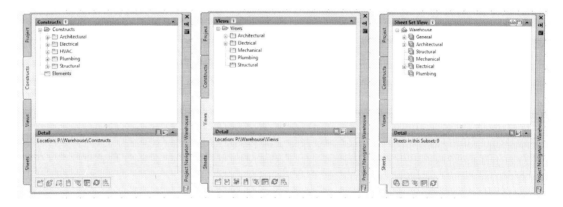

FIGURE 3.4
Each discipline will have its own folder on each Project Navigator tab

- As team members join the project, they simply load the project and begin working.
- Updates occur in real-time since all team members have access to the same set of files.

Variations on the procedure do occur, but the overall process typically mirrors the points noted here. In some firms, special security is enabled on the various folders to ensure that each discipline can change only the files that fall within its realm of responsibility. Such decisions are best made in consult with your IT department.

Let's do a short exercise emulating the process you would follow when joining an existing Project Navigator project.

INSTALL THE DATASET FILES AND LAUNCH PROJECT BROWSER

In this exercise, we use the Project Browser to load a project and then use Project Navigator to explore the various folders and files within the project structure.

1. If you have not already done so, install the book's dataset files.

   Refer to "Book Dataset Files" in the Preface for information on installing the sample files included with this book.

2. Launch AutoCAD MEP and then on the Quick Access Toolbar (QAT), click the Project Browser icon (see Figure 3.1 above).

3. In the "Project Browser" dialog, be sure that the Project Folder icon is depressed (this is item "h" in Figure 3.1 above).

4. Click to open the folder list and choose your *C:* drive. Double-click on the *MasterAME 2012* folder.

5. Double-click *MAMEP Commercial* to load the project. (You can also right-click on it and choose **Set Project Current.**).

At the top left corner of the "Project Browser" (items "a" and "b" in Figure 3.1 above), the project icon, name, number and description will change to reflect the current project. Also, on the right side of the dialog, a bulletin board will load. The bulletin board is just a small Web page. A default page can be used, or it can be different for each project. Any standard Web page can be used. On the left side, the name of the project in the list will now be bold indicating that it is active and loaded.

6. Click the Close button in the Project Browser.

IMPORTANT: If a message appears asking you to repath the project, click Yes. Refer to the "Repathing Projects" heading in the Preface for more information.

Note: You should only see a repathing message if you installed the dataset to a different location than the one recommended in the installation instructions in the preface.

When you close the Project Browser, the Project Navigator palette should appear automatically. If you already had this palette open, it will now change to reflect the contents of the newly loaded project. If for some reason the Project Navigator does not appear, click the icon on the QAT.

IMPORTANT: We will briefly explore this project as we continue to discuss the AE Firm sharing a single project scenario. Please do not make changes to any files. This project and its files will be used for many of the tutorials in the coming chapters. Feel free to open the files and explore, but please close files without saving to be sure the files remain in the state required for future lessons.

On a large project, there might be a single person responsible for project setup and overall maintenance tasks. This means others working in each discipline will be responsible only for their own files. Let's do a little exploration.

EXPLORE AN EXISTING PROJECT

Make sure the Project tab is active on the Project Navigator.

7. Study the three panels on the Project tab.

The name, number and description are listed at the top. In the Levels grouping, you will note that this building has four stores, a street level and a roof. There is only the one default Division. All these items can be edited with the icons in the right corner of the titlebars (as shown in Figure 3.2 above). We will explore some of these settings in the topics below.

8. Click the Constructs tab.

Two main folders appear here: *Constructs* and *Elements*. These are both required. Beneath these folders, several discipline folders have been included in this project as discussed above.

9. Expand the *Fire Protection* folder.

In this project, you will often see two versions of the files—the original file and a "complete" version. Since the project is provided to accompany the lessons in this book, a completed version of each file is typically provided so that you can check your work. Naturally there would be no "complete" versions in a real project.

10. Double-click to open the *01 Fire Protection Complete* Construct file.

Look at the titlebar at the top of your AMEP screen. Notice that *01 Fire Protection Complete.dwg* is listed there (see Figure 3.5). When you double-click a file in Project Navigator, it opens the corresponding drawing file from the project folder. This means that, unlike using the Open command, you do not have to browse to the location of the file first before opening it. Project Navigator knows where it is!

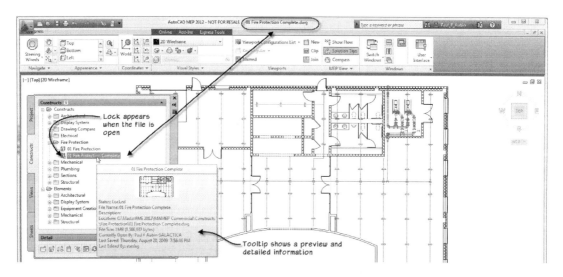

FIGURE 3.5

Open a Construct from Project Navigator

11. On the Project Navigator, right-click on the *01 Fire Protection Complete* Construct file and then choose **Properties**.

In the "Modify Construct" dialog, notice the checkmark in the Assignments area. The Assignments area lists all the Levels and Divisions in the project. This Construct is assigned to Level 1. This is why the prefix "01" is used in its file name. Whenever we ask Project Navigator to XREF this file, it will insert it relative to the Level 1 floor height.

12. Click Cancel to dismiss the dialog.

13. Close the *01 Fire Protection Complete* Construct file. When prompted to save, answer **No**.

14. Take a look at some of the other Construct folders.

Again, feel free to open any files you like in order to explore the project. Right-click and edit their properties as well. Just be sure not to change any assignments.

Note that on the Constructs tab is the *Element* folder. Some building elements are repeated in multiple locations throughout the project. In Project Navigator, Elements are used for this purpose. In this project, there are several examples. The *Core* file in the *Architectural* folder is one such example. Additional examples can be found in the *Structural* folder: *Column Grid* and *Typical Framing*. Each of these three files is XREFed to multiple locations in the building. This is because the building core and structural gird are the same on all floors. Using an Element allows changes to be made once in the Element file and then reflected throughout the building when the XREFs are updated.

Some additional Elements are also present: *Elevators* and *Third Floor Space Outlines*. The Elements folder is also a good location to store files that belong to the project but that you don't wish to have automatically XREFed anywhere. In this case, both of these files are stored here so that they will be easily accessible to the project team, but will not appear in any of the automatically generated files. Elements are not required to use Project Navigator. But as you can see they can have some useful benefits.

15. Click on the Views tab.

Use of View files varies with discipline and office standards/strategy. In this project, the Architectural folder has many view files. The other disciplines are not using that many views. Consult the "To View or not View" sidebar above for more discussion on this topic.

16. Click on the Sheets tab.

This tab provides access to all the Sheets in the project. The Sheets tab of Project Navigator fully integrates the AutoCAD Sheet Set functionality. When you click the Sheets tab, you will see a hierarchical series of items that look similar to a folder tree. These are the Sheet Set and its nested Subsets. A Sheet Set is used to organize drawing files and their Layout tabs for plotting. In the Project Navigator environment, the Sheets listed are saved in separate Sheet drawing files as well.

When you create a new project, a Sheet Set will automatically be created. A Sheet Set Template is used to create all of the initial Subsets. The entire collection of Subsets is completely customizable. It is likely that if you work in AE firm, the Sheet Set organization you see in a typical project will have been customized to include all the disciplines working in your firm.

CAD MANAGER NOTE: If your firm uses Sheet categories different from the ones in the out-of-the-box template, it can easily be modified to suit your firm's needs. To do this, open a project and then modify the Sheet Set by adding, modifying or deleting Subsets. Right-click each Subset and choose its Properties, such as the template file to use. Once you are satisfied with the Subset organization, copy the DST file from Windows Explorer to your AMEP templates folder. From the Application Menu, choose Options and click the AEC Project Defaults tab to set this DST as the default for new projects that are created without template projects. If you are planning to implement the project template feature (used at the start of this chapter to create the Commercial Project) you will want to load that template project (using Project Browser) and then modify its Sheet Set instead. This will give all future projects created from that template project a standard and consistent Sheet Set with office standard Subsets and Sheet templates.

To open and plot a Sheet, you simply locate it in the tree, double-click it to open like the files on the other tabs and then use the standard AutoCAD printing features. If you want to plot the entire set or a complete Subset, you can right-click on the item in the tree view and choose one of the options on the Publish sub-menu. Please refer to Chapter 17 for more information.

CLOSE OR SWITCH THE CURRENT PROJECT

As you go about your daily tasks, you may need to work in more than one project. Only one project can be active at a time in AMEP. This means that to perform work on a different project, you must return to Project Browser and switch the current project. To do this, simply repeat the process covered above in the "Install the CD Files and Launch Project Browser" topic starting with step 3. Since Project Navigator manages many linkages and relationships between files, it is important to close any open drawing files from one project before making another current. Fortunately, AMEP offers to do this for you automatically.

17. On the Constructs tab, double-click to open the *01 Fire Protection Complete* Construct file.

    If you left this file open above, this action will simply make this file active.

18. On the QAT, click the Project Browser icon.

19. In the "Project Browser" dialog, double-click the *Chapter00* folder.

20. Double-click the *Quick Start* project to load it. (You can also right-click on it and choose **Set Project Current**.).

21. Click Close in the "Project Browser" dialog.

A dialog box will appear indicating that you have open drawing files from the previous project. You are offered two options, including the option to close the files. In nearly all circumstances, this is the recommended choice. For a variety of reasons, it is ill advised to leave project files open from the previous project as you work in a different project.

22. In the "Project Browser – Close Project Files" dialog, click the Close all project files option (see Figure 3.6).

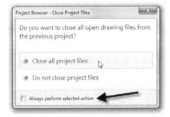

FIGURE 3.6
When you switch current projects, you are offered the option to close all open project files (recommended)

23. If prompted to save any files, just this one time answer No.

> Note: since we were just exploring and did not want to make changes, not saving is appropriate here. Normally, you will want to save your changes.

You will notice that there is also a checkbox in this dialog where you can make your choice permanent. Feel free to check this option if you wish, but it might be better not to check this box. This way you will always see a nice reminder when switching active projects with drawings open and you will not be left wondering what AMEP is up to as it begins closing all of your files.

In addition to switching projects, which closes the previous project in order to load the new one, you can elect to simply close the current project. There are times when you wish to close the active project and not load another one in its place. A good example might be if you are leaving the office for an extended period of time and do not want to accidentally leave any project files open where they will then be locked to your colleagues.

FIGURE 3.7

Close the current project without loading another one

> 24. On the Project Browser, click the Project tab.
>
> 25. At the bottom of the palette, click the Close Current Project icon (see Figure 3.7).

Like before, if there are open files, you will be prompted to close them. There are many more features to explore in the Project Navigator environment. We have yet to discuss how to create files (Constructs, Views and Sheets), how to XREF project files, or how to perform a number of other project related functions. We could certainly perform such tasks in either of the existing projects we explored here. However, let's move on to working with outside architectural firms and learn to perform those functions in the context of two sub-scenarios: loading a project from an outside architectural firm and creating a new project in-house.

ENGINEER RECEIVES PROJECT NAVIGATOR FILES FROM AN OUTSIDE ARCHITECT

In this scenario, the Architect is an outside firm that provides their Project Navigator files to your firm. Since they have already created the project, you can simply copy the entire project to a location on your server and load it with Project Browser in the same manner as you did above. When you first load the project, AMEP will report that the project has been moved from its original location. This is because it would be unlikely that your firm will have the same drive letters and folder structure on your server as the Architect has on theirs. This situation, however, reveals one of the major benefits of Project Navigator. You will be offered an opportunity to repath all of the reference files automatically! Let's take a first-hand look at this feature.

RECEIVING PROJECT FILES FROM AN OUTSIDE ARCHITECT

So your Architect has informed you that they are using AutoCAD Architecture and have their files setup in Project Navigator. Terrific! This will save you a lot of effort. All you need to do is get a copy of their project files and you can open them on your system and begin adding engineering data. Simple, right? Well, there is one very important request you must make. Make sure the Architect sends you *all* the project files, not just the DWG files. Ask them to ZIP the entire project directory into a single file. This is easiest to do in Windows Explorer with WinZip or the built-in Send to > Compressed (zipped) Folder options on the right-click menu.

It is important that you get all the project files because Project Navigator relies on more than just DWG files. Project Level and Division information (and in fact most Project Navigator data) are not stored in the DWG files. Level and Division information (and several other bits of global project data) are stored in a Project Information file with an APJ extension. Critical information about each Construct, Element, View and Sheet is saved in XML files that live in the same folders as the corresponding DWG files. Consider the contents of the *Plumbing* Constructs folder of the project shown in Figure 3.8. On the left side of the figure, you can see Project Navigator contains only one file: *Sanitary System.* On the right side, Windows Explorer is shown open to the same location. There you will note that there is a DWG file, and XML file of the same name, and in some cases you will even have BAK files as well.

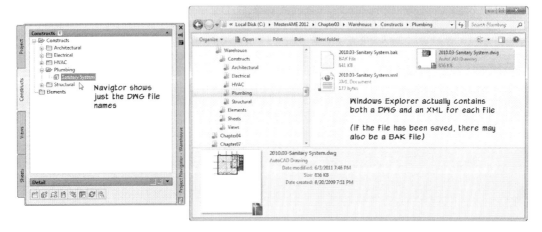

FIGURE 3.8

Project Navigator files have both a DWG and an XML for each file

In reality, Project Navigator is actually showing you the XML files. If you were to open one of these in an XML editor (not recommended) there is code in the file that instructs Project Navigator to open the appropriate drawing file. Furthermore, all of the project information associated with the file such as Level and Division for Constructs, is stored in the XML as well.

If you are starting to worry that you will now have to learn how to speak XML, don't. You *do not* have to do anything with the XML files. Project Navigator handles everything behind the scenes for you. There are only two reasons you even need to know they exist. The first is the issue we started discussing at the beginning of this topic: If you are receiving Project Navigator files from an outside firm, you need to be sure that they send both the DWG files and the XML files. If they only send you the drawings, Project Navigator will not recognize them and you will not be able to load the project with Project Browser or view and open the files in Project Navigator. The other reason you need to know that they exist (and that they are required) is so that you do not inadvertently delete them when doing file and folder cleanup. The icons in Windows Explorer look a little like BAK files (see Figure 3.8). While you can usually delete BAK files after the nightly backup is performed, you should not delete the XML files, ever!

If your Architect is kind, they will delete the BAK files before sending the project. This will help reduce the size of the ZIP file, and make the transmission via FTP or other file sharing quicker. Just make sure they send the APJ, the DST (Sheet Set file, see below) and all the XML files with the DWG files.

LOAD AND REPATH AN EXISTING PROJECT

After you have successfully received all the required files from the Architect, you are ready to load the project and begin working.

1. On the QAT, click the Project Browser icon.

2. In the "Project Browser" dialog, click to open the folder list and choose your *C:* drive.

3. Double-click on the *MasterAME 2012* folder and then the *Chapter03* folder.

4. Double-click *Warehouse* to load the project. (You can also right-click on it and choose **Set Project Current**.)

    The "Project Browser – Project Location Changed" dialog will appear (See Figure 3.9)

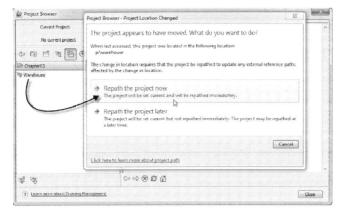

FIGURE 3.9
When a project is moved, AMEP will offer to repath all XREFs

5. Choose the "Repath the project now" option

This will instruct Project Browser to load the project and fix all XREF paths to match the new location on your server. Sit back and watch the progress bar as AMEP goes to work.

6. Click Close in Project Browser.

If the Project Navigator palette was not open, it will be displayed. If for some reason it does not, click the Project Navigator icon on the QAT.

7. Following the procedures above, click on each tab and open some of the files to look around.

This project is a copy of the project used in the Quick Start chapter, so if you completed that chapter's tutorials, you should already be familiar with the dataset.

UNDERSTANDING REPATH OPTIONS

When you make a project current, the software compares the original location of the project as saved in the project information file (APJ), with the current location of the project. The location stored is the location of the root folder of the project. By default, the root folder is named the same as the project. In this case, the name of the root folder is: *Warehouse* and is found in the *C:\MasterAME 2012\Chapter03* folder. (Your root path may vary if you installed the dataset to an alternate location.) The original root location of this project was *P:\Warehouse*. Since the current location does not match the saved location, AMEP prompted us to repath the project. When you do so, AMEP will execute a search and replace on all files in the project. All XREF paths stored in the root folder or lower will be repathed. Any XREFs pointing to files outside of the root path will not be changed.

There are also three options to how paths can be stored:

- **UNC paths:** *\\servername\share\Project folder\Constructs\File Name.dwg*
- **Mapped drives:** *P:\My Client\Project folder\Constructs\File Name.dwg*
- **Relative paths:** *..\Constructs\File Name.dwg*

Both of the first two options can be considered "Absolute" paths. There is not an explicit setting for the type of absolute path; rather this is determined by how you browse to the project. In the "Project Browser" dialog, if you navigate to the project through your local computer, mapped drive paths will be used. If you browse via network places, UNC paths will be used (see Figure 3.10).

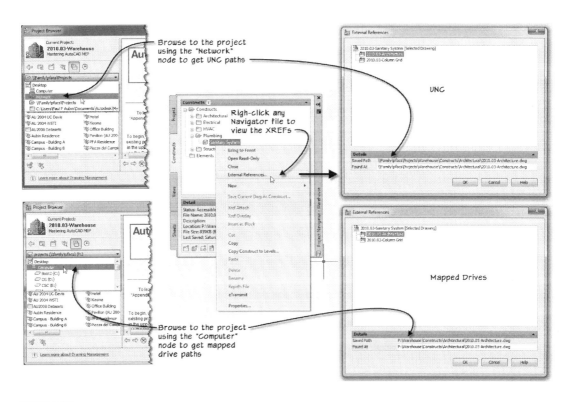

FIGURE 3.10
The way you browse to a project determines the kind of absolute path behavior

This is important because if you are using one option, and someone on the project team browses to the project the other way, AMEP will prompt that user to repath the project. If they answer yes, then all files will be updated to the other form of pathing. Naturally, it would be undesirable to have some members of the team applying mapped drive paths while others are using UNC. Therefore, it is important to discuss the desired method of mapping with all team members at the onset of the project and be sure that everyone is following only one method. Neither pathing method is inherently superior to the other. Check with your CAD Manager or IT support person for guidance on this issue. What is important is that one method is used consistently.

Pathing issues are not unique to the scenario currently under discussion. In other words, you will have to keep the same issues in mind if you work in an AE firm with an in-house Architect, have an out-of-house Architect using and sending you Project Navigator files or if you are creating the Project Navigator project in-house using non project backgrounds from the Architect. So please be sure to make the decision of path type carefully and discuss it with the team.

As noted above, there is a third type of path: Relative paths. Relative Paths can solve most of the potential pitfalls of using the other two. When Relative Paths are enabled for a project, AMEP will record only the part of the XREF path that occurs beyond the project root folder. So if the project lives on the P Drive in the *Projects\Client* A folder, a Construct called 01 Piping in the Plumbing folder would have an XREF path *..\Constructs\Plumbing\01 Piping.dwg* instead of *P:\Projects\Client\Constructs\Plumbing\01 Piping.dwg*. This is also the easiest way to prevent the need to repath the project each time the out-of-house Architect sends updated background files.

8. On the Project Navigator, click the Project tab.

9. At the top right corner of the palette, click the Edit Project icon (shown in Figure 3.2 above).

10. Within the Drawing Settings grouping, for "Use Relative Xref Paths," choose **Yes** and then click OK (see Figure 3.11).

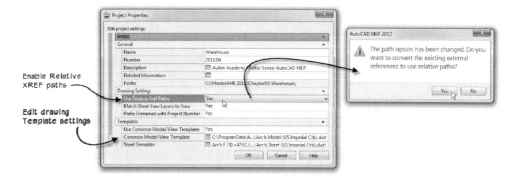

FIGURE 3.11
Enable Relative XREF Paths

When you click OK to close the worksheet, you will be prompted that the change requires the project to be repathed again. It is important to answer yes to this message as this will repath the project, once again replacing the absolute paths with relative paths.

11. In the alert dialog that appears, click Yes (see the right side of Figure 3.11).

From this point on, assuming you do not move the project again, you should not need to repath the project again.

12. On the Project Navigator palette, click the Constructs tab.

13. Expand the *Plumbing* folder and then right-click the *Sanitary System* Construct file.

14. From the right-click menu, choose **External References**.

15. Select the *Architecture* file in the list (see Figure 3.12).

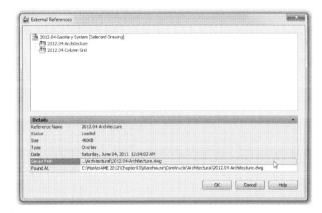

FIGURE 3.12
Check the resolved path of a relatively pathed XREF

At the bottom of the dialog, notice that the Save Path for this file is:

..\Architectural\2012.04-Architecture.dwg.

The file is found at:

C:\MasterAME 2012\Chapter03\Warehouse\Constructs\Architectural\2012.04-Architecture.dwg.

The Saved Path is the relative path. The Found At field always reads the full absolute resolved path to the file.

PROJECT TEMPLATE SETTINGS

The worksheet we opened in Figure 3.11 contains many useful project settings. We do not have the space to cover all of them here, but at least a few additional settings are worth a brief look. Look again at Figure 3.11. Notice the Templates grouping. In this location, you can choose one or more drawing template files (DWTs) that will be used automatically when you create files within the Project Navigator environment. Template files are very important to maintaining and ensuring compliance with established office standards. With the settings in this location, you can decide which DWT file will be used for each of the four Project Navigator files types (Constructs, Elements, Views and Sheets). If you do not want to designate different template for the three types of Model files: Constructs, Elements and Views, choose Yes from the "Use Common Model\View Templates" setting. This is the default setting. Regardless of the setting here, you will be able to specify a different DWT for Sheets. This is because Sheet templates are typically configured optimally for printing and include a titleblock and several layout tabs.

If you are working in a shared project in an AE firm, these settings will be managed by the project's project data coordinator. If you are working with an outside Architect and receiving Project Navigator project files, you may want to check these settings after you load and repath the project. The Architect's project will likely reference architectural template files. You will want to change the settings to point to your firm's office standard DWT files. This is a one-time setting at the beginning of the project.

CAD MANAGER NOTE: A/E firms that work with Architects in-house need to understand that Project Navigator only supports a single template file and does not allow for each discipline to specify their own template. We recommend that you do not specify a template, which will prompt each user to select the template that suits their needs. An alternative is to setup the template for the QNEW command for the project and use the SaveAS Construct functionality available on the Right Click menu as detailed in the section below titled Create a Construct from an Existing File.

Several other settings appear in this worksheet. Here is a brief description (consult the online help for complete details):

Match Sheet View Layers to View—With this setting, viewports on Sheets you create in the project will be using the layer settings of the associated View drawing. Existing sheet views will be synchronized the next time they are opened or when their external references are reloaded.

Note: Verify that VISRETAIN is set to 1 both in the View files and in the Sheet drawing files for the synchronization to be successful.

Prefix Filenames with Project Number—Some firms like to include the job number as a prefix to drawing file names. This can be accomplished automatically with this setting. When you enable it, the DWG file names in Windows Explorer are modified, but the job number prefix will *not* appear in Project Navigator. You can see an example of this in Figure 3.10. Look closely at the Sanitary System Construct file in the middle of the figure on the Project Navigator palette and compare it with the name shown in the External Reference dialogs on the right.

Project Standards—When you enable Project Standards you designate one or more drawing files to become project standard library files. These files contain master copies of styles and display settings that you wish to keep synchronized across all drawings in the project. Once standards are configured, you can synchronize project drawings at regular intervals to bring them up-to-date with the standard. When configured and used properly, this is a very powerful feature that can save the project team enormous time and effort, and help ensure more consistent and higher quality project files. To learn more about Project Standards, refer to the online help.

Project Browser—Items "a" and "I" in Figure 3.1 above discuss this feature. Use the settings in this grouping to point to a custom project image and/or bulletin board file.

Folders—By default, the *Constructs, Elements, Views* and *Sheets* folders are contained in the project root folder. In most cases, this is the best strategy. However, in some environments, it may be desirable to locate these folders in different physical locations. Use the settings here to change the path of project folders to different physical locations. Please note that you should make this change at the start of a project. Project Navigator will not be able to properly repath existing files to the new locations if you make this change after project files exist.

Detail Components—If you are using the Detail Component Manager feature in AMEP, this setting allows for custom project databases to be used in conjunction with that feature. Consult the online help for more information.

Tool Palettes—You can optionally enable custom tool palettes that are visible only when the project is active. Consult the online help for more information.

16. If you have opened the "Project Properties" worksheet, close it when you are finished studying the settings.

CREATE FOLDERS

Since you are starting your project with background files created by the Architect, there is a good chance that your next task will be to create discipline folders for each trade that is your firm's responsibility. This is an easy task to accomplish. You have two options. In Project Navigator, you can right-click any folder and choose **New > Category** to make a new folder. Alternatively, you can browse to the location in Windows Explorer and simply create a folder.

17. On the Project Navigator palette, click the Constructs tab.

18. Right-click the *Constructs* folder and choose **New > Category**.

19. Name the new folder: **Fire Protection** and then press ENTER.

> Note: The dataset provided here already has some discipline folders and files. However, as noted, typically you will only receive *Architectural* and maybe *Structural* from the outside Architect.

20. Click the Views tab and repeat the process.

21. Click the Sheets tab.

The procedure on the Sheets tab is nearly identical except that here you will create a Subset instead of a Category or Folder. This must be done within Project Navigator. You cannot make Subsets in Windows Explorer.

22. Right-click the Warehouse (top) node on the Sheet Set and choose **New > Subset**.

23. In the "Subset Properties" worksheet, type **Fire Protection** for the Subset Name.

There are some optional settings here. Subsets look similar to folders but are not folders. If you look in Windows Explorer, all the Sheet files are stored in the *Sheets* folder regardless of their Subset. However, if you wish to have folders created to match the Subset structure, you can do so with the "Create Folder Hierarchy" setting. If you do not wish to have the Subset you are creating published with the set, you can make this choice under the "Publish Sheets in Subset" option. To redirect the folder where Sheets for this Subset are stored, browse to a location in the "New Sheet Location" option. Each Subset can have its own Sheet template (DWT) file. This overrides the template choice made in the "Project Properties" worksheet above. You can also enable the "Prompt for Template" option if you need to choose a different Sheet Template for each Sheet. Unless you have a specific need to change any of these options, it is best to accept the defaults (see Figure 3.13).

FIGURE 3.13
Create a Subset and configure its options

> 24. Repeat the steps to create any additional discipline folders and Subsets required.

CAD MANAGER NOTE: Creating a well-planned folder (Category) structure for your projects can prove extremely beneficial. Once you have established a suitable folder structure, you can even re-use it in future projects by using the **Copy Project Structure** command (available as a right-click option in the Project Browser), which will copy all the sub-folders in the project to a new Project name and location that you specify. This will help you maintain consistency in project setup. Please note that **Copy Project Structure** does not copy any of the files. To create a project from another including all of its folders, Constructs, Views and Sheets, use the project as a project template when creating new projects.

Create Constructs

With the overall setup complete, the only task remaining is to create files. Begin with Constructs. Constructs are like building blocks for your project. Each Construct should uniquely represent some portion of the building. Constructs are required before you can do any meaningful work on Views and Sheets. Your Project Navigator Architect will have provided Constructs containing exterior Walls, interior Walls, Column Grids, Doors, Windows, Stairs and possibly Spaces, Slabs and Roofs. Such objects will be drawn in one or more Constructs per Level.

At a minimum, you should have one Construct per Level (and Division if the project has them), per discipline. In many cases you will have more than one Construct if the project is sufficiently complex enough to warrant it. For example, in a large or complex project, you may choose to have several Constructs for a particular discipline—perhaps having one construct per zone. This allows multiple team members to work in the project at the same time. So the quantity of Constructs your project will have is at least one per Level per discipline, and possibly several.

CREATE A CONSTRUCT FROM AN EXISTING FILE

If you have already begun preliminary design work in an independent AutoCAD file, you can easily convert such a file to a Construct and add it to the project.

1. On the QAT, click the Open icon (or choose **Open** from the Application Menu).

2. Navigate to the *C:\MasterAME 2012\Chapter03* folder and locate the file named *Sprinkler System.dwg.*

Independent files like this one can be added to the project. When you do so, you give it a Level and Division assignment so it knows to which part of the building it belongs.

3. On the Project Navigator palette, click the Constructs tab.

4. Right-click on the *Fire Protection* folder and choose **Save Current Dwg As Construct**.

   The Add Construct dialog box will appear.

5. Type **01 Fire Protection** in the Name field.

6. Click in the Description field and type **Sprinkler System**.

7. In the Assignments area, place a checkmark next to First Floor (see Figure 3.14).

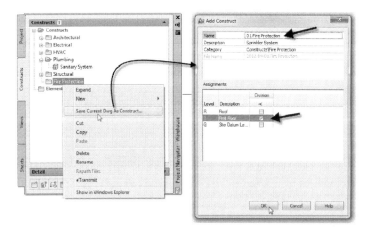

FIGURE 3.14

Save a Construct from an existing file and assign it to the First Floor

The Assignments area of the Add Construct dialog box is very powerful. It is here that you tell AMEP and the Project Management system which portion of the building this particular Construct represents. Since the Project Management system is aware of all the levels and divisions within a project, AMEP will be able to correctly XREF and locate this *Fire Protection* Construct drawing relative to all other drawings in the project.

8. Click OK to accept all values and create the new Construct.

Notice that there is now a new Construct in the Project Navigator (Constructs tab) named *01 Fire Protection* (see Figure 3.15).

FIGURE 3.15
The new Construct appears on the Project Navigator palette in the Fire Protection folder

9. Save and close the file.

> **Note:** It should be noted that if you use the **Save Current Dwg As Construct** option noted here, you will not benefit from the automated Template settings discussed in the "Project Template Settings" topic above. This is because a DWT file can only be applied at the time a drawing is created and not to an existing drawing.

CREATE A CONSTRUCT FROM SCRATCH

You can also create a new Construct from scratch. This is similar to creating a new drawing from the Application menu except that the new file will immediately become part of the project and will automatically use the template settings we looked at above.

10. Right-click the *HVAC* folder and choose **New > Construct**.

11. In the "Add Construct" dialog, type Energy Analysis for the name.

12. Input a description, if you wish, and check the First Floor box.

When you create a file in Project Navigator, you can have it open immediately so you can begin working in the file.

13. Check the "Open in drawing editor" checkbox (see the left side of Figure 3.16).

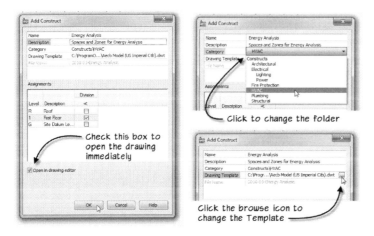

FIGURE 3.16

Create a new Construct from scratch

Notice in the figure that the Category (or folder) is *Constructs\HVAC*. Whatever folder you right-click is where the file will be placed. If you accidentally create a file in the wrong folder, you can actually click on the folder listed here next to Category and change it. You can also browse to a different Template file if necessary (see the right side of Figure 3.16). In order to ensure consistency with office standards, this should be done only in special circumstances. As already noted, you cannot change the Template after you click OK and create the file. But, again, to remain compliant with office standards, you will rarely need to do so.

14. Click OK to create the file.

After you click OK, if you realize that the Category/Folder is wrong, you can simply drag and drop the file to the correct folder within Project Navigator. When you do, AMEP will alert you that the action will require repathing of XREFs. You are given the option to repath immediately or postpone it until later (see the left side of Figure 3.17).

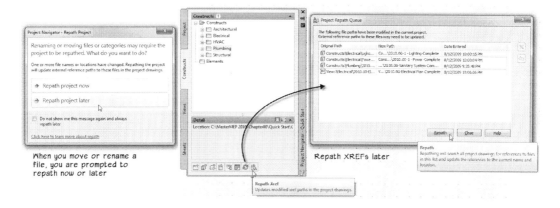

FIGURE 3.17

Name the Construct and assign it to Level 1

If you choose to repath later, you can click the icon on the Project Navigator palette when you are ready to view the Repath Queue and repath all files (see the right side of Figure 3.17).

ADD XREFS

Before you can begin your design work, you will need to load some files as backgrounds in your newly created Construct. Creating XREFs with Project Navigator is as simple as drag and drop.

15. Expand the *Architectural* folder.

16. Drag the *Architecture* Construct and drop it in the drawing window.

The architectural background will appear on screen in the correct location and with its AEC objects (like Walls and Doors) displayed screened. You may need to zoom or pan and possibly adjust some of the layer settings to get the display just right.

When you drag a Construct to a Construct as we have done here, the default XREF behavior is Overlay. This is because the assumption built into the system is that you are inserting other Constructs simply for reference. If you are intending to create and use Views, this behavior is suitable as the View will have its own XREFs. However, if you are planning to place your Constructs directly on Sheets, you can use the right-click menu to XREF instead of drag and drop.

17. Expand the *Structural* folder.

18. Right-click the *Column Grid* Construct and choose **Xref Attach** (see Figure 3.18).

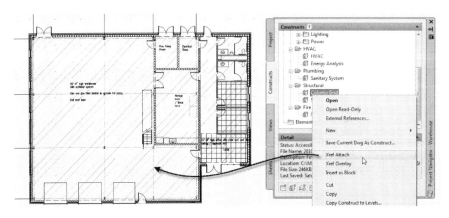

FIGURE 3.18
Xref commands on the right-click menu

You can create other Constructs following the same procedures. One additional tool on the right-click menu is worth mention. If you have a multi-storey building, you can right-click an existing Construct and choose Copy Construct to Levels. This command will show you a list of all the levels in your project. Check the boxes for the level(s) to which you wish to copy the Construct (see Figure 3.19).

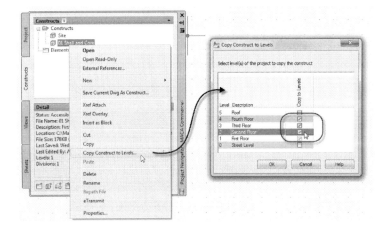

FIGURE 3.19
Copy Construct to Levels command

The copied files will end up with names like the original with a numerical suffix at the end. You can right-click the files and choose Rename to remove this suffix and change the name to something more suitable. Remember, after you make such a change, the "Project Navigator – Repath Project" dialog will appear. If you need to rename several files, postpone the repath by clicking the "Repath project later" option. When you are finished renaming files, you can repath the project manually as noted above. You can right-click any of the copied files and choose Properties. This will open the "Modify Construct" dialog where you can confirm that the assignments area correctly reflects the copied Level for the file.

> Note: It should be noted that copying Constructs to levels copies the files and their XREFs. So if you do not want the XREFs copied, perform the Copy to Levels before attaching the XREFs.

19. Repeat any of the above procedures for any additional Constructs that you need to create.

Spanning Constructs

In some cases you will have an element that spans across multiple Levels and/or Divisions. Building components that occupy more than one Level (or Division) in an AMEP Model are referred to in Project Navigator as "spanning" Constructs. Spanning Constructs will automatically be added to Views that reference them for each floor in which they span. To create a Spanning Construct, click more than one assignment checkbox (see Figure 3.20).

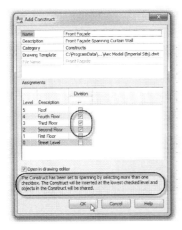

FIGURE 3.20

Assigning the Construct to multiple levels makes it a "spanning" Construct

Create Views

Views are an intermediate file between Constructs and Sheets. The intended workflow of Project Navigator is to create Constructs for all the physical geometry that represents real live elements. Constructs should only contain model elements drawn at their full, real world size and each Construct should uniquely portray some identifiable portion of the building. Next, Views are created to represent the variety of drawing types typically included in construction document sets. This includes plans, sections and schedules. Views automatically XREF the appropriate Constructs. Additional notes, annotations and embellishments are added on top of the XREFs in these files. Finally, Views are dragged to Sheets for plotting. This workflow works better for some disciplines than others. It works best for architectural and structural disciplines. For mechanical, plumbing, piping and electrical, this workflow presents certain challenges. There can be performance issues, and due to the nature of automated AMEP annotation routines, it is often not possible to create annotation in View files, which must instead be created with the model in the Constructs. If you decide that skipping the creation of Views is more beneficial to your workflow you should be aware of certain limitations. Automatic fields provided in the Sections and Elevation tags in AMEP will not work as designed. Adding constructs to sheets directly will require you to manually create the viewport in Paper space and manage multiple views by assigning the Xrefs to a unique layer so the XREF's visibility can be controlled using Layers. Chapter 13 uses the approach of creating a sheet directly from a construct.

Based on these issues, you will find some View types more useful than others. There are three types of View file: General, Section/Elevation and Detail. A Detail View is intended to be used for detail drawings that make use of the Detail Component Manager in AMEP. If you are creating Detail drawings and wish to begin your details with a cut from the model, the Detail View type is the best choice. When you create sections, use the Section/Elevation view type. It will automate the process of gathering all the Constructs required for cutting your section. Finally, a General View is used for everything else. If you want to create a 3D model view, or a plan view, then use the General option. Once you have chosen a View type to create, the remaining steps are the same for all three types.

BUILD A COMPOSITE MODEL VIEW

A composite model View will help us visualize what we have so far. The tools in Project Navigator make creating a View easy. By simply designating which portions of the building we wish to include in the View, we can have the system gather all the required XREFs and assemble them at the correct relative locations and heights. A composite model view is an excellent tool for studying the model and looking for clashes. (Refer to Chapter 15 for more information on this topic.)

1. In Project Navigator, click on the Views tab to activate it.

2. Select a folder where you would like to add the View.

   If you like, you can create a new folder first.

3. At the base of the palette, click the Add View icon (second from the left).

   The "Add View" dialog will appear with the three types of View listed as choices.

4. Select the "Add a General View" option (see Figure 3.21).

FIGURE 3.21
Add a General View

5. For the Name, type **G-CM01** and for the Description, type **Clash Detection Composite Building Model** (see Figure 3.22).

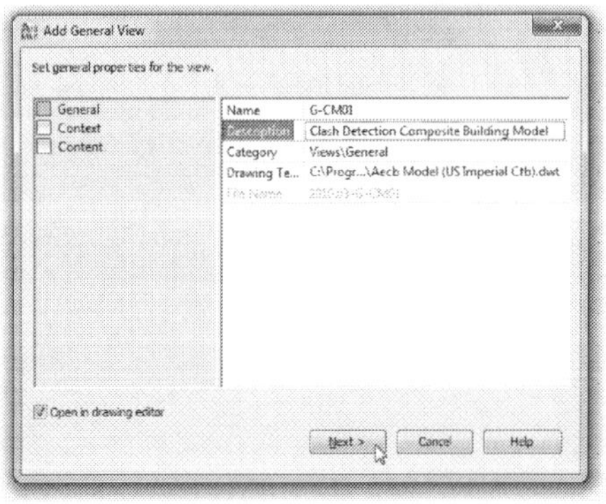

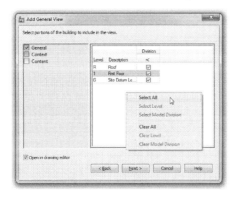

FIGURE 3.22

The first page of the Add View wizard

File naming conventions vary widely from one firm to the next. While it is possible to adapt existing file naming conventions to Project Navigator files when migrating to AMEP, it is often necessary to make some adjustments. In this book, we will use simple descriptive names for Constructs (as you saw earlier) and both descriptive names and US National CAD Standard (NCS) names for Views. See the "File Naming Strategy" heading above for more on file naming. Feel free to use your firm's file naming strategies rather than those recommended in this text. Changing the names of the files will not alter the tutorials in any way. In this case, for our Composite Model, "G" stands for "General," "CM" for "Composite Model," and "01" is simply a placeholder for enumeration. For a floor plan this would be a floor number designation, and for sections and elevations it is typically a simple sequential number.

6. Click OK to close the Description dialog and then click Next to move to the Context page.

FIGURE 3.23

Include all the Levels (and Divisions) in the View

7. Right-click anywhere in the Levels area and choose **Select All** (see Figure 3.23).

8. Click Next to move to the Content page.

9. On the Content page, verify that all Constructs are selected, that the "Open in drawing editor" checkbox is selected, and then click Finish (see Figure 3.24).

FIGURE 3.24

Complete the Add View wizard by verifying that all Constructs are included

10. On the View tab, on the Appearance panel, scroll through the list of preset views and choose SE Isometric.

11. Next to the view list, click the drop-down button on the Visual Styles tool and choose the Conceptual tool (See Figure 3.25).

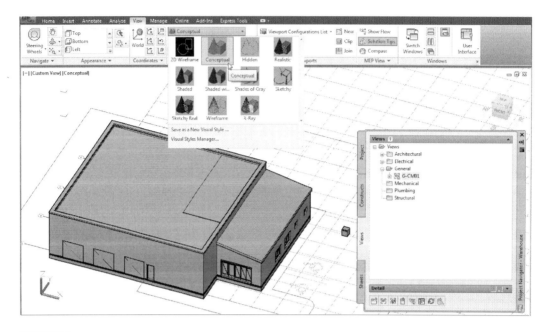

FIGURE 3.25
The composite model viewed from the southeast and displayed in Conceptual

Notice that all of the Construct files have been XREFed into this View file and inserted at their correct locations and heights respectively relative to the Levels settings on the Project tab of the Project Navigator. This is one of the major benefits of the Project Management system in AMEP. Once the basic parameters have been established, all the level and height information, as well as which files are required to assemble a particular View, are handled automatically by the system.

To study clearances and determine if any of the systems are clashing, you can use the interference detection tool. (Refer to the "Interference Detection" topic in Chapter 15 for more information.)

    12. Save and Close the file.

Visual Styles provide a wonderful way to visualize your designs in different graphical presentations such as hidden line or the Conceptual style shown here. However, if you do not have a suitable video card, performance can suffer when viewing a drawing in Visual Styles other than 2D Wireframe. To see if your video card is recommended by Autodesk for use in AutoCAD MEP, visit www.autodesk.com.

UNDERSTANDING FLOOR PLAN VIEWS

In the next series of steps, we'll learn how to create a Floor Plan View file. Like the Composite Model, this will also be a "General" View; however, it will only include one level of the project rather than all levels as in the composite model. Remember, when creating a View the software will automatically gather all of the correct Construct files required to make the View at a particular Level and Division combination. Thus the View file created will be ready to receive notes, dimensions, and other annotation appropriate to floor plans (or whatever specific type of drawing is intended). Later we can compose one or more of these Views (including their drawing specific annotations) onto a Sheet for printing. This is the intended workflow built into the Project Navigator toolset. Admittedly, however, this workflow is more suited to architectural projects than engineering. As such, some engineering firms choose not to create Views for plans at all, but work only in Constructs. There are three basic approaches to take with respect to the question of floor plan View files:

Create floor plan Views for all plans—As noted, this is the workflow followed by most architectural firms and many engineering firms. It is the "as designed" workflow of the software.

The advantages of this approach are that you can rely fully on the automated XREF behavior that Project Navigator offers. When working in Constructs simply drag and drop files to XREF (as we saw above). Since they will use Overlay, there will be no conflict or circular references when View files are created. When you create a plan View file, simply check the level and discipline folder(s) you want, and a View complete with all required XREFs will be created. Add any final embellishments such as view-specific notes, dimensions and in some cases detail linework to this View. When you drag this View to a Sheet (see below) it will automatically create a Viewport and scale properly. You can even take advantage of the cross-referenced callout features in AMEP, and have the titlebars and section callouts coordinate with the Sheet numbers via specially embedded field codes.

One disadvantage of this approach is that you may see degraded performance when creating separate Views, making them unworkable in many environments. Many users find the workflow of going from Constructs to Views to Sheets confusing and overly complex. As noted in the "To View or not View" sidebar above, most AMEP objects have integral annotation and labels that must be placed with the geometry in the Constructs. This diminishes the need and value of the separate View file and annotation championed by the "intended" workflow, and a lack of support for integral AMEP annotation which appears in the Construct as part of the AMEP objects.

Create plan Views for Sheet Benefits only—Even if you have decided that you wish to do all annotation tasks directly in the Constructs with your geometry, you may still want to experiment with creating plan View files particularly if you wish to take advantage of automated XREFs, automatic Sheet and viewport creation, and cross-referenced callout field codes. Even though the annotation would be part of the Constructs in such a scenario, having the separate floor plan View file would allow you to take the fullest advantage of the integrated Sheet Set in Project Navigator. In an architectural firm, most Constructs are XREFed into several other files (both Constructs and Views). This makes keeping them "pristine" and free of drawing-specific annotation an important requirement of using Project Navigator correctly and effectively. In an engineering environment, each discipline mostly will maintain its own files, and there will be less need for one discipline's work to be XREFed to others. Even when they are, AMEP offers display tools to control the display of automatic labels and annotation in ways not possible in traditional AutoCAD. So one discipline can reference the files of another at the Construct level and have a much higher degree of control over what is shown than an Architect using AutoCAD Architecture.

Do not create floor plan Views—In this approach, you would simply choose to do all floor plan work directly in Constructs, saving View files only for coordination type Views, like the composite model we created above, or automatically generated Views like the ones created when cutting sections (see Chapter 14). If you decide to take this approach, you will have to plan your XREF strategy carefully to avoid creating circular references (using Attach instead of Overlay). You will also have to create your Sheets and their viewports manually.

CREATE A FLOOR PLAN VIEW

The process to create a floor plan View is nearly identical to the process we used to create a composite model View.

13. On the Project Navigator palette, on the Views tab, click the Add View icon (second from the left at the bottom) and then choose the "Add a General View" option.

14. Name the file **P-FP01**, give it a description of **First Floor Plumbing Plan**, and then click Next.

15. On the Context page, place a check mark in the First Floor box only and then click Next.

On the Content page, notice that all of the discipline folders will be selected, but that only Constructs associated with the First Floor Level will be displayed.

16.Clear the checkboxes next to the *Electrical*, *Fire Protection* and *HVAC* folders.

Notice that this also clears any subfolders and the Constructs they contain. You can also simply deselect the Constructs themselves. The difference is, when you select or deselect a Category (folder) you are establishing this as a rule going forward. In other words, if another First Floor Construct is added later to one of the deselected folders, it will *not* be added to the View file even when that View is regenerated. If you add a Construct on the First Floor to one of the folders that is selected like *Architectural* or *Plumbing* in this case, that Construct will be added automatically to the View the next time it is regenerated. If you check or uncheck the Constructs themselves, the choice will affect only that Construct, and you will have to edit the View Properties later to determine the behavior of any Constructs added to the project after the View is created. This is an important benefit of careful category/folder planning in your projects.

17.Click Finish to complete the wizard and create and open the View.

The new View file has been created and opens onscreen (see Figure 3.26).

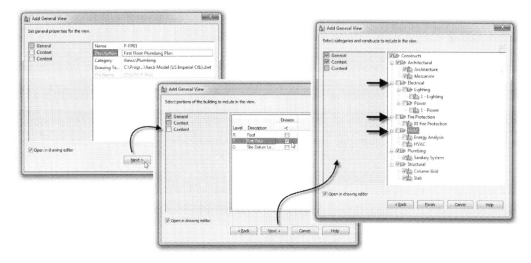

FIGURE 3.26

Create a Floor Plan View by selecting only one Level

Notice that all the first floor architectural and structural files have been XREFed into this file. Further, since we deselected most discipline categories, only the plumbing information has been XREFed.

18. On the Insert tab of the ribbon, click the External Reference dialog launcher icon on the Reference panel (see Figure 3.27).

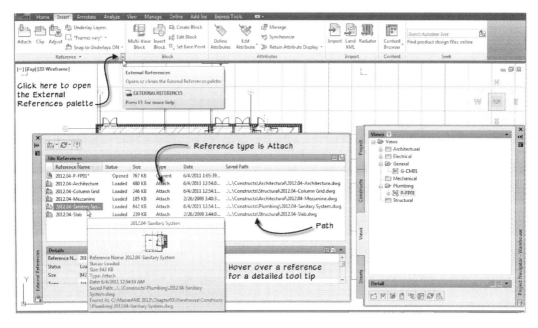

FIGURE 3.27

Study the External References in the newly created View file

The External References palette is the core AutoCAD tool for viewing and managing XREFs. We can see most of the same information directly from Project Navigator. However, Project Navigator's "External References" dialog (shown in Figure 3.12 above) does not show us whether the XREF is Attach or Overlay. Since a View file is intended as an intermediary file that will be dragged and dropped on a Sheet, all XREFs here use Attach automatically.

> Tip: XREF Attach creates a "nested" XREF. This means that if you XREF file A into file B, and then XREF file B into file C, file C will also see file A "nested" within. When you choose Overlay, the XREF is only one level deep. This means that if you Overlay file A into B, and then Overlay B into C, only B will appear in file C. File A will not nest through.

If you decide to create floor plan Views like the one here, your next task will be to add plan-specific notes, dimensions, detail items or other embellishments to make this file ready for placement on a Sheet.

ADD A FLOOR PLAN MODEL SPACE VIEW

If desired, you can make some decisions about how you would like this particular View to appear on a Sheet for plotting even before you create the Sheet. To do this, you can designate the portion of the model that you wish to appear within the viewport of our Sheet Layout. You can assign a display configuration, a title, and plotting scale ahead of time as well. This step is optional in your own projects, but it is being shown here because although this adds an extra step, it will be much easier to compose your Sheets later if you make this extra bit of pre-planning effort up front.

19. On the Project Navigator palette, right-click the *P-FP01* file, and choose **New Model Space View** (see the left side of Figure 3.28).

> Note: A Model Space View is simply the AutoCAD Named View that is essentially a "saved zoom." You could achieve almost the same result by going to the View tab of the ribbon on the Viewports panel, and clicking the Named tool.

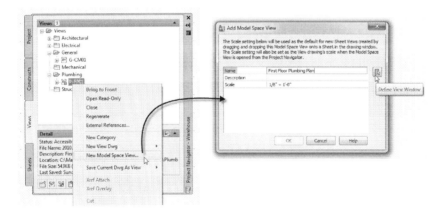

FIGURE 3.28
Create a New Model Space View in the P-FP01 file

20. In the Name field, type **First Floor Plumbing Plan**.

21. Verify that the Scale is set to 1/8" = 1'-0".

It is not necessary to type a description for this exercise.

22. Click the Define View Window icon on the right (see the right side of Figure 3.28).

23. At the "Specify first corner" prompt, click below and to the left of the floor plan.

24. At the "Specify opposite corner" prompt, click above and to the right of the floor plan (see Figure 3.29).

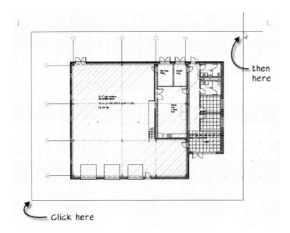

then
here

Click here

FIGURE 3.29
Designate the Model Space View boundaries

25. The "Add Model Space View" dialog will reappear. Click OK.

In Project Navigator, indented beneath *P-FP01*, an icon labeled "First Floor Plumbing Plan" will appear. AutoCAD Named Views (Model Space Views) appear in Project Navigator on both the Views and Sheets tabs. You can select these in Project Navigator to view detailed information or a preview at the bottom of the palette just as you can the actual drawing files. You can toggle between the Detail and Preview views with the icons on the preview pane (see bottom right corner of Figure 3.30). To open a file and zoom right to the Model Space View, double-click the Model Space View in Project Navigator. Since this is just an AutoCAD Named View, you can also see this new Model Space View on the View tab of the ribbon.

26. In Project Navigator, double-click First Floor Plumbing Plan beneath *P-FP01* (see Figure 3.30).

FIGURE 3.30
Name the View and then designate its boundaries—double-click to zoom to it

A temporary label will appear on screen with the Model Space View name (First Floor Plumbing Plan in this case). This will vanish the next time you zoom or pan.

ADD A TITLEMARK

In addition to providing a convenient way to zoom to a particular portion of a View file directly from Project Navigator and giving us a way to pre-assign the extents of a Sheet viewport (as we will see later), the name of the Model Space View is referenced automatically into drawing Titlemarks.

27. On the Tool Palettes click the Annotation tab, scroll to the bottom and then click the first Titlemark tool.

> Note: You will find the Annotation palette in most of the tool palette groups, so you shouldn't have to switch Workspaces.

Move the mouse around onscreen. You will notice that the First Floor Plumbing Plan Model Space View shows a temporary border, and whenever you move your mouse within it, the border highlights in red—this indicates that the callout you place will be associated to the Model Space View name.

28. At the "Specify location of symbol" prompt, click a point within the highlighted border and beneath the plan (see Figure 3.31).

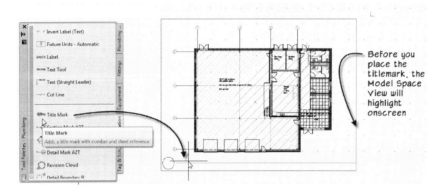

FIGURE 3.31

Add a Titlemark beneath the Plan, within the First Floor Plan Model Space View boundary

29. At the "Specify endpoint of line" prompt, drag to the right and click to designate the length of the title bar.

Notice that the label of the Title Bar has automatically picked up the name of the Model Space View. There is gray shading surrounding this label. This indicates that this is a field code. Field codes can be added to any piece of text including within Block Attributes (as is the case here) and then set to reference data from some other location. In this case, the field is configured to read the name of the Model Space View in which it is contained. Had we inserted it outside the Model Space View boundaries, it would have read the name of the drawing file instead. This is why it was important to insert the Titlemark within the boundaries of the Model Space View. Notice that the Scale has also been inserted as a field code and correctly reads the values we assigned earlier.

Finally, there is a third field code within this Titlemark: the number within the round bubble. Currently it is displaying a question mark (?). This is because it is tied to the actual drawing number for this plan from the Sheet. Since we have not yet built the Sheet, the field cannot yet display the correct number. Later when we build our Sheets, this question mark will automatically be replaced with the correct designation (see Figure 3.32).

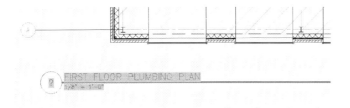

FIRST FLOOR PLUMBING PLAN

FIGURE 3.32
The Titlemark contains field codes that reference the drawing name and scale

30. When you are satisfied that the file has been created correctly, save and close the file.

You have now seen some of the features and potential benefits of creating floor plan Views. Below, in the "Create Sheets" topic, we will drag this View to a Sheet. After you have completed that and the other exercises in this chapter, please refer back to the "Understanding Floor Plan Views" topic above and consider the various approaches to dealing with floor plans before making a final decision on the approach you will take in your own projects.

CREATING SECTION VIEWS

AMEP provides several tools to view and document building sections. We can cut a "live section" view through the model that crops away the portion behind the Section line and reveals the sectioned portion in a live view. Depending on how these are cut and configured, live sections can even be printed under certain circumstances, but are really intended for coordination purposes and design review.

For most design development and construction document needs, the AMEP 2D Section/Elevation object gives us the required level of control by creating a separate two-dimensional drawing that remains linked to the original building model and can be updated when the original changes. To get just the right section requires a bit of careful configuration. (In Chapter 14, we will explore the 2D Section/Elevation object in detail.) In the "Add a Callout" topic in Chapter 14, detailed steps are given to create a section line, callout and associated section View file. For this reason, and due to the similarities to the steps already presented here, we will not outline the steps here. Please refer to that lesson and chapter if you wish to cut a section now.

Create Sheets

Sheet files are the final file type required in a Project Navigator project. The purpose of Sheet files is for printing documents. The Sheet file will have a titleblock border and one or more viewports. These viewports will display XREFed View or Construct files. In this exercise, we will create a Sheet file for our project.

CREATE A PLAN SHEET FILE

Let's continue with our First Floor Plumbing Plan.

    1. On the Project Navigator palette, click the Sheets tab.

        Be sure that the Sheet Set view is active. If it isn't, click the Sheet Set icon at the top-right corner of the Sheets tab. If Sheet Set View is active, but only Warehouse shows in the list, click the plus (+) sign icon to expand (see Figure 3.33).

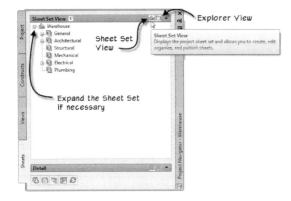

FIGURE 3.33

Be sure the Sheet Set view is active

    2. Select the *Plumbing* Subset.

    3. At the base of the palette, click the Add Sheet icon.

    4. For the Number, type **P-101** and for the Sheet Title, type **Plumbing Plan**.

Notice that the File Name is created automatically by concatenating these two fields. Also, since we are using the project number prefix option, the project number is also added automatically to the beginning of the file name. You can edit the File name field manually before clicking OK, but it is recommended that you accept the default naming behavior. Even though it may not match precisely the way you name Sheet files today, it is a logical system and happens automatically, making it much easier to ensure consistency from drawing to drawing and project to project.

5. Check the "Open in drawing editor" checkbox and then click OK (see Figure 3.34).

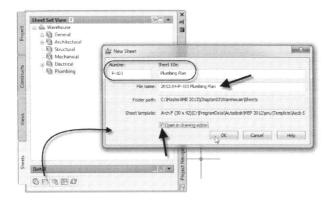

FIGURE 3.34
Create a new Sheet for the Plumbing Plan

> **Tip:** When you rename a Sheet file, you can choose whether the associated drawing file renames as well. When you do this, you will have the option to use the Name only, or the Name and Number in the new file name. To see these options, select a Sheet, right-click and choose **Rename and Renumber**.

Like other kinds of files in Project Navigator, there is also a checkbox here to open the drawing in the editor after creation.

> **CAD MANAGER NOTE:** You will notice that the template used for Sheets is different from the one used by the other file types. When you first set up a project, there are settings for this. You can assign the template used directly to the Sheet Set, or each individual Subset can have its own template—this can be very useful in multi-discipline firms. We are using the defaults here which loads the *AECB Sheet (Imperial ctb).dwt* template file to create Sheets. You can change these settings using the Edit Project icon on the Project tab and by right-clicking on the Sheet Set.

6. Zoom in on the titleblock and examine the fields. Notice that many of them have filled in automatically with project data field codes.

7. Zoom out to see the whole sheet.

8. On the Project Navigator palette, click the Views tab.

9. From the *Plumbing* folder, drag and drop the *P-FP01* View from the Project Navigator palette directly onto the Sheet Layout (see Figure 3.35).

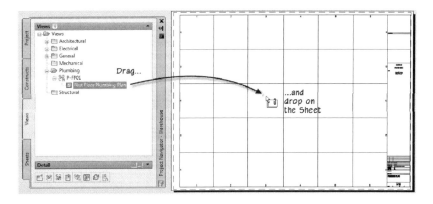

FIGURE 3.35
Drag the Plumbing Plan from the Project Navigator onto the drawing Sheet

The image of the file will appear onscreen with the lower left corner attached to the cursor.

10. Move the Viewport to position the Plan to an optimal position onscreen.

You may need to fine tune the position or size of the Viewport after placement. You can move it around and resize it with the grips.

Dragging the View onto the Sheet in this way has created a single Viewport scaled at 1/8" = 1'-0". (We can verify the scale by selecting the Viewport object. A small Quick Properties panel will appear showing the layer and scale of the Viewport.) Notice that the Viewport is also locked. This prevents the viewport scale from being changed accidentally when someone has the Viewport active and then zooms. Notice also that the Viewport is automatically placed on a non-plotting layer.

Notice that the bubble next to the Titlemark for the First Floor Plumbing Plan has filled in automatically with the number 1. This is because this is the first drawing on this Sheet. If you drag another view onto this Sheet, it will become number 2, and so on (see Figure 3.36).

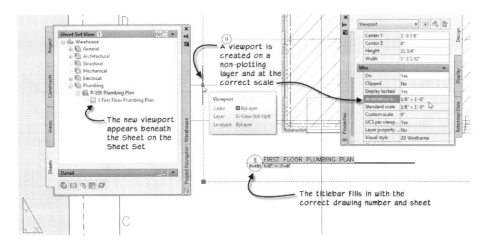

FIGURE 3.36
A locked, properly scaled and non-plotting viewport is created—the Titlemark callout updates as well

Take a look at the Sheet Set in Project Navigator. Notice that there is now a View indented beneath the Sheet name. These are more AutoCAD Named Views. (We have been calling them Model Space Views till now; however, that name is not applicable in Paper Space Layouts.) These Named Views represent the area of the Sheet that is associated to a particular number. Try double-clicking on the View. It will simply zoom to that location on the Sheet.

11. Save and close the *P-101 Plumbing Plan* Sheet.

> **CAD MANAGER NOTE:** Although your office standards may vary considerably from the file naming and XREF structure presented here, the more important issue is the strategy of using consistent file naming and XREF structures from one project to the next. It is also critical to set up the project files as early in the project life cycle as possible. Building the set early allows for easier setup and maintenance, and allows the project team members to follow an established standard.

ADDING CONSTRUCTS TO SHEETS

When you drag Views to Sheets, several steps are performed for you automatically. The View file (and its nested XREFs) is XREFed to model space at the correct coordinates. Thus, the XREF created is placed on its own layer (this facilitates viewport layer freezing). A viewport is created in paper space at the correct scale, locked and placed on a non-plotting layer. If they exist, callouts are updated throughout the project. Not bad for a simple drag and drop.

If you have chosen to forego Plan View files, you will need to XREF your Constructs directly to your plan Sheet files. Unfortunately this will take a few more steps than the simple drag-and-drop process outlined in the last sequence. However, it is still a little easier to perform the required steps using Project Navigator than it would be following the traditional manual AutoCAD procedures.

Here is a simple summary of the process. You will create the Sheet using the same steps as those covered above. Instead of dragging and dropping the Construct directly onto the Sheet (in paper space) as you were able to do with the View file, you will first need to switch to model space in the Sheet file. You can drag and drop the Construct from Project Navigator to model space. It will XREF the file as an attachment to the correct X and Y coordinates. You will need to adjust the Z coordinate manually if necessary. You will also need to place the XREF on its own layer manually if desired. Return to paper space, create a viewport, zoom and scale it to the inserted XREF. Adjust any layer settings and lock the viewport.

MAKING SHEETS MATCH ARCHITECTURAL SHEETS

When you receive the background files in Project Navigator as we have been discussing in this scenario, it is likely that the Architect has also configured Sheets of their own. If they have sent you these Sheets, you can save some work in configuring your own Sheets by "borrowing" their viewports. To do this, open the Architectural Sheet you wish to mimic. In paper space, select all the viewports (you can select one, right-click and choose **Select Similar**). On the Home tab of the ribbon, expand the Modify panel and choose Copy with Base Point from the Copy to Clipboard drop-down button. Type 0,0 for the base point and then press ENTER. Create your Sheet using the procedures covered here. Expand the Modify panel again, and choose the Paste to Original Coordinates tool. (If this does not place them correctly, try Paste and then type 0,0 again.) Switch to model space and perform the procedures to XREF the required Constructs into the file as noted in the previous topic.

UPDATES

When the Architect sends you updated files, you can simply overwrite the previous versions of the project files with the new ones they send. Naturally you will want to backup the project directory before overwriting any files. Make sure that the Architect has not sent you copies of your own files before doing this. Communication with the Architect and running a few tests will ensure seamless updates and prevent loss of work. Remember to backup often!

If your Architect is not using relative paths as outlined above in the "Understanding Repath Options" topic, encourage them to do so. It will make updates much easier as you will not have to mimic the Architect's drive letters and folder structure. If they insist on absolute paths, your next best option is to create a custom drive mapping that mimics the Architect's drive and folder structure.

12. Close the current project and, when prompted, close (and save) all project files.

ENGINEER IS USING PROJECT NAVIGATOR, ARCHITECT IS NOT

Our third sub-scenario: The first option under Scenario 2 occurs when you have decided to use Project Navigator, but your Architect has not. This would apply if they are using plain AutoCAD or some other CAD program.

The only major difference in this scenario from the other two covered so far is that you will have to create the project yourself. The basic process involves creating the project, setting up the Levels and Divisions (if being used), adding folders and then adding files. Since we have already covered how to add files (Constructs, Elements, Views and Sheets) we will focus on the initial setup tasks in this topic.

CREATE A NEW PROJECT

1. On the QAT, click the Project Browser icon.

Browse to the location (usually on the project server) where you want to save the project. This is the location from which all team members will access it. For this example, we'll continue to work in the *Chapter03* folder.

2. With the *Chapter03* folder showing in the drop-down list, click the New Project icon (see item j in Figure 3.1 above) at the bottom of the "Project Browser" dialog.

> Note: Be sure to browse to the folder first, then click the New Project icon. Review items c through h in Figure 3.1 above for the tools used to navigate within Project Browser.

The Add Project worksheet will appear.

3. In the Project Number field, type **2012.05**.

4. In the Project Name field, type **Chapter 3**.

5. For the Project Description type **AAMS AutoCAD MEP 2012**.

6. Clear the "Create from template project" checkbox (see Figure 3.37).

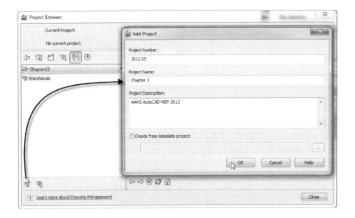

FIGURE 3.37

Input the name, number and description of the new project

7. Click OK to create the project.

CAD MANAGER NOTE: The bulletin board and image in this project are blank. Any HTM or HTML file can be used as a project bulletin board. You can use virtually any word processor, text editor, or HTML editor to create or edit project bulletin board files. A project bulletin board can be used by a Project Coordinator or CAD Manager to keep project team members informed on project news, or as a way to provide links to project standards, project tools or links to related Web sites and resources. The Project Web page displays in its own integrated Web Browser window within the Project Browser, complete with its own Back, Forward, Home and Refresh icons (see item i and item l in Figure 3.1 above). In this way, the project Bulletin Board can actually reference a home page to an entire intranet project Web site. If you prefer not to use the Bulletin Board feature in projects, simply leave it blank.

The Project Image file is a logo for the project. Any image file can be used for this image, but it must be saved in BMP format. Use Photoshop, Windows Paint or any other image editor to save the image file. If your project has its own logo, you can load it here. Otherwise, you can use your company logo, the client's logo, or you can leave it blank.

In this example, we are creating the project from scratch (we have cleared the template project checkbox). However, you have the option to create projects from a template project. A template project is much like a drawing template (DWT): It provides a pre-configured starting point for a new project that can potentially include settings, drawing template file references, tool palettes and even premade drawing files. There are no good MEP samples provided with the software. There is an architectural example named *Commercial Template Project* that provides a good example of the potential of template projects. It includes pre-configured levels and a collection of premade Constructs, Views and Sheets and other settings as well. You can also find a detailed description of the Commercial Template Project in the help file.

> The new **Chapter 3** Project will appear in the Project Browser (highlighted in bold to indicate that it is current).

> 8. Right-click on the *Chapter 3* project in Project Browser and choose **Project Properties** (see Figure 3.38).

We have already discussed the settings in this dialog above. Make the following changes:

> 9. For "Use Relative Xref Paths," choose **Yes**.

> 10. For "Match Sheet View Layers to View," choose **Yes**.

> 11. For "Prefix Filenames with Project Number," choose **Yes** (see Figure 3.38).

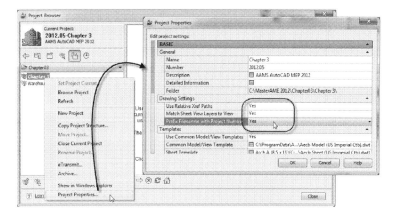

FIGURE 3.38

The Project Browser—editing properties of the newly created project

When you create a new project, a Tool Palette Group (refer to Chapter 1) can optionally be associated with the project. If you have Tool Palettes within this group, they will automatically load in project team members' workspaces. We will not use Project-based Tool Palette Groups in this book. Therefore, let's turn this off.

12. In the "Project Properties" dialog, scroll down to the Tool Palettes item.

13. For "Project Tool Palette Group" choose: **None** and then click OK.

14. Confirm any warnings about changing path types and if messages regarding the bulletin board and project image appear, simply close them.

15. Click Close to close the Project Browser and return to AMEP.

The Project Navigator palette will appear onscreen with the new project loaded.

> Tip: You can also load your project directly from Windows Explorer before even launching AMEP. Simply browse to the directory where the AMEP project is found and double-click on the APJ file located directly in the project's folder. AMEP will launch a new session and make the selected project current. If AMEP is already running, launching a project through Windows Explorer will open a second session.

SET UP PROJECT LEVELS

The first thing we need to set up in our new project is the Levels.

16. On the Project tab, click the Edit Levels icon (see Figure 3.39).

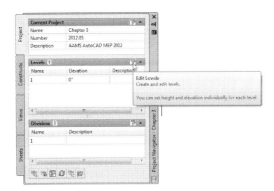

FIGURE 3.39
Edit the Levels from the Project tab on the Project Navigator

Let's create a simple three-storey building. We will need a Level for each. We also typically want to add a Level for the site conditions and the roof. This will give our new project five Levels.

The default Level is named "1."

17. Click in the Floor Elevation column and edit the value to **1'-0"**.

18. Change the Floor to Floor Height to **12'-0"**.

19. Change the Description to read **First Floor** (see Figure 3.40).

FIGURE 3.40
Edit the parameters of the First Floor Level

20. At the top right corner, click the Add Level icon three times.

Notice that the new Levels have automatically been named sequentially, and the Floor Elevations begin at height of the Level below. This is due to the "Auto-Adjust Elevation" setting, which is on by default (see the bottom left corner of Figure 3.40).

21. Edit the Descriptions to read: **Second Floor**, **Third Floor** and **Roof**.

22. Rename Level 4 to R and change its ID as well (see the left side of Figure 3.41).

FIGURE 3.41
Add and edit several Levels

23. In the Name column, select 1, right-click, and choose **Add Level Below** (see the right side of Figure 3.41).

    A Level is added below the current level; it is named "5." This name and the Floor Elevation of this Level require adjustment.

24. Click in the Name column (on the number "5"), pause a moment, and then click again.

    This should activate the rename mode for Level 5. You can also right-click and choose **Rename Level**.

25. Change the Name to **G**.

26. Click in the Floor Elevation column and edit the value to **0**.

    This will distort all the other Floor Elevation values. Don't worry—with Auto-Adjust on, it is easy to fix.

27. Change the Floor to Floor Height to **1'-0"**.

28. Change the ID to **G** (for "Grade") and the Description to **Grade Level** (see Figure 3.42).

FIGURE 3.42

Completing the edits to the Project Level Structure

29. Click OK to accept the values and dismiss the Levels dialog.

30. A prompt will appear asking you to "Update all Project Views." Click the Yes button (see Figure 3.43).

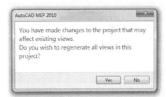

FIGURE 3.43

Completing the edits to the Project Level Structure

Answering "Yes" to this prompt will update all drawings that reference the Project Levels to incorporate the new values just entered. There are not any drawings in this project that would require such updating, but you should typically answer "Yes" to this prompt regardless. This will ensure the integrity of your project files and their relationships to one another.

Let's assume that the project we are configuring is large enough to warrant more than one Division. Divisions break the project into sections. You can use them as a way to split the work on a single Level between two or more individuals. Most projects will not use Divisions, but in case you have one that justifies their use, let's walk through the steps here.

31. On the Project tab, click the Edit Divisions icon

32. In the "Divisions" dialog, make the edits shown in Figure 3.44.

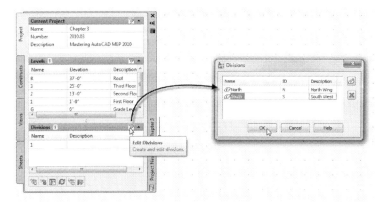

FIGURE 3.44

The Project Navigator complete with Divisions and Levels

33. Click OK to complete the Divisions.

At this point, we have completed the general project parameters and established that the building will be divided into five Levels (including the Grade and Roof Levels) and two Divisions. Keep in mind that we can revise this structure later if necessary and adjust the Level and Division structure as changing project needs dictate. Naturally, we would want to try to avoid drastic changes to the Level and Division structure wherever possible. The important point to note here is that this structure *can* be adjusted later if required.

The next step is to build the files called Constructs and Elements that will populate this building framework. These will be comprised of drawing files that represent various portions of the building information model as defined in the next topic.

IMPORT ARCHITECTURAL BACKGROUND FILES

In this scenario, we are receiving basic AutoCAD background files from our Architect. These are not Project Navigator files and in fact are simple 2D line drawings that do not use AEC objects at all. You can XREF the files directly into your files, but you might want to consider creating a "buffer" file first. When you get files from an outside source, you have no control over the contents or quality. Furthermore, even if you spend time to "cleanup" a received file, you will lose the benefits of your efforts the next time the Architect sends an update. Therefore, consider creating an empty file using your office standard template (DWT) file first. XREF the Architect's background into this file. Make any modifications to layers and other settings in this host file. You will be modifying the XREF layers, but the settings will be saved and maintained with the host file. You can even move or rotate the XREF if necessary, and those transformations will be saved with the file as well. When the Architect sends their update, overwrite their original file and then reload your XREFs. All of your cleanup settings will be preserved! The *Elements* folder in Project Navigator provides an excellent staging area for such procedures.

34. On the Project Navigator palette, click the Constructs tab.

35. Right-click the *Elements* folder and choose **New > Category**.

36. Name the new folder: **Architectural Backgrounds**.

You can save each file individually to the Elements folder by opening them in AMEP, right-click the Architectural Backgrounds folder and choose Save Current Dwg as Element. Confirm the name in the dialog that appears and then click OK. This is fine if you have only a few drawings to process. However, if there are many, you might want to try this alternative:

37. Open a Windows Explorer window and browse to the location of the files you received from the Architect.

   You will have to unzip the files to a folder on your system first. This location will most likely be a folder in the project directory on the server. Check with your project data coordinator to be sure.

38. Select the *Architecture* and *Column Grid* files, drag them from Explorer and drop them on the *Architectural Backgrounds* folder directly in Project Navigator (see Figure 3.45).

FIGURE 3.45

Drag the Architectural backgrounds from Windows Explorer to Project Navigator

An "Add Element" dialog will appear for each file dragged and dropped.

39. Simply click OK in each dialog to confirm them (see Figure 3.46).

FIGURE 3.46

Confirm the importing of each Architectural file as an Element

You now have a list of Element files matching the names of the files you received from the Architect. These are copies of the files you received saved in the *Architectural Backgrounds* folder. In addition, each one now has an accompanying XML file like the other project files as discussed earlier in this chapter. Our next step is to create a Construct for our Architectural files. You have a few options here. You can create a Construct for each file, or create a single Construct per floor Level. For example, if the Architect sent a core and shell file, an interior walls file, and a column grid file for each floor, you may want to merge all three of these into a single Construct. Simply create a Construct per Level and then drag and drop in the appropriate Element files.

40. On the Construct tab, right-click the *Constructs* folder and choose **New** > **Category**.

41. Name the new Category: **Architectural**.

42. Right-click the *Architectural* folder and choose **New** > **Construct**.

    If you wish, add a Description.

43. In the "Add Construct" dialog, name the file according to your standards, perhaps: **01 Architecture**.

44. If you created Divisions above, check both North and South for Level 1, check "Open in drawing editor" and then click OK (see Figure 3.47).

FIGURE 3.47
Confirm the importing of each Architectural file as an Element

45. From the *Elements\Architectural Backgrounds* folder, drag and drop the *Column Grid* file and then repeat for the *Architecture* file.

46. Make any layer adjustments necessary and then save and close the file.

When your Architect sends an update, simply browse in Windows Explorer to the *Elements\Architectural Backgrounds* folder and copy the new versions over the originals. Only the DWG files will be replaced; the XML files will be unaffected. The next time you open your *01 Architecture* Construct or update its XREFs, it will show the latest changes.

CREATING FILES

The rest of the procedure is the same as the steps covered above in the "Create folders," "Create Constructs," "Create Views" and "Create Sheets" topics. Create a folder for each discipline in the *Constructs* and *Views* folders. Create Subsets on the Sheet tab. Add all the Constructs, Views and Sheets that your project requires.

Save everything and make a backup of the project when your setup is complete.

TEMPLATE PROJECTS

Template Projects have already been noted above. If you find that you are performing many of the same setup tasks over and over on each new project, it is time to consider creating a Template Project. Any project can become a Template Project. Simply set up a project that adheres as closely as possible to your office standards, and then copy it to the server to a convenient location. The next time you create a project, check the "Create from template project" checkbox in the "Add Project" dialog (see Figure 3.37 above). Browse to the Template Project and AMEP will copy the Template Project to the location of the new project, rename it, and repath all XREFs. Your new project will begin where the template left off.

CONGRATULATIONS!

You have discovered many of the benefits of working in the Project Navigator environment. While the workflow may be different than you are used to today, you are encouraged to spend the time required to get comfortable with this powerful tool and all its features. Once you start reaping the benefits of a well structured Project Navigator project, you will wonder how you ever got by without it.

SUMMARY

✓ Thorough project setup can help give a good sense of project drawing requirements early in the project cycle.

✓ Using the Project Browser and Project Navigator tools makes setting up a project quick and easy.

✓ AMEP Drawing Management tools make use of XREFs to relate files to one another.

✓ XREF Overlay is used when you want the XREF to go only one level deep.

✓ XREF Attach creates nested references, which create a hierarchical reference structure.

✓ Model files are full-scale drawings used to generate actual project data on a daily basis.

✓ Constructs and Elements are Model files representing individual pieces of a complete building model.

✓ Constructs have a unique physical location within the Building Model; Elements do not.

✓ Views are used to gather a collection of Constructs (and any nested Elements that they may contain) for a specific viewing purpose.

✓ Sheet files are used for setting up "ready to plot" sheets for printing document sets.

✓ There are two overall project startup scenarios each with two sub-scenarios.

Scenario 1—Project Navigator is already set up. This occurs either with an in-house Architectural department, if you are part of an AE firm, or when your outside Architect sends you Project Navigator files.

Scenario 2—Project Navigator is not set up. Here you can still set up Project Navigator yourself for your engineering work on the project.

✓ Each scenario has its own unique setup issues and plenty of areas of overlap with the other scenarios. Become familiar with all tools, techniques and options available in Project Navigator to get the most out of the tool.

WHAT'S IN THIS SECTION?

This section includes a chapter for each major discipline included in AutoCAD MEP. The goal of each chapter is to introduce the overall concepts and procedures for using the various MEP objects. In Chapter 4, you will learn how to perform energy analysis using Spaces, Zones and gbXML export. In Chapter 5, the various mechanical objects will be introduced. Piping is the subject of Chapter 6, electrical systems will be explored in Chapter 7 and Chapter 8 looks at conduit systems.

If you wish to gain expertise in all disciplines, feel free to read the entire section. If you are only interested in a particular discipline, you may skip right to the chapter for your professional expertise.

SECTION II IS ORGANIZED AS FOLLOWS:

Section II

Chapter 4

Energy Analysis

INTRODUCTION

Most building projects require the mechanical design professional to calculate square footages of rooms, exterior walls, exterior wall openings, roof, and roof openings. This data is needed to help determine the appropriate heating and cooling requirements for the rooms and/or areas of a project. This task can consume a considerable amount of time, even before we factor in any project-related changes to the building design after the mechanical design professional has performed the initial calculations. The AutoCAD MEP (AMEP) Space object can help minimize the overall time required to gather a project's square footage data as well as assist in managing project related changes. When applied correctly, the use of Space objects can be the beginning of substantial time savings and greater accuracy of calculations.

OBJECTIVES

In this chapter we will focus on the Space and Zone object and the gbXML tools in an effort to learn how these tools can help the mechanical design professional share information from AMEP with third-party energy analysis programs. The following topics will be explored in detail:

- Workflow concepts for Spaces.
- Spaces setup, Styles, behaviors, and how Spaces relate to one another.
- How to manipulate the Space boundaries.
- Understanding plenum space and when you might need one.
- Understanding Zones and their use in single or multiple drawings.
- Using the Space/Zone Manager.
- Exporting and importing gbXML files.

SPACES

A Space object is typically used to represent a single room or a contiguous area in an open plan assigned to a particular function. If the room is enclosed by Walls on all sides, it is easy to make the Space. In an open plan, we sometimes need to draw the Space manually or add linework, using a Space Separator tool to assist in the auto-generation process. Spaces have many useful characteristics. They can be queried for area and volume, be assigned a room name and number, have many useful properties applied such as occupant loads and analysis values, and can be displayed graphically in almost limitless ways. We will explore many of these features throughout the lessons in this chapter.

As the mechanical design professional, you could work with different architectural firms and therefore receive background data in a variety of ways. In some cases, only a 2D AutoCAD file will be provided. In other cases, the Architect will be using AutoCAD Architecture and will provide Walls, Doors, and Windows. In still other cases, the Architect, using AutoCAD Architecture, will also lay out Spaces for the project. In such a scenario, this theoretically means that you can use the Spaces the Architect created to perform your mechanical design analysis. This can be a huge time saver. However, the Architect's goal in creating Spaces may vary, but typically they use Spaces to calculate the square footage of the project and assign room names and numbers. The boundaries defined for Spaces serving those goals may not always match the needs of the mechanical design. There might be times in which the Space created by the Architect does not meet your needs. A simple example might be an open office area. The Architect might create several small Spaces within such a space for the purposes of labelling each cubicle or for tracking square footage by occupant to charge to certain departments. For purposes of determining heating and cooling loads, such divisions in the overall space may not prove meaningful. Furthermore, since you as the mechanical design professional carry your own project liability, you may choose not to use the Architect's Spaces even if they are apportioned the way you require. Regardless of the route chosen, use the Architect's Spaces or create your own, using Spaces will enhance your workflow and typically yield a significant reduction in project time compared to similar traditional methods.

SPACE TYPES

Spaces can be represented in three distinct geometrical ways: 2D, Extrusion, or Freeform (see Figure 4.1). Prior to assigning a Space to a particular room or area, you can use your Properties palette to select one of the three available geometry types. You can also preset the Space Create type on a Space tool (see the "Create a Tool Palette" topic below).

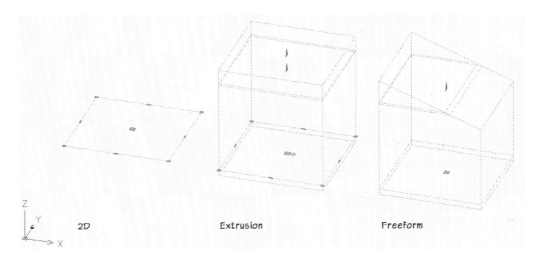

FIGURE 4.1

Spaces can be represented in three distinct geometrical ways

A 2D Space will merely generate the square footage information for a particular room or area. If no volumetric data is required, the 2D Space is appropriate. A 2D Space will generate the following information once placed in a room: length, width, base area and base perimeter (Actual Dimensions).

The Extrusion Space will not only generate the square footage information for a particular room, but will also generate the square footages of exterior walls and roofs and their respective openings. The Extrusion geometry type adds a three-dimensional characteristic to the Space and exposes the floor, ceiling, and volume components of the Model display representation. An Extrusion geometry type of Space will generate the following Component Dimensions once placed: overall space height, ceiling height, floor thickness, ceiling thickness, height above ceiling and height below floor. In addition to the Component Dimensions that are generated, the Actual Dimensions of length, width, base area, base perimeter and base volume are generated.

The Freeform geometry type of space will generate the same information as the Extrusion Space. When choosing Freeform you gain the ability to modify the three-dimensional form of the volume component. A Freeform geometry type of Space will generate the following

Component Dimensions once placed: ceiling height, floor thickness, ceiling thickness, height above ceiling and height below floor. In addition to the Component Dimensions that are generated, the Actual Dimensions of length, width, base area, base perimeter and base volume are generated. When an Extrusion type of Space is converted to a Freeform type of Space all the openings will be retained with the same location, shape and area.

The three different types of Spaces control how the Space is represented graphically. When any one of the three types of Space is placed, the actual area and perimeter can be seen on the Properties palette (see FIGURE 4.2). When an Extrusion or Freeform type of Space is placed, the base area, perimeter, and volume, etc., are representative of the actual Space geometry.

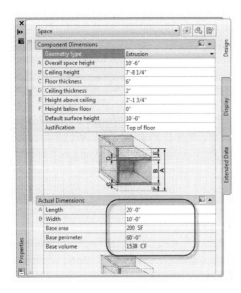

FIGURE 4.2

Base dimensions of all three geometry types are the actual dimensions of the Space and not based on calculations or interpretations

SPACE CREATION TYPES

Another important component of the Space is the Create Type. There are four creation types for Spaces: Insert, Rectangle, Polygon and Generate. Insert places a Space with an area determined by the style, and you can define the length and width during placement. You can also edit it after placement. The rectangle option allows you to pick two opposite corners to create a rectangular-shaped Space any size you require. The Polygon option allows several vertices to be placed in order to draw a non-rectangular shape. The process is like drawing a polyline. After placing Spaces with any of these three methods, the Space grips can be used to modify the Space boundary shape (see Figure 4.3).

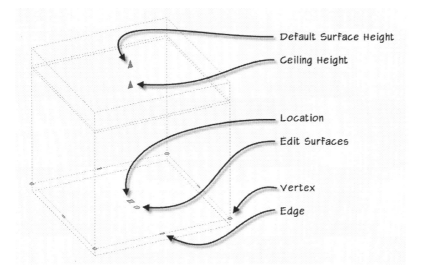

FIGURE 4.3
Understanding Space object grips

For the Insert, Rectangle and Polygon Create type, the Associative setting is always No. This means that if Walls or other room boundaries change, you will need to manually update such Spaces to match the new plan layout.

The final Create type is Generate. A Generated Space can be either Associative (automatically updates for any changes to the objects it is associated to, such as wall, floors, etc.) or Non-Associative (requires manual update). Generate is the quickest way to create Spaces when Wall or other bounding geometry is present in the file.

SPACE BOUNDARIES

If the mechanical design professional would like to have even more control or provide more detail with the Space object, there are several alternative boundaries available with each Space object. To expose the Space boundaries, click on a Space and when the Properties palette appears change the Offset boundaries to Manual. Click the Space again to see additional Space grips at the center of the Space. Hover over each of the grips and you will see that you can enable and disable one of four different boundary types (see Figure 4.4).

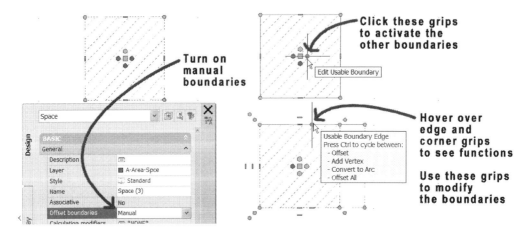

FIGURE 4.4
Manual boundaries can be used to modify the Base, Net, Usable and Gross areas

Spaces have four alternative boundary types. Each boundary type displays a different aspect of the Space and can be edited individually. Here is a brief summary of the Space boundary types:

Base boundary—normally representing the inner area of a room covered by a space. This is the area generated by bounding objects in an associative space. In most cases, the base boundary is identical to the net boundary.

Net boundary—this boundary is offset from the base boundary and can be used for planning and detailed design. The net boundary can also be used for special applications when the calculated area of a space is smaller than the base boundary.

Usable boundary—this boundary is offset from the base boundary. The usable boundaries typically extend from the inside of the exterior walls to the middle of the interior walls (or a specified distance into the interior walls).

Gross boundary—the gross boundary is offset from the base boundary. Normally, the gross boundary is measured from the outside of the exterior walls to the middle of the interior walls.

Note: The preceding definitions were excerpted from the AutoCAD Architecture online help.

UNDERSTANDING THE WORKFLOW

Space objects provide an easy and sophisticated means of calculating room data. They also give the mechanical design professional the ability to share that information with other applications outside of AMEP through the gbXML functionality built into AMEP. The Space tool on the HVAC Analysis palette is predefined to Generate spaces bound by Walls or other such objects. Using this tool will automatically generate Space boundaries by making them conform to the shape of the surrounding geometry. Space objects typically display on screen like an AutoCAD hatch pattern; however, the graphical display can be adjusted. Spaces contain room-specific data like the square footage of the room, the square footage of the room's exterior walls and wall openings and the square footage of the roof and its openings. In addition, a Space can also house other data like the ceiling height, plenum height and equipment loads in a particular room based on user input. Spaces, and their accompanying data, can then be associated to a Zone object. A Zone object is a group of one or more Spaces, typically served by an air distribution device (like an air-handling unit, VAV box, etc.). In a nutshell, a Zone object in AMEP has the same meaning as a zone in the mechanical engineering industry. After the Spaces have been associated to a Zone (or Zones) the Space information can be exported out of AMEP to third-party energy analysis programs. An energy analysis can be performed in the third-party application from the exported data. Depending on the analysis tool, the results may be imported directly back into AMEP and stored within the individual Spaces from which the data was previously exported.

Space objects serve many different purposes for several disciplines. Architects can use them in AutoCAD Architecture to house room data such as name, number, department and other data. They can also be used in conjunction with certain callout tools to assist in the creation of interior elevations. Even the electrical design professional can save a considerable amount of time utilizing the Space tool to place their electrical devices around a room (see Chapter 7 to learn more).

CHOOSING YOUR WORKFLOW PATH (2D OR 3D)

The key component that you must identify before you can be successful with Space, Zone and gbXML is an understanding of what you are trying to accomplish with the Space, Zone and gbXML tools. You must set your objective before you place your first Space in your project. For example, if you are merely seeking room square footage information for your project, that will take you down one path. However, if you are seeking room square footage information and the square footage information of all the room's associated components (i.e. walls, door, windows, roofs, etc.) that you intend to utilize in a third-party energy analysis program that will take you down a different but parallel path. Although the two are very similar in nature they will produce extremely different data.

Space objects can be either 2D or 3D. The most likely determinant in your decision over which mode to use will be the kind of data you wish to generate. If the architectural files are traditional 2D CAD files, you might choose simply to create 2D Spaces. However, in this situation you could also use 3D Spaces like the Extrusion or Freeform space. Either one of those spaces will allow you to add opening information (like doors and windows) to your Space object (using grips) even though the architectural floor plan is a traditional 2D file (see Figure 4.5). It might be important to you to manually generate this opening information, especially if you are strongly considering using the gbXML tool in AMEP.

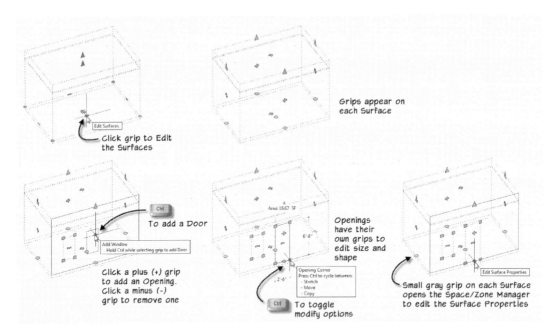

FIGURE 4.5
If your architectural reference drawing is in a 2D format, editing the Surfaces of your Spaces allows you to add opening information

If you receive AutoCAD Architecture (ACA) backgrounds containing Walls and other 3D objects, it makes the most sense to use 3D Spaces as you will not need to manually add the openings. Both modes offer square footage calculation benefits. Utilizing Spaces is an extremely quick and efficient way of determining room square footages for your project.

There are added benefits to utilizing the Space tool within an ACA file. Not only can the room square footage be quickly determined, but since ACA uses objects (Walls, Doors, Windows, Roofs, etc), the square footage of the walls and all their openings, and the square footage of the roof and all its openings can be determined as well. This is one of the many time saving benefits that 3D project modeling has over 2D electronic drafting.

Even though there can be a considerable difference between the 2D and the 3D information generated with the Space tool, you should not be discouraged from utilizing the Space tool in a 2D environment (lines, arcs and circles) where appropriate. Even in a 2D environment, the square footage of a project can be determined quickly and effectively—much faster than by using the traditional methods that have gripped this industry since the advent of electronic drafting. This book will focus primarily on the modeling (3D) aspect of projects as the industry appears to be moving in the direction of project model creation. However, the authors believe that it is important to mention the 2D aspect of Spaces as many design professionals will be dealing with 2D legacy files for many years to come.

WHERE TO BEGIN

Since most projects will include some sort of background file created by an Architect, you will need to assess the data such backgrounds contain and then make a decision about the workflow you wish to pursue. If the Architect is using ACA, they will most likely use the Space tool to generate the square footage information and the room names and numbers for the project. If provided with such a file, you can elect to use all the Space information that the Architect generated, some of the Space information the Architect generated, or none of the Space information at all. Since you likely carry project liability as the mechanical design professional, you may choose to ignore the Space information generated by your architectural partner and add your own Space information in an effort to feel more comfortable with the data generated. You do have complete flexibility in this situation and can make this decision as you see fit. The workflow concepts discussed in this chapter will outline both scenarios.

SPACE WORKFLOWS

As noted above, it is typical to begin a project with background files received from an architectural firm. These files are typically XREFed into your files to form the background. AMEP does not require this workflow to change. However, if the Architect's file already contains Space objects, and you have decided to use them, you must understand how you can use these architectural Spaces in your work. Let's quickly examine a few possible workflows.

- If you are satisfied with the way the Architect has assigned the Spaces to all the rooms in the project then you can simply reference the Architect's project files into your drawing file, insert a Zone(s) and then assign the Space(s) to the Zone(s). This is the simplest workflow process.
- If only some of the Spaces that the Architect created meet your needs, you have the option to add new Spaces over the top of just the architectural Spaces that are not suitable.
- If none of the Spaces that the Architect created meet your needs, you will have to add all new Spaces over the top of all the Architect's Spaces.

Now that we have reviewed a few workflow concepts for Spaces, let's adjust some Space options before we begin placing Spaces.

OPTIMIZING SPACE OBJECT SETTINGS

One of the first things that should be done before placing Space objects in the project is to configure some options that will allow you to layout and manage your Spaces more efficiently. These settings are located in a variety of places in the software, including the "Options" dialog, the "Drawing Setup" dialog and the "Style Manager."

INSTALL THE DATASET FILES AND LAUNCH PROJECT BROWSER

In this exercise, we use the Project Browser to load a project and then use Project Navigator to explore the various folders and files within the project structure.

1. If you have not already done so, install the book's dataset files.

   Refer to "Book Dataset Files" in the Preface for information on installing the sample files included with this book.

2. Launch AutoCAD MEP 2012 and then on the Quick Access Toolbar (QAT), click the Project Browser icon.

3. In the "Project Browser" dialog, be sure that the Project Folder icon is depressed, open the folder list and choose your *C:* drive.

4. Double-click on the *MasterAME 2012* folder.

5. Double-click *MAMEP Commercial* to load the project. (You can also right-click on it and choose **Set Project Current.**).

IMPORTANT: If a message appears asking you to repath the project, click Yes. Refer to the "Repathing Projects" heading in the Preface for more information.

Note: You should only see a repathing message if you installed the dataset to a different location than the one recommended in the installation instructions in the preface.

6. Click the Close button in the Project Browser.

When you close the Project Browser, the Project Navigator palette should appear automatically. If you already had this palette open, it will now change to reflect the contents of the newly loaded project. If for some reason, the Project Navigator did not appear, click the icon on the QAT.

CONFIGURE SPACE OPTIONS

Let's begin with settings in the Options dialog.

7. On the Project Navigator, click the Constructs tab. (If the Project Navigator did not open automatically when you closed Project Browser, press CTRL + 5 to open it now or select the Project Navigator tool on the QAT.)

8. Under the Constructs node, expand the *Mechanical* folder and then double-click on the 03 Spaces file to open it.

9. From the Application menu, choose Save As > AutoCAD Drawing.

10. Name the copy: **03 Spaces 2D** and then click Save.

> **Note:** Using Save As will not add the drawing to the project in Project Navigator. To do so, with the drawing closed, right click on the drawing name in Project Navigator, and select Save Current Dwg as Construct.

Notice that the architectural floor plan (XREFed here as a background) already has a few Spaces (displayed as diagonal hatching in the rooms) for the two toilet rooms and a few adjacent rooms. Recalling the discussion on workflows, you have two options: You can use those Spaces or generate new ones over the top of the architectural ones.

11. From the Application menu, choose **Options**.

12. Using the scroll buttons at the top right, scroll to and then click the AEC Object Settings tab.

13. Verify and/or change the following settings in the Space settings grouping:

⇨ Turn on: "Automatically update associative spaces".

⇨ Change the "Maximum Gap Size" to **4'-0"**.

⇨ Verify that the "Maximum Automatic Adjacency" is: 1'-0".

⇨ Make sure that the "Calculation Standard" is **None** (see Figure 4.6).

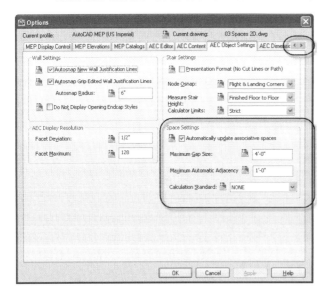

FIGURE 4.6
AEC Object Settings in the Options dialog box

14. Click OK to dismiss the dialog.

15. If not already active, load the HVAC Workspace with the Workspace Switching tool on the Application Status Bar.

If you need to switch your Workspace, use the tool on the Application Status Bar as discussed in Chapter 1.

An associative Space is shaped by the objects that define it (i.e. Walls, Floor, Roof or even lines). If the shape or the sizes of the objects that define the shape of the room are modified, the associative Space will automatically be updated. Configuring Spaces to automatically update can be a huge time saver. With the Maximum Gap Size set to 4'-0", we allow for openings in the walls between Spaces. These settings will help you layout and manage your Spaces more efficiently.

If your project contains walls that are thicker than 1'-0" you will need to set your Maximum Automatic Adjacency to match the largest wall thickness of your project. Space objects are inherently intelligent and can detect the surface of adjacent Spaces.

> **Tip:** At any point after placing your Spaces in your Drawing you can graphically see the relationship between adjacent spaces by typing `AecShowSpaceAdjacencies` *at the Command prompt.*

Regarding Space adjacencies and the gbXML functionality within AMEP, let's review the programmed intelligence of the Space objects to help you gain a better understanding of Space adjacencies. A single Space within a drawing assumes that since it is "all alone in the world" its walls are exterior facing (in lieu of interior) and that it has a roof above since there is no other Space above it (see the left side of Figure 4.7). If another Space is placed directly next to the Space in the discussion, the adjoining walls will now become interior walls (see the middle of Figure 4.7). The others will be exterior walls. Going one step further, if another Space is placed directly above the original Space in discussion the original Space will no longer reflect a roof above (see the right side of Figure 4.7). Having this basic understanding of Space adjacencies and how Spaces interact with each other is a must in order to understand how Spaces get exported from AMEP to gbXML, and subsequently to a third-party heating and cooling load analysis program. The Space/Zone Manager can be used to manipulate surface types. We will cover the Space/Zone Manager later in this chapter.

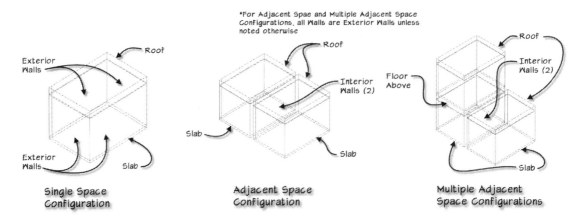

FIGURE 4.7
Adjacencies are automatically determined and treat the surfaces of Spaces differently depending on the other nearby Spaces

DRAWING SET-UP FOR SPACES

If either a 2D or a 3D drawing will be utilized as the background to place the Space objects, the referenced drawing must be set up to be Space bounding. This is a fairly simple process.

1. XREF the architectural drawing into the drawing where you will be placing Spaces.

2. Select the XREF drawing, right-click and choose **Properties**.

3. On the Properties palette, expand the Advanced grouping and set the "Bound spaces" setting to **Yes**.

This setting will allow the XREF to be used for Space bounding objects. If the objects or external references in your Drawing have not been set to Space Bounding, then you will not be able to generate Spaces because Spaces must have bounding objects in order to be generated (see Figure 4.8).

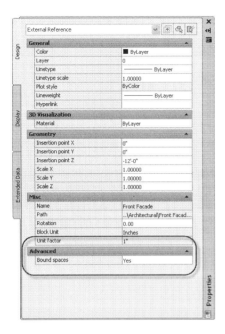

FIGURE 4.8

XReferences and linework must be set to be Space bounding or you will not be able to generate Spaces automatically

Let's create some Space Styles and begin placing Spaces.

ADDING SPACES TO A DRAWING

Like other AEC objects, Spaces are style-based. A style provides a way to configure a collection of parameters and settings that apply to all objects using the style. If the style is edited, the change affects all objects in the drawing assigned to that style. For example, you could easily select all objects in a drawing belonging to the same style and change the watts per square foot, or you could assign a value for the number of people in a Space of this style, or set the ceiling height that is typical for the Space style, etc. As is true for all objects, the default Space style is called Standard.

> **Note:** Even though a Space style or a Space name has been assigned to a Space (i.e. Office) a third-party heating and cooling load analysis program may or may not read that Space style name in its respective program.

However, if your primary goal is to calculate square footage information because you have no desire to export the Space data from AMEP to a third-party analysis program, it may not be necessary to create any Space styles for the project at all. You have complete flexibility here. Furthermore, if you begin the project using only the Standard style, you can always reassign Spaces to other styles as you create them later in the process.

Let's explore how we can quickly generate square footage information for our project using the Space tool and the auto generation functionality built into the Space tool.

1. On the Analyze tab of the ribbon, on the Space & Zone panel, click the Space tool.

2. On the Properties palette, for Tag, choose: **Aecb_Space_Name_Tag**.

3. Beneath "Generate Space" change Allow overlapping spaces to **Yes.**

4. Beneath "Component Dimensions" change the Geometry type to 2D.

5. Zoom in on the upper middle area of the plan. Move your mouse into the empty area directly to the left of the Spaces that the architect generated in their plan. Do *not* click yet.

   Notice how the room boundary highlights when the cursor is within it (see Figure 4.9).

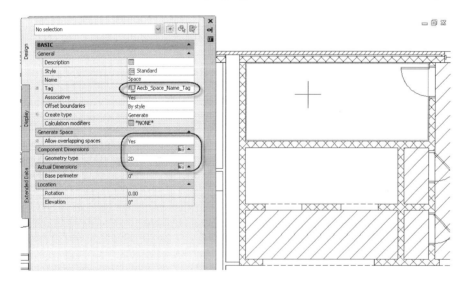

FIGURE 4.9

With the generate option enabled, the boundary and shape of the room is determined by the surrounding objects

Before you click the mouse, move to the area directly below the area shown in Figure 4.9 and the outline that room will highlight in red. This is typically the case for all rooms that have not had Spaces generated for them.

Conversely, if you try to place a Space in a room where a Space has already been generated (even if it is not visible due to layer or other visibility settings), you will get a Tool Tip notification that a Space has already been created (see Figure 4.10).

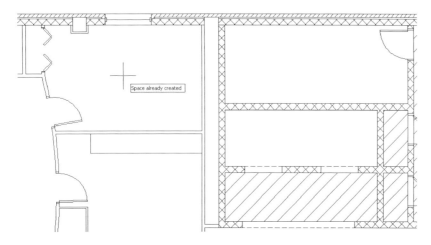

FIGURE 4.10
If Spaces have already been created for rooms AMEP will notify you through the Tool Tip

Tip: If you have to pan or zoom during Space generation, you may see "No space found" in the ToolTip at the cursor and a "Valid boundary not found." warning message on the command line. Type **V** (reset Visible boundaries) and press ENTER at the command line to have previously off-screen objects included in the boundary set.

6. Move the mouse back to the original area as shown in Figure 4.9, click within the area to place your Space and then press ENTER.

Notice that a Space object (represented graphically with diagonal hatching) has been generated in the shape of this room and has been named and numbered automatically.

7. Select the new Space and, if not already displayed, right-click and choose **Properties** to make the Properties palette appear.

Beneath the "Actual Dimensions" grouping, notice that several calculated values are displayed, including the Base area (square footage) of the Space. This information was generated far more quickly and efficiently than would be possible with the traditional methods typically employed by engineering professionals using CAD tools.

8. Zoom Extents and then, on the Analyze tab of the ribbon, click the Space tool again and move your mouse to the corridor in the left of the plan. Do not click yet.

Notice that the red boundary is highlighting not only the corridor, but also the reception and conference room areas (see Figure 4.11). This is because there is no clear boundary between these areas. We can use Walls,lines, polylines and other geometry. A file with such lines has been provided in the *Elements* folder.

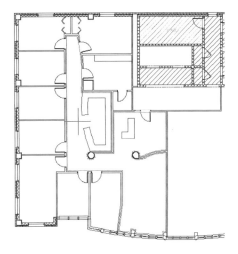

FIGURE 4.11
When Walls do not separate the rooms, a single continuous Space will be generated

Tip: Before placing Spaces in your drawing it is recommended that you freeze all layers that do not contain boundaries. This will help avoid Spaces being generated around objects like casework, toilet partitions and other similar objects. This will help ensure that your square footage calculation for such rooms is not distorted.

9. Press the ESC key to cancel the command without creating any Spaces.

Let's use the polylines to generate a more defined area. A file containing polylines further delineating the Spaces has been provided in the *Elements* folder.

10. On the Project Navigator, on the Constructs tab, expand the *Elements\Mechanical* folder.

11. Right-click the *03 Space Outlines* element and choose **Insert as Block**.

12. In the Insert dialog box select the Explode checkbox and then click OK.

13. At the insertion point prompt, type **0,0** and press ENTER.

This will insert the *Third Floor Space Outlines* file into the *03 Spaces* drawing. A set of magenta lines have been added to the drawing file. These represent the boundaries we need for the Spaces.

14. On the Analyze tab, click the Space tool again. Move your mouse into the corridor again.

15. Notice that there has been a clear delineation to the Corridor Space with the inclusion of the lines. Click to create the Space.

> **Note:** If you need to create additional boundaries, this can be done with the lines or polylines. On the Properties palette, under Advanced, set Bound spaces to Yes.

Generating Spaces on large projects can be a time consuming task. Let's expedite this process by having AMEP automatically generate all of the Spaces for us.

16. Right-click and choose **Generate all**.

17. Press ENTER to complete the command.

Zoom into the Toilet Room area and notice that Spaces have been generated over the top of the defined Spaces in the architectural plan. As the mechanical design professional, you will need to decide which Space to use. If you are comfortable with the Space the Architect generated, you can simply erase the Space you just generated. If you are more comfortable with the Space you just generated, you can freeze the layer containing the architectural Spaces. In the same area, notice that a Space has been generated for the plumbing chase. The Space tool is only intelligent enough to determine bounding objects for Spaces. It is not intelligent enough to determine which Spaces are valid and which are not. You should review all Spaces generated to determine which ones are valid and simply erase the ones that are not.

For the purposes of this exercise, let's assume that we are satisfied with the Spaces that we just generated. How can we summarize the Space and the Space square footage information? The easiest approach is to generate a Schedule. AMEP has Schedule Table style located in the Content Browser preconfigured for such a task. We will discuss how to access the Content Browser and the tools in the Content Browser later in this chapter. For the purposes of this exercise the schedule has been preloaded in our *03 Spaces 2D* drawing. Let's insert this Schedule now to get a quick snapshot of both the individual Space square footage values and the sum of the Space square footages.

18. On the Annotate tab, on the Scheduling panel, click the Schedule drop-down button and then select the Space Engineering Schedule (see Figure 4.12).

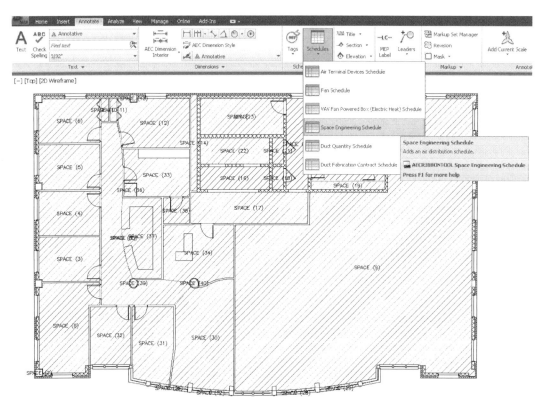

FIGURE 4.12

Using a predefined Schedules style, we can get a quick report of all Spaces in the drawing

Note: The Space Engineering Schedule defined in the *03 Spaces 2D* drawing is different from the style that may be preloaded into other templates or that may be loaded using the tool on the ribbon if not already present in the drawing.

19. At the "Select objects" prompt, make a window selection around the entire floor plan and then press ENTER.

20. Pan the drawing to allow room next to the plan and then click a location for the Schedule in the drawing.

21. Press ENTER to accept the default scale.

Your schedule, summarizing your Space square footage information, has been added to your drawing. Study the individual entries and note the total at the bottom (see Figure 4.13). Your results might be slightly different.

SPACE ENGINEERING SCHEDULE		
Room #	Space Type	Area
SPACE	STANDARD	160.6
SPACE (10)	STANDARD	8.7
SPACE (11)	STANDARD	8.7
SPACE (12)	STANDARD	181.2
SPACE (13)	STANDARD	2.3
SPACE (14)	STANDARD	32.4
SPACE (16)	STANDARD	17.9
SPACE (35)	STANDARD	293.
SPACE (36)	STANDARD	8.7
SPACE (37)	STANDARD	159.4
SPACE (38)	STANDARD	35.4
SPACE (39)	STANDARD	2.8
SPACE (4)	STANDARD	139.5
SPACE (40)	STANDARD	3.0
SPACE (5)	STANDARD	143.2
SPACE (6)	STANDARD	140.2
SPACE (7)	STANDARD	2.3
SPACE (8)	STANDARD	242.3
SPACE (9)	STANDARD	2,596.1
		6,539.8

FIGURE 4.13
The Space Engineering Schedule presents a simple list of the Spaces and their areas

We quickly learned how to generate 2D Spaces for square footage information for our project. Remember we identified that as our objective at the beginning of this exercise. The key component that must be identified before you can be successful with Space, Zone and gbXML is an understanding of what you are trying to accomplish with the Space, Zone and gbXML.

To achieve the best results, this objective should be set before you place your first Space in your project.

> For our next exercise, we will reuse the *03 Spaces* Construct provided with the CD ROM files.

22. From the Application menu, choose **Close**.

23. When prompted to save the *03 Spaces 2D* drawing, choose Yes.

WORKING WITH SPACE STYLES AND TOOLS

Now let's explore how we can generate different types of Spaces through the creation of Space styles and Space tools in an effort to obtain more information from the Space object and to have more control over the Space object as well.

Let's look at how Space styles can be created. There are two processes that we will cover. We will look at the manual creation process of Space styles through the Style Manager and copying predefined Spaces from the Content Browser, and modifying that information once it has been placed on a tool palette. With our Space styles created we will use the Space tool to place the Spaces in the rooms of our drawing.

CREATE A SPACE STYLE

1. On the Project Navigator, click the Constructs tab. (If you closed the Project Navigator click the Project Navigator icon on the QAT).

2. Expand the *Mechanical* folder and then double-click the *03 Spaces* file to open.

Let's learn how to create Space Styles to provide more definition to the Spaces in our project. Make sure that the HVAC Workspace is active and that your Tool Palettes are displayed for this exercise.

3. On the Manage tab of the ribbon, on the Style & Display panel, click the Style Manager button.

4. Expand the *Architectural Objects* folder and then select the *Space Styles* item.

5. In the right-hand pane, right-click and choose **New**.

6. For the name of the new Space style, type **Lease Space,** and then press ENTER (see Figure 4.14).

FIGURE 4.14

Space Style created through the Style Manager

When first created, all Space styles typically will have a cross-hatching pattern as their graphic representation. Space styles can be configured to display differently from one another to provide a clear differentiation between the different types of Spaces in a project. Let's create a solid pattern that will clearly identify all of our Lease Spaces in the project so we can visually locate those Spaces more easily.

7. Expand the Space Styles item in the left pane, and then select the Lease Space style beneath it.

8. In the right pane, click the Display Properties tab.

The active Display Representation will be bold in the list. (It should be Plan Screened in this case). In the Display Property Source column, each representation is currently controlled by the Drawing Default. We want to apply a new hatch pattern to Spaces of this style only. To do this, we'll add a style override.

9. Place a checkmark in the Style Override checkbox next to Plan Screened.

10. In the "Display Properties" dialog box, click the Hatching tab.

11. Next to the Base Hatch Display Component, click on User Single in the Pattern column.

12. In the "Hatch Pattern" dialog box, for the Type select **Solid Fill** from the drop down menu and then click OK (see Figure 4.15).

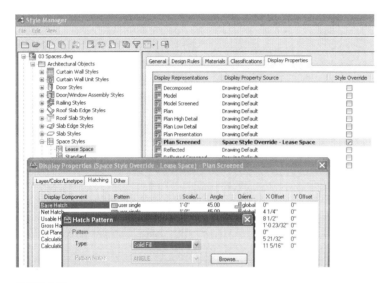

FIGURE 4.15

The graphical representation of Space Styles can be modified to meet user preferences.

13. Click the Layer/Color/Linetype tab.

14. For the Base Hatch Display Component, change the color to Color 1 (Red) and then press OK.

15. Click OK in the "Display Properties" dialog box, and then click OK again to dismiss the Style Manager.

ADD A STYLED SPACE

We have now created a new Lease Space style that will appear with a solid red pattern in our project drawings. Now let's place our newly created Space Style in our project.

16. Zoom Extents so that we can see the complete drawing.

You should notice a large open area below the two Toilet Rooms in our project drawing. This is our Lease Space area where we want to apply our new Lease Space Style. You can add Spaces using the same method we did above, but there are also other methods as well. In this sequence, we will use the Tool Palettes.

17. On the Tool Palettes, click the Analysis tab.

18. On the Analysis tab, click the Space tool.

19. On the Properties palette, beneath the Basic grouping, choose Lease Space for the Style.

20. For "Create type" select **Generate**.

21. For the "Geometry type" select **Extrusion**.

22. Place your cursor in the open area to the right, click to create the Space and then press ENTER to terminate the command (see Figure 4.16).

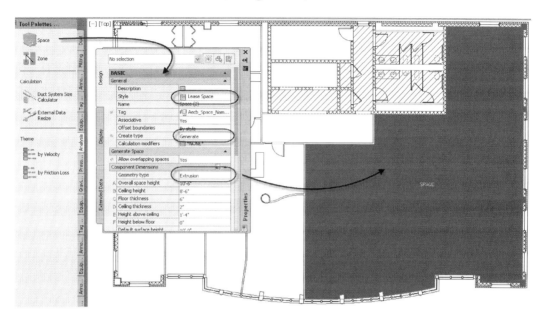

FIGURE 4.16

Space generation through the tool palette

23. Select the Space that you just placed, right-click and choose **Properties**.

Scroll through the Design tab in the Properties palette to review the square footage information that you just generated in a matter of a few mouse clicks. Notice that we now have Component Dimensions like Ceiling height that can now be modified; and also notice that we have Base volume (cubic feet) information available in the Actual Dimensions. These components were not available with the 2D Spaces we originally placed in our drawing in the previous exercise (see Figure 4.17).

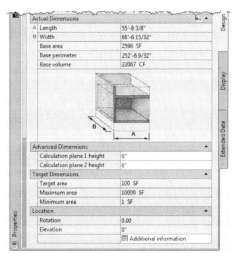

FIGURE 4.17
More information is available for extruded Spaces than was for 2D Spaces

CREATE A TOOL PALETTE

You can also place a Space through the ribbon as we did in the "Adding Spaces to a Drawing" topic above. The exact procedure you use is a matter of personal preference. Now let's look at a different way to create Spaces and edit Space Styles by using the Content Browser.

24. Right-click on the Tool Palettes titlebar and choose **New Palette**.

25. Type **MAMEP Spaces** for the name of the new palette and press ENTER.

You now have a blank tool palette to begin storing different Space Styles as you create them. There are Space styles provided with the software. You can access them using the Content Browser. Let's copy a few Space tools from the Content Browser and place them on our newly created tool palette.

CAD MANAGER NOTE: This tool palette will reside locally on your computer only. If you wish to share your palette with other users, you will need to copy the entire palette to a network location. Palettes stored on the network can be accessed using Content Browser or an Enterprise CUI.

ACCESSING PRE-MADE SPACE STYLES

26. On the Home tab, on the Build panel, click the Tools drop-down button, and choose **Content Browser**.

    Content Browser is a separate application and will open in its own window.

27. On the main panel of the Content Browser, click the Design Tool Catalog – Imperial catalog.

    A list of categories appears on the left.

28. Click to select Spaces.

29. On the right side, click the link for Commercial and then select Page 3 of Commercial.

30. On page 3 you should see Office (Large), Office (Medium), and Office (Small).

31. Hold down the CTRL key and click each of these three Spaces (see Figure 4.18).

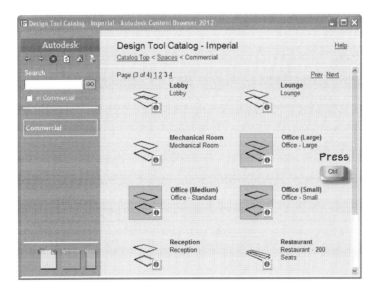

FIGURE 4.18

Space tools in Content Browser. Use CTRL to select multiple items

32. Right-click one of the highlighted items and choose **Copy**.

33. Close the Content Browser.

34. Back in AMEP, right-click the MAMEP Spaces palette and choose **Paste**.

IMPORTING GBXML PROPERTIES

You have now added three new Office Space Styles to the MAMEP Space tool palette. Adding tools to a tool palette makes them more convenient to access and gives us the ability to pre-configure options for tools. For example, the style of each of these tools is already pre-configured. We can change styles just by clicking a different tool without the need to change the properties palette. Any of the other settings like description, create type and basic dimensions can also be pre-configured in the tool. Another very powerful feature that we can take advantage of with our tools is pre-assignment of Property Sets. In this case, let's add some gbXML Property Sets to our drawing to increase the power of the Spaces that we are going to generate.

35. On the Insert tab of the ribbon, on the Block panel, click the Insert Block tool.

36. In the "Insert" dialog box, click the Browse button.

37. In the "Select Drawing File" dialog box, click the *Content* folder icon in the left pane.

38. Double-click the *Styles* folder and then double-click the *Imperial* folder.

39. Select the *gbxml Property Set Definitions (Imperial).dwg* file, click Open. When the Insert dialog box appears, make sure that the Specify On-screen selection box in Insertion Point is deselected, and Explode is selected, and then click OK in the "Insert" dialog box (see Figure 4.19).

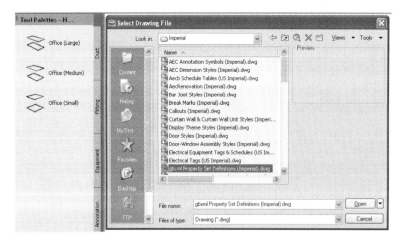

FIGURE 4.19

Insert the gbXML Property Set Definitions for Spaces

40. Even though we exploded the block on insertion, and the block contains no geometry, the style contents of the file have been added to the current drawing. The important part of the process is that we now have the gbXML Property Sets available in the current drawing. We will discuss the gbXML Property Sets further as we continue this exercise.

You may be wondering, "What is a Property Set?" Here is a brief summary of Property Set Definitions. A Property Set Definition is a list of Property Set Data available to the objects and schedules in a drawing file. Property Set Data attached to objects is comprised of one or more object properties defined by one or more Property Set Definitions. Property Sets establish links between objects and the schedules that report them. A Property Set Definition determines how a Property Set will be applied (object-based or style-based), what properties it contains, and how the properties are configured. To gain a better understanding of Property Set Definitions, please refer to Chapter 16.

It should be noted that the SpaceEngineeringObjects Property Set that we will be adding to our drawing plays a key role for load calculations for all the Spaces you wish to analyze. Whenever you would like to analyze Spaces, at a minimum you must have the following Property Sets in your drawing:

SpaceEngineeringObjects—applies to Spaces in AMEP and will allow you to assign occupancy factors, lighting and miscellaneous equipment load factors, and certain airflow values to the Space.

ThermalProperties—allows you to apply U-values to Wall Styles within AMEP.

ZoneEngineeringObjects—allows you to set the design heating and cooling temperatures for the Spaces.

Each of these Property Sets is found in the *gbxml Property Set Definitions (Imperial).dwg* file inserted above.

CAD MANAGER NOTE: You have the ability to manually populate additional Property Set information in the Space for a particular room in order to make this information available to your third-party analysis program. This additional information can be lighting values, total number of people, equipment loads, etc. However, the authors of the book understand that most mechanical design professionals will have this type of room information defined in their third-party heating and cooling load analysis program. Based on the establishment of space type information in the third-party analysis program populating data within AMEP might be a duplication of effort. Whether you add such data to AMEP or not is a matter of personal preference. Although the authors of the book understand the traditional workflows of the mechanical design professional and their third-party analysis programs they believe that they will be doing the mechanical design professional a great disservice by not mentioning the additional interoperability between AMEP and other products herein. Interoperability results will vary based on the third-party load analysis application.

EDIT A SPACE STYLE

An advantage of using tool palette tools is that we can pre-assign many properties and settings to the tools. This means that objects created from the tool will automatically have all the desired settings without our being required to apply them manually. Let's modify one of the Office Space tools to meet our needs for this project.

41. On the MAMEP Spaces tool palette, click the Office (Small) space tool.

42. Press the ESC key after the Analyzing Potential Spaces dialog disappears.

    Running the tool and then canceling it imports the Office_Small Space style into your drawing. You can also right click the tool to Import or Re-import the associated style.

43. On the MAMEP Spaces tool palette, right-click the Office (Small) tool and choose **Space Styles.**

    The Style Manager will appear and Space Styles will be selected.

44. On the left side, select the Office_Small Space style. On the right, click the Display Properties tab.

45. Place a checkmark in the Style Override checkbox next to Plan Screened.

46. In the "Display Properties" dialog box, click the Layer/Color/Linetype tab.

47. Change the color for the Base Hatch Display Component to Color 3 (Green) (see Figure 4.20).

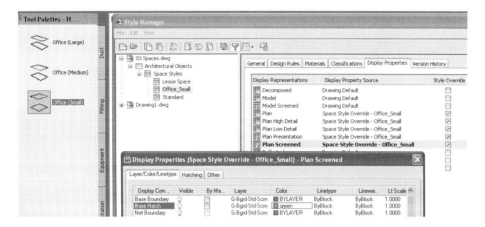

FIGURE 4.20

The graphical representation of Space Styles can be modified to meet user preferences

    48. Click the Hatching tab.

    49. Next to the Base Hatch Display Component, click on User Single in the Pattern column.

    50. In the "Hatch Pattern" dialog box, for the Type, select **Solid Fill** from the drop-down menu and then click OK.

    51. Click OK twice to return to the drawing.

You have now modified the Office (Small) Space style and it will appear with a solid green pattern when used in this drawing.

> **CAD MANAGER NOTE:** By modifying the Space Style in this manner, the changes apply only to the current drawing file. If you want the modified Style available from the project, you need to place the modified version in your office standard template (DWT) file, edit the original library file (the one accessed from Content Browser) and make the modified version available on the server to all users.

ADDING PROPERTIES TO SPACE TOOLS

Now let's modify some of the Space tool information.

    52. On the MAMEP Spaces Tool Palette, right-click the Office (Small) tool and then choose Properties.

53. Edit the following items in the Basic > General Grouping:

⇨   In the "Tool Properties" dialog box, click in the Description field.

⇨   In the "Description" dialog that appears, type **Office Space,** and then click OK.

⇨   From the Name pop-up menu choose **Office Area**.

⇨   From the Tag pop-up menu, choose **Aec7_Room_Tag**.

Note: Tags are Multi-view Blocks. The list of styles that appears is already available in the current drawing. Keep this in mind in your own projects. If the Tag you want is not on the list, you will need to exit the dialog and then import it into the drawing using a Tag tool or the Style Manager first. Alternatively, you can specify a Tag location (a .dwg file) that contains the desired tag style, and have the specified style imported when necessary.

⇨   From the Create type pop-up menu, choose **Generate**.

⇨   In the Component Dimensions grouping, from the Geometry type pop-up menu choose **Extrusion** and then click OK (see Figure 4.21).

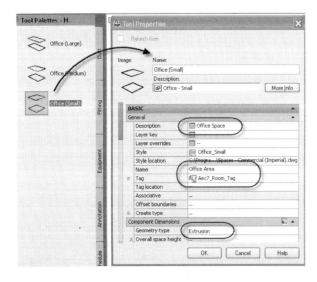

FIGURE 4.21

Configure the Tool Properties for the Office_Small tool

PLACE SPACES FROM A CUSTOMIZED TOOL

We have now defined specific information for this Space tool that will be applied to every Space that is placed using the tool. We will cover additional properties that we can modify Space by Space later in this chapter. For now let's use the modified Space tool to place some spaces.

54. Zoom Extents so that you can see the complete drawing.

    You should notice five rooms on the left side of the plan. Those five rooms are your office Spaces to which you'll apply the modified Office (Small) Space Style.

55. On the MAMEP Spaces Tool Palette, click the Office (Small) tool.

The Properties palette should appear. If you study the settings on the palette, you will note that you do not need to provide the description, Style type, name, tagging information, etc., as we preset this information in the Space tool.

56. Place your cursor in the office at the top of the drawing, and begin placing the Office Spaces and working your way down the bank of rooms.

57. Once you have placed all five rooms, press ENTER to terminate the command (see Figure 4.22).

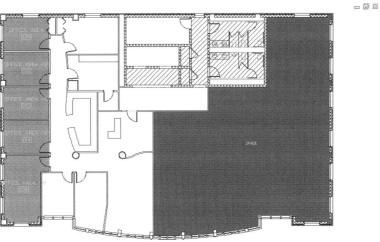

FIGURE 4.22

Place five new Spaces using our modified Space tool.

You should note that a room tag has appeared in the Space. Note that the room numbers associated with the tags have automatically incremented sequentially beginning at 101 and ending at 105.

> **Note:** The starting number for the room tag is defined in the SpaceObjects property set definition and can be revised through the Style Manager. In this exercise we could have elected to start with a 300 series number to correspond to the floor where we were placing our Spaces. To gain a better understanding of Property Set Definitions, please refer to Chapter 16.

You should also be aware of the fact that the room tag and number will not match information in the architectural background. While this fact is potentially disappointing, there are a few things that you should consider. First, it is not necessarily bad that the names and numbers do not match. Typically, the mechanical design professional will do a whole project analysis during the schematic design and design development phase of a project. More than likely, the Architect's room names and numbers have not "settled down" at this point in the project. In addition, there are likely to be changes to the room names and numbers during the early phases of the project. Having a different room name and number tag for your Spaces gives you the ability to create "a roadmap" of their project Space information that will allow you to move forward without the need for the Architect to generate this information. Finally, as we will see in the next sequence, you may need to separate your Spaces differently from the architectural plan to account for perimeter zones or other design issues. In this case, it would be difficult to maintain a correlation between the names and numbers as well.

Now that we have placed some Spaces in our project, let's dissect one of those Spaces. What happens if there is a need to provide further clarification to a Space(s) that we have already added? For example, perhaps there is a perimeter zone we would like to account for. Certainly we could use lines or polylines to help identify this area, but AMEP includes a Space Separator tool that can be used for this. Let's look at this tool.

58. Zoom in on the office in the lower left-hand corner of the drawing.

DIVIDE A SPACE

Let's create a perimeter zone in this room on the west (left) side of the room.

59. Select the lower left hand corner office Space.

Tip: Try clicking it by the doorway to select the Space instead of the XREF.

60. On the Space tab of the ribbon, on the Modify panel, click the Divide Space tool.

61. At the "Specify first point" prompt, click the left endpoint of the window on the south side (bottom) of the room.

62. At the "Specify next point" prompt, extend the separation line up to the north wall in the room, and click a perpendicular point on that wall (see Figure 4.23).

63. Press ENTER to terminate the command.

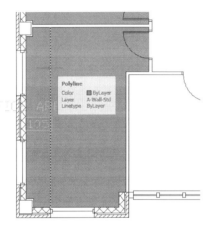

FIGURE 4.23
The Space Separation tool is merely a polyline that divides the Space further

The Space has now been separated into two (see Figure 4.24) and a separation line has also been placed. Both Spaces are the same Space style, but there are now two distinct Spaces in this room. Optionally, a "Perimeter Zone" style can be created with different settings than the office style. In this case, after you divided the Space, you would simply select the Space at the perimeter zone and change its Space style to Perimeter Zone.

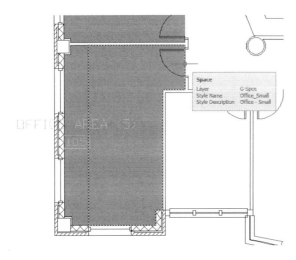

FIGURE 4.24

After the Space Separation tool is applied the Space is separated into two distinct Space.

We have quickly learned how to create Space styles, change their graphic representation and modify existing Space tools from the Content Browser. We have also learned how to place Spaces using autogenerate. Now let's gain a better understanding of Spaces and their related components.

> Note: The Space Separator tool draws polylines that will have the Bound spaces property preset to Yes. This tool also uses a separate layer key. This will allow you to change the layer key to a non-plotting layer if you do not want the polylines to plot.

MODIFYING SPACES

There are a few ways we can modify existing Spaces. We can make a global modification, a style-level modification or even modify each Space individually. This is not unique to Spaces. Most style-based objects can be modified in a similar fashion. In general, you will want to attempt global modifications first, and then style-level modifications, and only resort to object by object modifications if the other methods will not achieve your desired goal. We have already seen style-level modifications in the topics above. Let's take a look at how we can accomplish a global modification.

1. With the *03 Spaces* drawing still open, Zoom Extents.

2. Using a window selection, select all objects in the drawing.

   If the Properties palette is not displayed already, right-click and choose **Properties**.

At the top of the Properties palette, the drop-down list should read something to the effect of "All (17)." "All" indicates that the selection includes a mix of object types, and the number is the quantity of objects selected. If you click on the list and open it, each type of object will be listed by name with its own quantity next to it. For example, it might read: Space (7), Polyline (1), etc. Your totals may vary from those indicated here.

3. From the object list (reading: All (17) or similar) choose Space (7).

   Your quantities may vary.

4. The Space information should now appear in the Design tab of the Properties dialog box (see Figure 4.25).

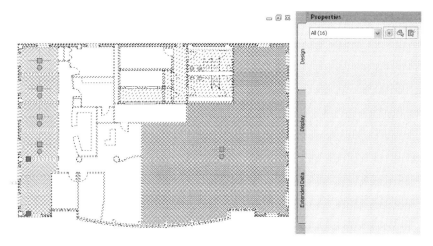

FIGURE 4.25
Selecting all Spaces to modify globally

From here you can change any of the Properties of the selected spaces. For example, you could change the Floor Boundary thickness and/or Ceiling boundary thickness for all Spaces even though they are different styles. You will want to use caution with a selection like this. Notice that values like the Name and Style say: Varies. This is because the objects in the selection belong to different styles.

5. Change the Ceiling Thickness to 1/2".

Naturally we cannot see the effect of this change in plan view. If you wish, you can change to a front or side view to see the new thickness, but this is not necessary.

6. Press the ESC key to deselect all objects.

Another way to change many Spaces is by selecting similar Spaces. We did this in one of the previous topics, but look again at how Space styles can help isolate objects further. If you changed to an elevation view, return to plan view before continuing.

7. Select the Space in the lower left-hand corner of the building.

8. On the Space tab of the ribbon, click the Select Similar button.

   The Properties palette should read Space (6) this time (five offices and the one perimeter Space you divided above).

This time, the style on the Properties palette will read Office_Small, but the Name will still read: Varies. From here we can change any of the properties of the Office_Small Space instances. For example, we could change the Ceiling height of all the offices to 9'-0".

9. Press the ESC key to deselect all objects.

OVERLAPPING SPACES

It will sometimes be desirable to generate Spaces over the top of existing Spaces. One such example would be when XREFing the Architect's background drawing. If the Architect's file contains Spaces you can certainly use them as is. However, if the Spaces created by the Architect are not satisfactory to meet your needs, you can create new ones on top. To accomplish this, the "Allow overlapping spaces" setting on the Properties palette has to be enabled prior to inserting the Space over the top of another Space. The Allow overlapping spaces setting can also be enabled in the Space tool. Let's take a look.

1. Zoom into the Toilet Room area in the building core.

As you can see, the Architect has created Spaces in the core area that we have referenced into our drawing. We are going to place an overlapping Space over the architect's Space. First we need an appropriate style.

2. From the Home tab of the ribbon, from the Tools drop-down button, choose Content Browser.

3. Click on the Design Tool Catalog – Imperial catalog to open it.

4. In the left pane of the Content Browser under the Design Tool Catalog – Imperial, select Spaces.

   Depending on the size of your Content Browser window you might need to scroll down to find Spaces.

5. Click on the Commercial link, and then click on the Page 3 link.

6. Right-click on Restroom – Men (Small) and then choose **Copy**.

7. Close the Content Browser.

8. Back in AMEP, on the MAMEP Spaces tool palette, right-click and choose **Paste**.

9. Right-click the new Restroom - Men (Small) tool and choose **Properties**.

10. In the "Tool Properties" worksheet, make the following changes:

⇨ For the Description input **Men's Toilet Room**.

⇨ For the Tag choose Aec7_Room_Tag.

⇨ For the Create type choose **Generate**.

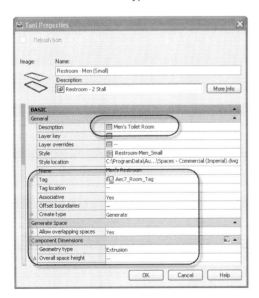

FIGURE 4.26

Configure the Properties of the Restroom – Men (Small) tool

⇨   For Associative choose **Yes**.

⇨   For Allow overlapping spaces, choose **Yes**.

⇨   For Geometry type, choose **Extrusion**, and then click OK (see Figure 4.26).

The key tricks to overlapping spaces are to ignore what you see and what you don't see on your computer screen. What do we mean by this? When you attempt to place a Space in a room that already has a defined Space you will get a Tool Tip at your cursor telling you that a Space has already been created. You should ignore that. What you will not see on your screen is the typical red highlight box indicating the outline of the Space that you are about to place. You will have to take a leap of faith and just place your Space. So, in both cases, the visual feedback will not help you in placing the overlapping Space. Let's look at creating an overlapping Space.

11. Click the Restroom – Men (Small) Space tool.

12. Place your cursor in the bottom (lower) Toilet Room and then click your left mouse button to place the Space.

13. Press ENTER to terminate the command (see Figure 4.27).

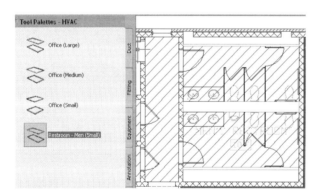

FIGURE 4.27
Create Overlapping Spaces.

An overlapping Space has now been generated. It should be noted that since we did not modify the fill pattern it will graphically appear like the Spaces that the Architect placed in their Drawing file. The only graphical difference between the two was the Space tag that we placed when we placed the Space itself. Our intent for the steps above was to show how we could generate overlapping Spaces. For the purposes of this exercise let's assume that we will be using the Spaces the Architect generated for the project thus far, and we no longer need the overlapping Space that we just generated.

14. Enter the command U, and press the ENTER key.

15. Zoom Extents.

For the Zones exercises we will use the Spaces the Architect has generated in conjunction with the Lease and Office Spaces we generated in this exercise

PLENUM SPACE

Third-party heating and cooling load analysis programs handle plenum spaces differently. Depending on the analysis program, it might be necessary to model the plenum space independent of the Space. The plenum space would then need to be associated with a Zone(s) for the load analysis program to calculate the heating and cooling requirements of the project correctly. A Space Style for the plenum could even be created that has occupancy of "0" and is set to "Unconditioned."

Now that we know more about Spaces let's learn about Zones and look at how we can associate a Space to a Zone.

ZONES

Zones are used to define which Spaces are served by an air-handling unit, VAV box, etc. Zones can even be attached to other Zones. If your goal for the Spaces in your project is only to calculate the square footage information, then you do not need to create Zones. Furthermore, if you do not want to share AMEP data with a third-party analysis program, it is not necessary to use Zones. However, if there is a desire to transfer square footage information of the room and all its associated components to a third-party heating and cooling load analysis program, then you will need to add a Zone(s) to your project and associate Spaces to that Zone(s).

> Note: You must have a Zone in order to transfer your project information from AMEP to a load analysis program. Without the Spaces being associated to a Zone(s) the mechanical design professional will not be able to reap the benefits of the gbXML export/import tool within AMEP.

AMEP allows you to have several Zones in your project, and Zones can be named in AMEP. For example AHU-1 might be a zone that serves the first floor of a project. Zones can be configured to display differently to provide a clear graphical differentiation between the Zones.

Let's establish a Zone in our drawing and then associate some Spaces to it. In our *03 Spaces* drawing we have Spaces XREFed from the architectural drawing as well as Spaces we have added to the drawing ourselves. This example will exploit both.

1. On the Analyze tab, on the Space & Zone panel, click the Zone button.

2. At the Command prompt type **N** (for Name) and then press ENTER.

3. Type **AHU-1** at the command prompt and press ENTER.

We now have a Zone named AHU-1. Next we have to place the Zone object in our drawing. It does not matter where you place the Zone object. However, you might want to place it away from your floor plan, but not too far away so you can avoid having to pan from the Zone to the Spaces.

4. Place the Zone to the lower right-hand corner of the Lease Space next to the floor plan.

5. Press ENTER to complete the command (see Figure 4.28).

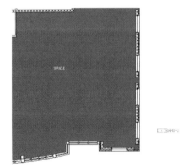

FIGURE 4.28
A Zone object placed in the Drawing

A simple rectangular icon with a label is used to represent the Zone object. The next step is to add Spaces to our Zone.

6. Zoom extents and then select the Zone object.

7. Select the plus (+) grip which appears on the Zone object.

8. At the "Select space and/or zones to attach" click below and to the right of the plan, and then click above and to the left to make a crossing window selecting the entire drawing.

You should note that although a large group of objects were selected, only the Spaces were included in the selection set, as only Spaces or Zones can be attached to Zones. All other objects were filtered out.

9. Press ENTER to complete the selection.

You should see several distinct lines from the Zone icon to each of the attached Spaces. Notice that all the Spaces placed in the previous exercise—in addition to the all the Spaces the Architect placed in their drawing file that was XREFed into our drawing—have been attached to the Zone object. You can see how many Spaces you have attached by selecting the Zone.

10. Select the Zone and then look at the Properties palette. (The Zone may still be selected after the previous step.)

Both the "Number of Spaces," "Total Number of Spaces," and dimensional data will be indicated on the palette (see Figure 4.29).

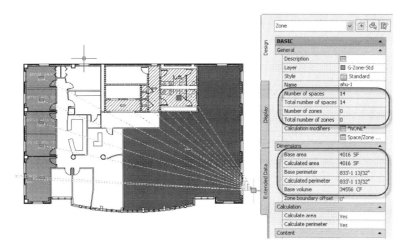

FIGURE 4.29

Spaces associated to a Zone object

By default, a Zone style has a double cross-hatching pattern as it graphic representation. Like other styles, Zone styles can be configured to display differently from one another to provide a clear distinction between the different Zones in a project. This will help greatly if you are creating a zone plan to review with your client. Therefore, to make it easier to clearly identify what area of the building the Zone is serving, we can configure our Zone to display differently.

11. Select the Zone and on the Zone tab of the ribbon, click the Edit Style button.

12. In the "Zone Style Properties" dialog box, click the Display Properties tab.

13. Select the Plan Display Representation and then check the Style Override checkbox.

14. In the "Display Properties" dialog, click the Hatching tab.

15. Click on User Double next to the Hatch Display Component.

16. Choose Solid Fill for the pattern type and then click OK.

17. Click the Layer/Color/Linetype tab.

18. Turn on the Zone Boundary component and change its color and the Hatch component's color to Color 6 (Magenta).

   Make sure that the Zone Boundary and Hatch Display Components are Visible.

19. Click OK twice to return to the drawing (see Figure 4.30).

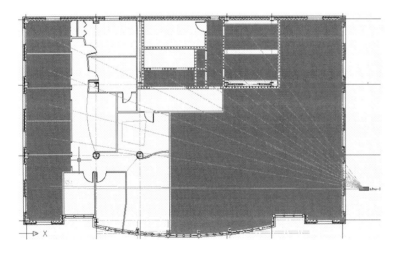

FIGURE 4.30
Change the graphical representation of a Zone object

You should now see the box associated with your Zone having a solid fill of Magenta and the Spaces associated with your Zones should reflect a solid fill pattern. This has not modified the fill pattern of the Spaces. The Zone pattern is simply lying over the top of the Spaces. Let's freeze the Zone layer to show our Space patterns once again.

20. In the Layer Properties Manager (Home ribbon, Layers panel, button in upper left corner), select G-Zone-Std and freeze it.

With the ability to freeze and thaw the Space and Zone objects, you can create colored plots for the client's review clearly showing the thermostat zones or the air-handling unit zones.

ZONE DESIGN RULES

Zones can also have restrictions placed on them. The first type of Zone restriction is "Space Exclusive". This type of restriction ensures that a space is only associated with one zone instance of the particular zone style. For example, you may want to use this to ensure that each space is only associated with a Zone style named "HVAC".

The other type of Zone restriction is called a "Zone Exclusive." A Zone Exclusive restriction will only allow Zones of similar styles to be attached to each other. For example, a Zone restriction might be that only laboratory air-handling units Zone styles can be associated to each other. In essence, "birds of a feather flock together."

To create Zone restrictions, use the Design Rules tab of the Zone Style.

1.   On the Manage Tab, on the Style & Display Panel, click the Style Manager button.

2. Expand Documentation Objects.

3. Expand Zone Styles.

4. Select the Zone Style you wish to restrict, and then click the Design Rules tab (see Figure 4.31).

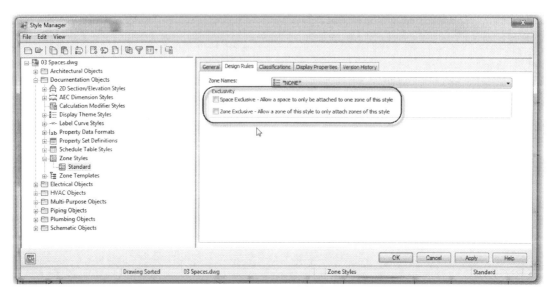

FIGURE 4.31
To change the restrictions on a Zone style, edit the Design Rules

WORKFLOW CONSIDERATIONS FOR SPACES AND ZONES

Let's briefly discuss how Zones behave in single drawing files and how a project model will need to be created for a whole building analysis of projects with multiple drawing files.

If all project Spaces are located in a single drawing file the workflow is simple. The Zone(s) can be added to the same file and the project Spaces can then be associated to the Zone(s). Once all of the Spaces have been associated to the Zone(s) the gbXML export tool can be used to "harvest" the Space data for export to a third-party analysis program. We will discuss gbXML below.

If the project has multiple drawing files that contain the project Spaces, those drawing files will have to be assembled (via XREF) to create an overall project model. This is where Project Navigator can be a helpful tool (please refer to Chapter 3 for additional information on the Project Navigator). Once the project model has been assembled, the Zone(s) can be added and the Spaces can be associated with the Zone(s). All of these tools recognize Spaces in XREF files, so there is no need to try to create everything in one file in larger and more complex projects. When creating the project model, individual areas or floors could also be assigned to different layers. This is optional, but doing so would allow those layers to be used to help keep the primary focus on one particular floor or area. It should be noted that once a referenced file is detached from the project model, the Spaces and Zones in the detached file will no longer be included in the project model. If the file is reattached, the Spaces and Zones will have to be reattached manually to the objects they were previously attached to. Therefore, it is recommended that you not detach XREFs, but rather unload them or use layers when you wish to hide portions of the model.

SPACE/ZONE MANAGER

The Space/Zone Manager organizes all the Space and Zone information in a centralized location. This information is displayed in a tree view format. Information about each Space and its related surfaces and openings can be viewed. You can also see the Zone with which a Space is associated. It is also possible to use drag-and-drop functionality in the tree view to associated Spaces to Zones.

You can access the Space/Zone Manager from the ribbon. To do so, select either a Space or a Zone in the drawing, and then on the ribbon (on either the Space or Zone tab) on the Helpers panel, click the Space/Zone Manager button.

> **Note:** To use the Space/Zone Manager to its fullest potential, it is recommended that you freeze the Base Hatch Pattern of the Spaces and Zones whenever you are using a solid fill pattern. Doing this will allow you to see the highlighted components from the Space/Zone Manager better.

Since the fill pattern from our Zone layer has already been frozen, let's freeze our Office style Base Hatch so we can better understand the power of the Space/Zone Manager tool.

1. Zoom into the lower left corner of the drawing and select the Office in that corner.

2. On General panel, of the Space tab of the ribbon, click the Edit Style button.

3. In the "Space Style Properties" dialog box, on the Display Properties tab, click the Edit Display Properties icon.

4. Click the Layer/Color/Linetype tab.

5. Turn off the visibility of the Base Hatch and click OK twice.

The solid Base Hatch pattern is now turned off, leaving only the outline of the Space boundary.

6. The Space should still be highlighted (if it is not, select it again).

7. On the Space tab, on the Helpers panel, click the Space/Zone Manager button.

8. In the lower left-hand corner of the "Space/Zone Manager" dialog, check the Show All Zones and Spaces and Show Space Surfaces checkboxes (see Figure 4.32).

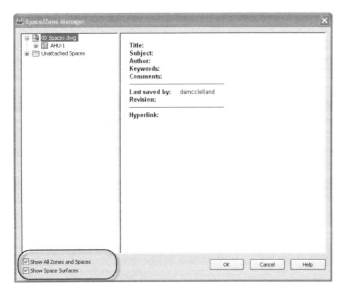

FIGURE 4.32

Open the Space/Zone Manager and display all Zones, Spaces and Space Surfaces

9. Expand the AHU-1 Zone and then expand the Office Area (6) Space listed.

Each of the surfaces of the Space object will be listed. Even though most of the rooms are essentially rectangles, several of them were created around columns and other small obtrusions. This accounts for there being more than four surfaces for many of the Spaces.

10. Select the Ceiling surface of the Office Area Space and review the information associated with the Ceiling surface on the right.

11. Select Surface 1 and note that this is a wall surface. Expand this surface and note that a window is associated with the wall surface (see Figure 4.33).

You will notice that objects like windows have been accounted for in the surface face. This information is also included in a gbXML export to a third-party load analysis program.

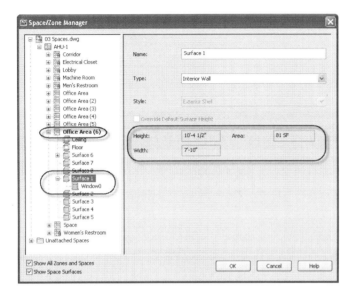

FIGURE 4.33

The Space/Zone Manager houses surface data regarding the Spaces, and indicates which Space is associated to which Zone

If you position your Space/Zone Manager dialog box in such a manner that you can see the Office Space you selected, you should notice that as you select these surfaces, the Space surface itself will be highlighted in red in the project drawing. You also might notice that you can modify the type of surface. For example, you can change the surface type from an interior wall to an exterior wall. Please remember our earlier discussion regarding the intelligent behavior of the Spaces when they are exported to a gbXML file. Although you modify the wall type in the Space/Zone Manager, the intelligent behavior of the Space will override your modification once the gbXML export tool is used to export the data from AMEP to a third-party analysis program.

> **Note:** Any Unattached Spaces that appear in the Space/Zone Manager can be dragged to the Zone to which you would like them attached. This is another alternative to attach Spaces in lieu of the Zone (+) grip. The association created through the Space/Zone Manager will be shown graphically in the drawing once the Space/Zone Manager dialog box is closed.

12. Click OK to exit the "Space/Zone Manger" dialog box.

Finally, there is a minus (-) grip at the Space and a plus (+) grip at the Zone. If you have attached the wrong Space to a Zone, use the minus (-) grip at the Space to detach the Space from the Zone.

Now that we have a basic understanding of Spaces and Zones, let's dive deeper into how this information can be made available to third-party heating and cooling load analysis packages through the gbXML export/import tools inside AMEP.

GBXML EXPORT

The gbXML export tool will compile all of the Space and Zone information into one single XML file. This XML file can then be read by many other third-party heating and cooling load analysis programs. All of the Space and Zone information that we have been discussing in this chapter can be exported to the gbXML file. This includes the name and square footage information of the exported Spaces, the Zone information, and the surface and opening information. Other information, if provided through user input, can also be compiled in the XML file. This includes information on lighting, miscellaneous equipment or people values to name a few. However, it should be noted that not all third-party heating and cooling load analysis programs will import all the exported information from AMEP. For example, if an Office Space style has been created, the analysis program may or may not read the style information into the program. In addition to this, not all analysis programs will import information like zip codes or building types. That does not mean that this information is not present in the XML file; it only means that the analysis program does not import that information from the XML file. As with gaining an understanding of how Spaces and Zones interact with each other, you will need to gain an understanding of how an XML file will interact with your preferred analysis program and adjust your workflow accordingly. Unfortunately, time and space do not permit us to cover the many analysis programs available in this text, so such explorations are left to the reader.

> Note: Refer to http://www.gbxml.org for more information on the gbXML schema and a link to various software applications that are gbXML capable.

Let's export our project information to a gbXML file.

1. We will first need to thaw our Zone layer. On the Home tab, click the Layer Properties button and then thaw G-Zone-Std.

2. From the Application menu choose Export > gbXML.

3. In the "gbXML Export" dialog box, next to item 1, click Browse.

4. Navigate to the *C:\MasterAME 2012* folder, accept the default file name and then click Save.

5. Next to item 2, click the select objects icon.

6. This returns you to the drawing to make a selection. Select the AHU-1 Zone object onscreen and then press ENTER.

Certain third-party load analysis programs require a building type and zip code to be input to perform the heating and cooling analysis (like Green Building Studios). Other programs do not (like Trane Trace 700). Assuming that you will typically have this information readily available, it will be best to input it. If your analysis program does not use that information on the import of the gbXML file, it will simply ignore this data. If you are certain that your program does not need this data, you can opt to skip it.

7. For this example, for item 3, select Office as the Building Type and use 60611 for the zip code (you can substitute your own zip code if you like) (see Figure 4.34).

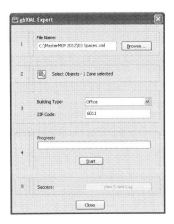

FIGURE 4.34

Exporting your AMEP Space data to a gbXML file allows you to easily exchange your project information with a third-party analysis program

8. Next to item 4, press the Start button.

The "Event Log" will show any error messages you receive with through the export. Several of our Spaces will be listed with the error: "No space type specified" (see Figure 4.35).

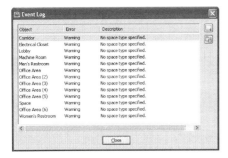

FIGURE 4.35

The Event Log dialog box will indicate any errors encountered during the export to a gbXML file

9. Click Close twice.

After reviewing the Spaces associated with the Zone that was exported, you should note that a Space type was not provided for the Spaces (see Figure 4.36).

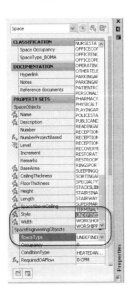

FIGURE 4.36

Not providing a Space type will provide an error in the Event Log. Becoming familiar with your third-party analysis program will help you determine if this or other errors are cause for concern

For the purposes of this exercise let's assume that we knew exactly how the data would be exchanged between AMEP and our third-party analysis program and that is one of the reason that we elected not to provide this data in AMEP since our third-party analysis program would not import this data. Therefore these particular errors in the Event Log can be ignored.

Any errors in the export will appear in the "Event Log" dialog box. In the "Event Log" dialog box you can highlight a Space and select the "Highlight and zoom selected objects" button to zoom to the particular Space(s) exhibiting the error. In some cases, some of the errors can be ignored as the load analysis program that you are using does not require the information requested. For example, suppose you are using Trane Trace 700 for their project load analysis. Furthermore, assume that you did not provide the building type or zip code information. The "event Log" dialog will appear with an error message indicating that the building type and zip code is missing. In this case, these error messages can be ignored as Trace 700 will not import this information from the gbXML file. If you already have your preferred tools in place, a few simple trial exports from AMEP and imports into your analysis package should be all you need to determine what information is required and thereby focus your process on providing such data to your model. If you are evaluating many potential analysis packages, you will have to employ such trial and error across each product you are considering.

What can be done after the mechanical design professional performs their load analysis of the project? Let's review how completed load calculations can be imported from the gbXML file back into AMEP.

CAD MANAGER NOTE: When exporting to gbXML true North will always be up in your drawing file. Your project might have to be rotated to true North if it is not before you export. Some third-party analysis programs allow you to rotate a project inside their program, so it might not be necessary to rotate your project inside AMEP. It should also be noted that the current gbXML schema that is exported by AutoCAD MEP 2012 is 0.37.

GBXML IMPORT

Some third-party heating and cooling load analysis programs can export the completed analysis back to the original XML file. This allows you to "round trip" the data, and avoids potential piles of paper output from the analysis program. The information from the completed analysis stored in the XML file can be imported back into AMEP. The gbXML import tool stores the calculation results as property set data for the particular Space and Zone object. (See the "Importing gbXML Properties" topic above for more information on the three property sets that should be associated with all Spaces for the import process to be successful.) It should be noted that not all calculations will be imported back into AMEP. As with the gbXML export, you will need to understand how an XML file interacts with their preferred analysis program, and what data makes the round trip and what does not. After the completed calculation results have been imported back into AMEP those results can be viewed through the Properties palette for any selected Space or Zone object. For "round trip" to function properly, you must import the *same* gbXML file into the same drawing file from which it was exported.

To help you see the "round trip" potential of the gbXML export and import, we have provided a file generated from a third-party analysis program. Using Trane Trace 700, a heating and cooling load calculation analysis has been performed from the building model data of our project. Now let's import the completed load analysis back into our completed Space drawing.

> **Note:** Typically, the data exchange between AMEP and third-party load analysis programs will want to keep the same file naming structure. This is the best workflow for the data exchange between the products.

1. On the Project Navigator, click the Constructs tab. (If you closed the Project Navigator, click the Project Navigator icon on the QAT.)

2. Expand the *Mechanical* folder and then double-click the *03 Spaces 3D Complete* file to open.

3. Open the Layer Properties Manager and freeze the G-Zone-Std and layer.

4. Zoom Extents.

5. Using a window selection, select all objects in the drawing.

   If the Properties palette is not displayed already, right-click and choose **Properties**.

At the top of the Properties palette, the drop-down list should read something to the effect of "All (17)." "All" indicates that the selection includes a mix of object types, and the number is the quantity of objects selected. If you click on the list and open it, each type of object will be listed by name with its own quantity next to it. For example, it might read: Space (7), Polyline (1), etc. Your totals may vary from those indicated here.

6. From the object list (reading: All (17) or similar) choose Space (7).

7. Click the Extended Data tab.

8. At the bottom of the Properties palette, click the Add Property Sets icon.

9. Deselect the **RoomFinishObjects** Property Set and then click OK (see Figure 4.37).

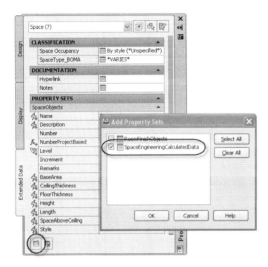

FIGURE 4.37
Adding Property Set Data to your Spaces will allow you to associate completed calculation results from third-party load analysis programs to your Spaces

Note: The "Add Property Sets" dialog can be a little confusing. It lists all available Property Sets (available to add, not already added) in the current drawing for the object type selected. So in this case, the RoomFinishObjects and SpaceEngineeringCalcualtionData Property Sets are available in this drawing and apply to Space objects. This does not mean that they are already applied to the selection of Spaces. Whatever Property Sets you select (check) in this dialog, will be added to the selection of objects in the drawing. Since we are only interested in adding SpaceEngineeringCalcualtionData to our seven Spaces in this case, we only selected it and not the other Property Set.

10. Scroll down to view the SpaceEngineeringCalculatedData property set data and notice that the information has a zero result (see Figure 4.38).

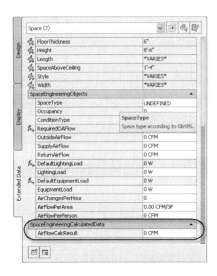

FIGURE 4.38

Until populated, Property Set Data can have a zero value

11. Deselect all objects in the drawing.

12. On the Analyze tab of the ribbon, on the HVAC panel, click the Import gbXML button.

13. Navigate to the *C:\MasterAME 2012\Chapter04* folder and choose the *03 Spaces 3D Completed.xml* file, and then click Open.

14. Zoom in on the upper left area of the plan, in particular Office Area 102.

15. Select the Office Area 102 Space, right-click and choose **Properties**.

16. Scroll down to the SpaceEngineeringCalculatedData and review the data populated from your imported calculation results for this Space (see Figure 4.39).

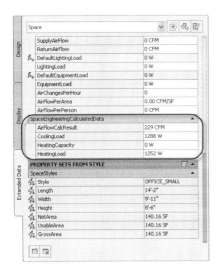

FIGURE 4.39
Imported data from a third-party load analysis program can be stored with the Space

The airflow, heating and cooling values have now been associated with the space in the project file. This allows the project team to share project knowledge electronically in lieu of sorting through potential reams of paper for the data. To have this information at your fingertips can be a huge time saver when you are beginning to place your ductwork systems. Continue to query other Spaces in the project drawing to review the imported load analysis data if you wish before continuing.

LEGACY 2D DRAWINGS

We have covered many procedures in this chapter. Many of the topics have implied a workflow appropriate to new projects. However, you may be wondering if you can use these powerful tools on your existing projects. Frequently a project will begin and then be stalled for whatever reason, only to reemerge later. Other common situations include phased projects or projects with a long-term client performing regular modifications to existing facilities. You may also inherit a project from another firm where work has already begun. In such situations (and several others) you may find yourself with existing AutoCAD files that do not have Spaces and Zones, but rather a collection of polylines. Can you use such polylines to calculate square footage, link to Zones, and export to gbXML for analysis? Not directly, but it turns out that "converting" existing polylines to Spaces is quite simple. All you need is a tool on a tool palette like the ones we created in the "Create a Tool Palette" heading above. Once you have one or more Space tools, simply right-click the tool you wish to use and choose **Apply Tool Properties to > Linework and AEC Objects**. Follow the prompts to select the polylines you wish to convert. If you will no longer need the polylines, you can delete them when prompted. The remainder of the procedures can then be performed as indicated throughout this chapter. You may need to adjust some of the settings in the Zone Manager to fine-tune the model before gbXML export, but overall you should find the process similar to the procedures outlined herein.

SUMMARY

Throughout this chapter we reviewed many key aspects of Space, Zone and gbXML tools in AutoCAD MEP. We have applied concepts to our sample commercial office building. You should now have a good feeling for how the Space, Zone and gbXML tools work, as well as having a broader understanding on how to apply those tools in your projects. In this chapter we learned:

✓ Basic drawing set-up and Space object options required to "start off on the right foot" with Spaces.

✓ You can use simple 2D Spaces to quickly generate area takeoffs from almost any plan.

✓ To generate square footage information, simply add the Spaces, then query them on the Properties palette or add a quick Schedule Table.

✓ Create Space styles in Style Manager and optionally add them to tool palettes.

✓ There are many advantages to creating Space styles.

✓ Place Spaces using Tool palette tools for more flexibility.

✓ Modify Spaces to fit the specific project needs on the Properties palette.

✓ Zones allow you to group Spaces and other Zones in logical ways.

✓ Use the Space/Zone Manager to review Space and Zone information.

✓ Export data from AutoCAD MEP to third-party energy analysis programs using gbXML export.

✓ Once the energy calculations are complete, you can import that data back into AutoCAD MEP.

Chapter 5

Mechanical Systems

INTRODUCTION

After using Spaces to help you perform your energy analysis of your project, it is time to begin designing your ductwork system(s). AutoCAD MEP (AMEP) allows you to easily design and document your project's ductwork systems.

In this chapter we will cover the fundamentals of how ductwork works in AMEP. We will discuss the settings that control the ductwork and the preferred workflow approaches when placing ductwork systems in your project drawings. In addition to this, we will discuss the creation of new ductwork system definitions, the fittings that can be used in your ductwork systems, how ductwork systems are displayed, and how the routing tools work when you are placing your ductwork systems.

OBJECTIVES

In this chapter we will focus on ductwork and related tools in an effort to learn how these tools can help the mechanical design professional design ductwork systems for their projects. We will discuss:

- Basic ductwork options for placing and displaying your ductwork and their associated components.

- Additional settings for HVAC Objects through the Style Manager.

- Ductwork placement and display behavior.

- Tools for automatic sizing and resizing of your ductwork system or selected objects in your ductwork system.

- Display Themes and their ability to help you better understand the systems in your projects.

INITIAL SETUP IS VITAL TO YOUR SUCCESS

Before placing ductwork in your drawing you should familiarize yourself with many settings and configure your preferences before drawing your ductwork system(s). Many settings have a direct impact on the parts chosen and solutions offered by the software as you draw AMEP objects. Configuring your settings first and then saving such preferences in a template file will help you avoid wholesale substitutions later in your project. In other words, if you do not carefully consider such things as the graphical representation of your ductwork and the factors for sizing your ductwork at the onset, you might need to adjust many (or all) of the ductwork systems that you have placed in your drawing. However, since projects change on a daily basis, you might not know how to set some of these Options initially. Don't let that stop you from making your best attempt to configure the various options and settings. That will serve you well as you begin your project. As you place your ductwork systems in your drawing, you will begin to find the right balance between optimal initial setup, enforcing your office standards, and providing the flexibility to modify systems as ongoing project needs dictate.

Once you have determined the best configuration for your ductwork options, preferences, and styles it is recommended that you save those settings in a template file (DWT). This will ensure that you preserve your work and allow all your drawings to be started from a common point. The first several pages of this chapter will cover the settings that you should consider as part of your initial setup.

It is highly recommended that you follow along through the initial setup topics in an effort to better understand how you might create a template file for your company based on your office standards. At the completion of the initial topics, we will save our work in a new drawing template file (DWT) that we will use to complete the remaining tutorial lessons in the chapter.

> Note: In the following topics, every attempt has been made to establish best practice recommendations for the available ductwork options and settings. Considering that every office has its own set of standards, this is not always an easy task. You are encouraged to determine how best to set up your ductwork options and styles in a way that incorporates your company's standards and procedures with the recommendations made herein. Please consult with your CAD Manager and/or IT support person for assistance.

INSTALL THE DATASET FILES AND CREATE A DRAWING

The lessons that follow require the dataset files provided for download with this book.

1. If you have not already done so, install the book's dataset files.

   Refer to "Book Dataset Files" in the Preface for information on installing the sample files included with this book.

2. Launch AutoCAD MEP (US Imperial) from your desktop.

For the first part of this chapter, we will explore the ductwork settings in an effort to create a new ductwork template file. As noted, your office standard template may vary from the out-of-the-box template that we are beginning with here. It is recommended that you start with the out-of-the-box template and from there you can choose to load (copy) your office standards to the out-of-the-box template and then resave that as your new company standard template file once your company standards have been copied. Starting with your company standard template file might cause your screens not to exactly match the screen captures herein. If you have time, try exploring the lessons with the out-of-the-box template first, and then repeat them incorporating your office standards into the template. In this way you can make a good comparison between the two and reconcile any differences. As noted, please consult with your CAD Manager and/or IT support person for assistance.

3. From the Application menu, choose **New > Drawing**.

4. In the "Select template" dialog that appears, select the *Aecb Model (US Imperial Ctb).dwt* template file and then click Open (see Figure 5.1).

FIGURE 5.1

Create a new file from the out-of-the-box template

If you do not have this template file installed, a copy has been provided with the book dataset. Browse to *C:\MasterAME 2012\Template* folder to locate it.

Note: Please remember to set your workspace to HVAC. For more information on how to set workspaces refer to the Quick Start chapter.

DUCTWORK OPTIONS

Configuring your ductwork Options allows you to provide information to help define your ductwork system(s). For example, ensuring that your ductwork systems connect correctly, controlling the center line display of the ductwork system, creating elevations at which to place your ductwork systems, and ensuring that the correct part catalogs are referenced when parts are placed in your drawing are among the options available.

CONFIGURE MEP OPTIONS

1. From the Application menu choose **Options** (see Figure 5.2).

FIGURE 5.2
Options is available on the Application menu

The "Options" dialog could fill almost an entire book in itself. Across the top of the Options dialog are multiple tabs. The first several tabs, on the left side, are from base AutoCAD. Tabs containing settings unique to AutoCAD Architecture, upon which AMEP is built, appear at the right side. Their names are prefaced by "AEC." In the middle, between the AutoCAD and AEC tabs, is a collection of four tabs prefaced by "MEP." Those are MEP Layout Rules, MEP Display Control, MEP Elevations, and MEP catalogs. We will focus on these four tabs in this discussion as these settings are unique to AMEP. Use the scroll buttons at the top right of the dialog to scroll through the tabs.

Take a look at Figure 5.3and notice the AutoCAD drawing icon at the top of the dialog and next to some of the items in the "Options" dialog box. This icon indicates a setting that is saved only in the current drawing file. As we explore and configure the settings on the MEP tabs below, you will note that all settings on the "MEP Layout Rules," "MEP Display Control" and "MEP Elevations" tabs have this icon next to them. This means that most MEP specific settings are saved with the drawing file, and that you must use drawing template files to manage these settings. If you or other members of your team begin drawing with the wrong template file, you may not have the correct MEP options enabled. This is the primary purpose for the inclusion of the first several topics in this chapter.

MEP LAYOUT RULES

2. Scroll to and click the MEP Layout Rules tab.

The "Connection Test Mismatch" setting is used for objects that are connecting to one another. The default setting is "Prompt for user input" so we can leave that setting as is. In the "Part Selection" area, the default setting, Prompt when Non-Standard Parts are Needed, will prompt you before placing non-industry standard parts. You have several selection options here. To limit yourself to using only parts out of the catalog, you should select Use Catalog Parts Only. To allow non-standard parts to be used, but to have AMEP advise you when this is being done, you should select Prompt when Non-Standard Parts are Needed. To have AMEP place non-standard parts that will allow you to continue placing your ductwork system as you wish, you should select Use Non-Standard Parts. Both of these defaults are recommended and should already be configured in your template, but it is always a good idea to double-check (see Figure 5.3).

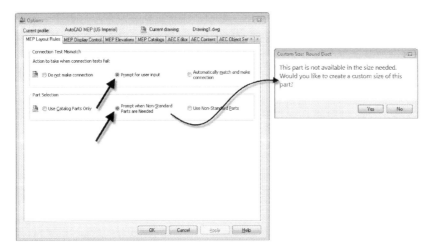

FIGURE 5.3
Ensuring that your MEP Layout Rules are set correctly can help notify you of ductwork connection problems

MEP DISPLAY CONTROL

Let's move next to the MEP Display Control tab. This tab allows you to set how hidden lines (Crossing Objects Display) are displayed for AMEP objects when they cross each other, and is the first step to displaying your center lines for your ductwork systems. The final step for displaying your duct center lines will be through the Layer Manager by thawing the center line layer for the ductwork. If you do not have your duct center lines set to display through this dialog, they will not be displayed on your drawing, even if the layers have been thawed for the particular Duct Style.

At the top of this tab are the Crossing Object Display settings. We have already discussed these settings in Chapter 2. To gain a better understanding on the behavior of crossing object display, refer to "Understanding Hidden line" topic in that chapter. For our purposes here, let's make our settings match the recommendations in that chapter.

3. Enable the "Apply Annotation Scale to Gap" setting and set the "B - Gap Paper Width:" to **1/16"** (see Figure 5.4).

Changing this setting allows the Gap to adjust based on your drawing scale factor. You will need to set this in all templates, including sheet templates to ensure consistency throughout your project. It is recommended that you adjust this gap to find a suitable distance to meet your office standards.

Leave the remaining settings as they are. Most of the default settings are suitable unless your office standards dictate otherwise. We will make just one other adjustment here.

4. Check the "Extend Center Line for Takeoffs" checkbox (see Figure 5.4).

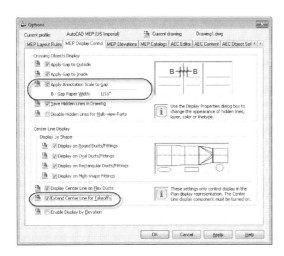

FIGURE 5.4

Use the MEP Display Control as the first step to show your ductwork center lines

This setting will extend the center line of the branch ductwork to the center line of the main ductwork, and is only needed if you typically display center lines for your ductwork. You can also deselect the shape that you do not want to display center lines for by simply removing the check next to the appropriate shape in the Center Line Display section. The display to the right will update accordingly to indicate which shapes will display center lines when the center line display component is on.

Finally, to gain a better understating of the "Enable Display by Elevation" checkbox, refer to the "Edit Display Properties" topic in Chapter 2.

MEP ELEVATIONS

The MEP Elevations tab allows you to create named elevations. When placing ductwork, you can use these named elevations in lieu of manually typing in elevation values. Although you might not know final elevations for you project at the start of it, you can establish preliminary elevations and set your final elevations once they are known.

For example, early on in a project you might set your preliminary elevation for all the ceilings in the project at 8'-6" A.F.F. This might allow you to begin placing ceiling diffusers with the understanding that you will need to modify the elevation of those diffusers at a later date once the final ceiling heights have been determined. As the project becomes more defined, the ceilings might actually be at a final elevation of 9'-0". In the "Options" dialog, you can return to the MEP Elevations tab and adjust your ceiling elevation accordingly. You can then reapply this elevation to all the ceiling diffusers and they will automatically be adjusted to the new elevation.

In the exercises later in the chapter we see how we can use the ceiling grid that the Architect placed to automatically set the elevation of our ceiling devices.

5. Click the Defining System Elevations icon in the lower left-hand corner of the MEP Elevations tab to create a new elevation.

   When you click this icon, "New Elevation" will appear at a height of 0" (see Figure 5.5).

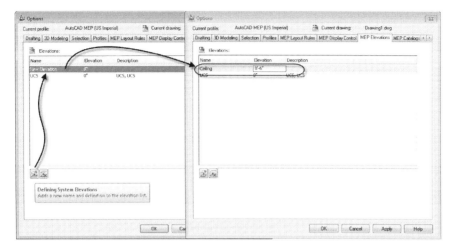

FIGURE 5.5
Creating unique elevations gives you the ability to define the location(s) in the model where your system(s) and their associated components will reside

6. Click directly on the name to edit it, and type **Ceiling**.

7. Do the same to edit the height making it **8'-6"** (see Figure 5.5).

8. Continue to add elevations until you have defined all of the heights that you might use in your project (See Figure 5.6).

Although you are creating common elevation names and setting preliminary elevations in your template file, you can modify the elevations at any point in time in your project. As your project becomes further defined, you can adjust the height of the preset elevations by simply picking the elevation and typing in your new elevation height. Creating initial elevations allows you to predefine the elevation of the center line or the bottom of duct for each system. This does not restrict the maximum height of the associated ducts for each system.

FIGURE 5.6
Different elevation names and heights can be created

Note: You can sort your project elevations by Name or by Elevation. This is done by simply clicking on either the Name or Elevation column heading.

MEP CATALOGS

Since we are already exploring and verifying settings in the "Options" dialog, it is a good idea to take a quick look at the MEP Catalogs tab and make sure that all is in order. At the top of this tab are listed several out-of-the-box AMEP catalogs. A catalog stores the Multi-view parts (content) that are available to be used within AMEP. Here you can incorporate a custom catalog containing the content that your company has created, or will be creating. Doing so will make such content available in the list of available parts for selection and placement in your files.

To gain a better understating of Catalogs and Content creation, refer to Chapters 9 and 10. The lower portion of the tab includes paths to style-based content. Each of these can be edited if necessary. To learn more about style-based content, please refer to Chapter 9.

When you are finished exploring and modifying settings in the "Options" dialog, you can exit the dialog. In this exercise, we will not explore any of the non-MEP tabs, but please feel free to explore them on your own.

9. Click the OK button to close the "Options" dialog and return to the drawing window.

That completes our work in "Options." We have a few other settings to explore. If you wish to save your work to this point, from the Application menu, choose **Save As > AutoCAD Drawing** and give the file a temporary name (perhaps your company name, since we are creating a template to incorporate your company standards) and location. When we are finished with all settings below, we will save the drawing as a template file.

Note: When modifying any of the out-of-the-box components within AMEP, it is strongly recommended that you save the modified item to a different file name. This will preserve the original item in case the item you are modifying does not work according to your expectations, or if for any reason you need the original.

HVAC OBJECTS IN THE STYLE MANAGER

The settings we configured above are overall settings that typically apply across the drawing until you change them. With styles, you can modify settings that apply only to objects belonging to that style. Should you later edit the style, all objects using the style will benefit from the change. You can further define your ductwork system(s) by configuring various styles. For example, you can identify the symbol(s) that will be used to represent your ductwork rises and drops, create duct system definitions that will help identify your ductwork systems from one another as you place them in the drawing, and more.

1. On the Manage tab, on the Style & Display panel, click the Style Manager button (see Figure 5.7).

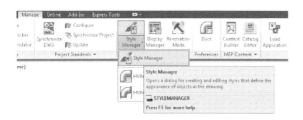

FIGURE 5.7

Use the Style Manager to modify HVAC objects

DUCT SYSTEM DEFINITIONS

Next we'll look at Duct System Definitions. Like most styles, we can modify any existing definitions or add new duct system definitions to meet our specific project needs.

2. In the Style Manager, beneath *HVAC Objects*, expand *Duct System Definitions*.

Here you will find several preconfigured System Definitions (see Figure 5.8).

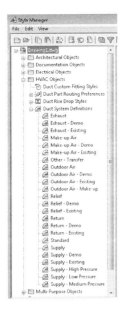

FIGURE 5.8

AMEP ships with multiple System Definitions to get you started

Let's assume that you would like to capitalize on the delineation of the supply air systems between low, medium, and high pressure systems. Doing this will give you the ability to use other tools, like the Duct System Size Calculator tool, to size your ductwork systems as well. Let's look at System Definitions.

Like other styles in Style Manager, you can create a new one by selecting the appropriate branch of the tree on the left and then clicking the New Style icon, or you can find one that matches the type of system you would like to create, copy, and rename it.

For this exercise, we will focus on a single System Definition: Supply – Low Pressure in this case. Keep in mind that the concepts covered in our discussion will apply to any System Definitions.

3. Beneath *Duct System Definitions*, select the Supply – Low Pressure System Definition.

4. Click the General tab.

The General tab allows you to change the Name and Description to this System Definition (see Figure 5.9).

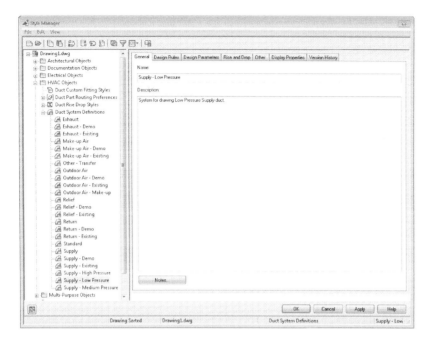

FIGURE 5.9

The General tab includes the Name and Description

At the bottom of the General tab you can click the Notes button. This opens a simple dialog with two tabs. Type in text on the Notes tab and use the Reference Docs tab to add links to documents on your system. Items you add to the Reference Docs page will be links to the original document. Double-click an item to launch the associated program and open the linked file. You should note that the Reference Docs is only associated with the System Definition in the Style Manager and not associated with every piece of Supply – Low Pressure ductwork you place in your project. This means that you will only be able to access the referenced document through the Style Manager (see Figure 5.10).

FIGURE 5.10

Reference Docs inside Style Manager give you the ability to create hyperlinks to important documents directly from the System Definition

5. Click the Design Rules tab.

On this tab, you can assign an Abbreviation and Layer Key to the System Definition. A Layer Key is already assigned to most System Definitions. The Layer Key will assign, and create if necessary, a layer from the Layer Key Style when this System Definition is used. To further differentiate the Layer assignment, overrides can be assigned to any or all layer fields (see Figure 5.11). See the online help for information on Layer Keys.

The abbreviation is used by the Labelling mentioned in Duct Preferences above. The System Group determines what systems should be allowed to connect. Only systems belonging to the same group or no group (blank) are allowed to connect. The default template assigns all Duct Systems to the DUCT group. To prevent interconnection between systems create a new System Group by selecting in the Systems Group box and enter the name of your new System.

FIGURE 5.11

The Design Rules tab allows you to change the Abbreviation and set the Layer Key

The Design Parameters tab allows you to set the Velocity and Friction loss of your ductwork system.

6. Click the Design Parameters tab.

Choosing the Velocity radio button will tell AMEP to calculate ductwork size for this system based on the Velocity value in the FPM box. Choosing the Friction radio button will tell AMEP to calculate the ductwork size for this system based on the friction loss in the ductwork system in W.G. per 100 feet. The Roughness of the ductwork and the density of the air in the ductwork can also be added to the Design Parameters tab (see Figure 5.12).

FIGURE 5.12

The Design Parameters tab of a particular System Definition allows you to set the Velocity, Friction loss, Roughness, and the density of the air in the ductwork

The settings on this tab are extremely important if you plan to use the sizing tools within AMEP to automatically size your ductwork systems. The ductwork will either be sized based on Velocity or based on Friction. However, if you use the Duct System Size Calculator to automatically size your ductwork, the ductwork will be sized on a maximum Friction value and/or a maximum Velocity value. See the "Duct System Size Calculator" topic below for additional information.

> **IMPORTANT:** It should be noted that the duct sizing tool within AMEP is strictly based on the ductwork sizing formulas located in ASHRAE Fundamentals Handbook. The ductwork sizes calculated by AMEP should be similar in nature to sizes obtained by industry standard ductwork calculators. It is extremely important that correct data be provided in this tab before the ductwork sizing tools within AMEP are utilized. Failure to do this could cause adverse results.

7. Click the Rise and Drop tab.

The Rise and Drop tab allows you to choose which Rise and Drop Style you will be using for this particular System Definition (see Figure 5.13). The choices available on the list here are the ones provided in AMEP. If you create any custom styles, they will appear here as well.

FIGURE 5.13

Specific Rise and Drop Styles can be assigned to a System Definition

If you wish to exclude the objects belonging to this System Definition from the shrinkwrap of 2D sections, you can enable the setting on the Other tab.

    8. Click the Other tab.

2D Section/Elevation objects are 2D drawings linked to the 3D model. You create a section line to indicate from where the section should be cut, and how wide and deep it should be. When the section is generated, AMEP will perform a hidden line removal on the 3D geometry to remove object in the background concealed behind objects in the foreground. Any object intersecting the cut line will be "shrinkwrapped." Shrinkwrap is a bold outline generated around any object that is cut through (see the left side of Figure 5.14). If you do not want objects in the current system to render in bold outlines when cut, you can exclude them from the 2D Shrinkwrap (see the right side of Figure 5.14). For more information on 2D Sections, refer to Chapter 13.

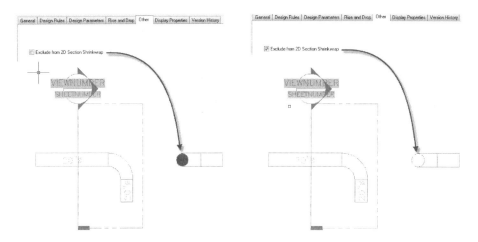

FIGURE 5.14
System Definitions can be excluded from 2D Section Shrinkwrap

    9. Click the Display Properties tab.

Display Properties allow you to control how you want this System Definition to appear graphically in your project drawings (see Figure 5.15). The Display System was discussed in Chapter 2, and you can also refer to Chapter 12 for more advanced coverage of the Display System. As you can see, there are several Display Representations which are used to change the way objects display in order to portray different types of commonly required drawings.

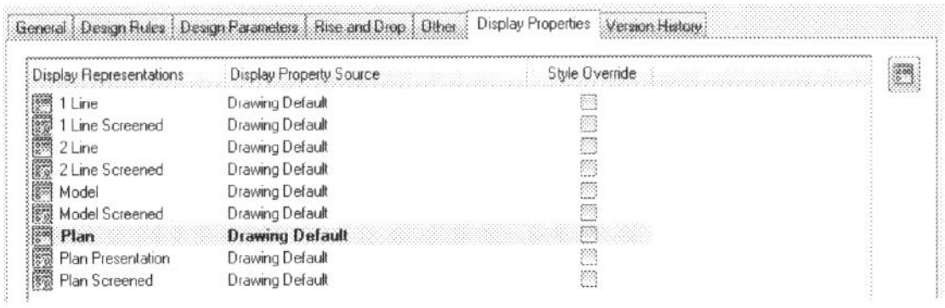

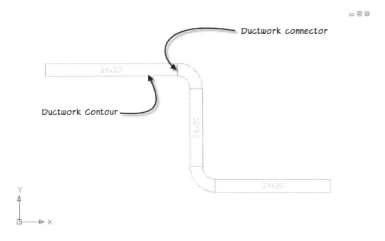

FIGURE 5.15

Display Properties allow you to define how this particular System Definition will be displayed in your project drawings

The basics of the Display System have been covered in detail already in previous chapters. When you configure a style, such as the System Definition in consideration here, you have the option to assign style-level display properties to it. Settings on the Display Properties tab are used to control the appearance of the style, in this case the System Definition. Before making any edits to the display properties of the System Definition, be sure you are comfortable with the display system hierarchy and definitions. Refer to the "Overview and Key Display System Features" heading in Chapter 2 and the "Display System Definitions" topic in Chapter 12 for definitions of the key Display Control terms. You will also find detailed tutorials on working with the Display System in Chapter 12.

Display Properties can be used to control the line color of ductwork objects. For example, you might not want the ductwork connector to show heavy like the contour of the ductwork when it is plotted (see Figure 5.16).

FIGURE 5.16

Display Properties allow modification to individual components of ductwork

The current active Display Configuration is Plan. You can tell this because Plan is bold in the list. If you simply click the Edit Display Properties icon (on the right) you will be editing the Drawing Default settings, which means that your change would apply to the Plan Display Representation of all Duct System Definitions. In some cases, this may be what you want. If you want the change to apply only to the Supply – Low Pressure style that we are currently editing, then you will want to apply a Style Override first. Let's assume that is what we want to do in this case.

10. On the Display Properties tab, check the Style Override box next to the Plan Display Representation.

11. In the "Display Properties" dialog that appears, change the color of the Connector component (see Figure 5.17).

FIGURE 5.17

A Style Override applied to the ductwork connector in order to meet an office standard

Assuming that you chose a color that plots lighter in your company standards, the result will be something like that shown in Figure 5.18. In this example, we have changed the color of the component. You can pick any color that plots in a light pen weight in your office standards. You will notice that some of the items in this dialog are assigned first to layers, and then the color is set to ByLayer. This is an appropriate option as well. The advantage of choosing a layer instead of a color directly is that you can assign the same layer to multiple styles and then change the layer settings later if required, effectively modifying several styles without having to modify them individually. The choice is up to you and is a matter of office standards.

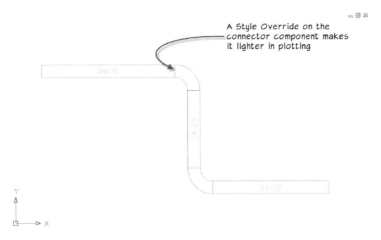

A Style Override on the connector component makes it lighter in plotting

FIGURE 5.18
The connectors will appear lighter when plotted by choosing a "lighter" pen color

The Version History tab is the final tab and part of the Project Standards feature in AMEP. If you are using Project Standards, you can apply versions here and view a history of previous synchronizations. To learn more about Project Standards, search the online help.

If you have set out to configure an office standard template file for use in AMEP, then you will likely spend quite a bit more time in the Style Manager tweaking styles and their settings. The more you can pre-configure in your template file, the fewer configurations you and your team will need to do later.

12. Click OK to close the Style Manager.

SAVE A TEMPLATE FILE

When you are satisfied with your settings and configurations, you are ready to save a template file to preserve your work. The process is nearly the same as saving any drawing file.

13. From the Application menu, choose Save As > AutoCAD Drawing Template.

In the "Save Drawing As" dialog, AMEP will open your default *Template* folder. While you could save our custom template here, to be sure we don't violate any office standards, let's save ours in the dataset folder for now. If you decide to use this template for real projects later, you can move it to the *Template* folder later.

14. Browse to the *C:\MasterAME 2012\Template* folder.

15. In the File name field, type **MAMEP Model.dwt,** and then click Save.

16. In the dialog that appears, type **Mastering AutoCAD MEP 2012 Mechanical Systems Template,** and then click OK.

17. Close the template file.

CAD MANAGER NOTE: When editing Templates, or creating new ones, you should PURGELAYERKEYSTYLES before saving and run BLDSYSPURGE to remove and content styles. This is important to allow for the Layer Key to be imported from the Layer Key drawing instead of the one from the template. Also purging the styles removes copies of the content that exists in the catalog.

DUCTWORK

Our next task is to learn how to place ductwork. Before we do that let's quickly look at how ductwork behaves in AMEP. We will start by reviewing some of the fundamentals of the AMEP ductwork tools.

DUCTWORK BEHAVIOR

There are 2 types of routing behaviors: unconstrained and constrained. Unconstrained applies when you are drawing ductwork in a "freeform" manner and not attempting to connect to another piece of ductwork or ductwork fitting (see Figure 5.19).

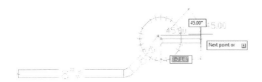

FIGURE 5.19
Unconstrained duct routing

Going one step further AMEP can automatically create a ductwork layout using an internal auto routing algorithm. This algorithm allows the software to utilize the information stored in the Parts tab of the Duct Layout Preferences to determine layout options and make connections for you.

A constrained layout method applies when you are drawing ductwork and you want to connect it to either another piece of ductwork or a duct fitting in the same system (see Figure 5.20).

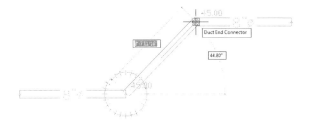

FIGURE 5.20
Constrained duct routing

During the constrained layout process you can get a preview of the different connection options. The Command prompt will indicate how many possible solutions are available based on the routing you picked. At the command prompt you can choose Next to review all the options before you decide which one to accept (see Figure 5.21).

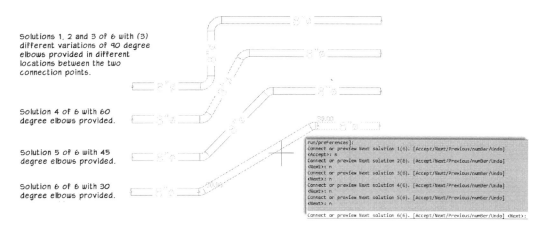

Solutions 1, 2 and 3 of 6 with (3) different variations of 90 degree elbows provided in different locations between the two connection points.

Solution 4 of 6 with 60 degree elbows provided.

Solution 5 of 6 with 45 degree elbows provided.

Solution 6 of 6 with 30 degree elbows provided.

FIGURE 5.21
A constrained auto routing solution with seven possible layout possibilities

The Figure above shows six possible routing solutions. The first three solutions have two 90° elbows at different locations between the two connection points, one solution has two 60° elbows, one solution has two 45° elbows, and the final solution has two 30° elbows.

Auto routing will create ductwork solutions based on parts stored in the Parts of the Duct Layout Preferences and will add any required fittings and change elevations. The auto routing algorithm works by creating a plane based on the initial ductwork location and another plane at the connecting ductwork location. The software will then calculate possible routes based on the available elbow angles stored within the Part preferences.

TYPES OF DUCTWORK

There are two distinct types of ductwork in AMEP. The types of ductwork are not defined by their shape, i.e., Oval, Rectangular or Round; but by their graphical representations. The two types of ductwork are 1-Line and 2-Line ductwork. The 1-Line ductwork can be broken down further to be displayed as Undefined or Defined 1-Line ductwork.

Let's review the two different types of ductwork.

1-Line Undefined Ductwork—1-Line undefined ductwork should not be confused with 1-Line defined ductwork. Although there might be graphical similarities between the two, these are very different representations of ductwork within AMEP. The best way to think about 1-Line undefined ductwork is to look at it as a placeholder within AMEP. You will define this placeholder later in your project.

A 1-Line duct tool for the 1-Line undefined ductwork can be found on the Duct tool palette. In order to create a 1-Line Duct through the tool on the Ribbon, the Routing Preference and Shape must both be set to Undefined in the Properties dialog box. Choosing the 1-Line duct tool from the Duct tool palette does this automatically.

The intent of the 1-Line undefined duct is to give you the ability to begin placing your ductwork in the predesign, schematic design, or design development phases. Using the 1-Line undefined duct will allow you to convey ductwork routings and equipment layouts very early on in your project without your needing to specify a lot of unknown details (see Figure 5.22).

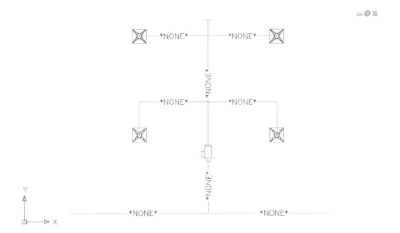

FIGURE 5.22
Medium and low pressure supply air ductwork drawn with the 1-Line duct tool.
The shape and size can be defined later on in the project

Notice that the duct size tag indicates NONE as no size, yet is associated with the 1-Line duct. Engineering design professionals have been using simple lines in plain AutoCAD for years to convey early design intent. The 1-Line undefined ductwork workflow process is in direct correlation with this. The major difference here is that you will not need to erase the 1-Line undefined duct that you placed in your AMEP drawing once you are ready to show double line ductwork. You can simply modify the 1-Line ductwork and define the duct connection type, shape, and size when you are ready. Specifying the 1-Line undefined ductwork sizes and shapes can also be done automatically through the Calculate Duct System Sizes tool. See the "Duct System Size Calculator" topic later in this chapter.

1-Line Defined Ductwork—1-Line defined ductwork should not be confused with 1-Line undefined ductwork. Although there might be graphical similarities between the two, these are very distinct representations of ductwork within AMEP. In simple terms, 1-Line defined ductwork is a modified Display Representation of 2-Line ductwork. The Display Representation of 2-Line ductwork has been changed through the Display Configuration to display the 2-Line ductwork as a simple 1-Line display (see Figure 5.23).

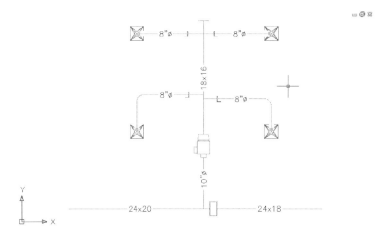

FIGURE 5.23

1-line ductwork is a modified Display Representation of 2-Line ductwork

In a 1-Line Display Configuration all ductwork fittings, accessories, and equipment will also display in a 2-Line format for clarity purposes. Although the Display Configuration of the ductwork has been modified to show as 1-Line, the 1-Line ductwork still thinks, acts, and behaves like a 2-Line duct. For example, in Figure 5.24, a piece of ductwork was placed next to the 10" branch ductwork going to the VAV box in such a fashion that interference would be created.

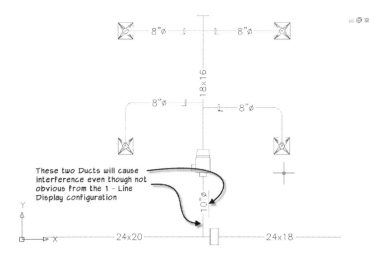

FIGURE 5.24

Your current Display Configuration may conceal that you actually have a conflict

Although the two ducts do not show any type of collision in a 1-Line Display Configuration, when an Interference Detection is run, the interference will be noted in AMEP (see Figure 5.25). To learn more about Interference Detection, refer to Chapter 14.

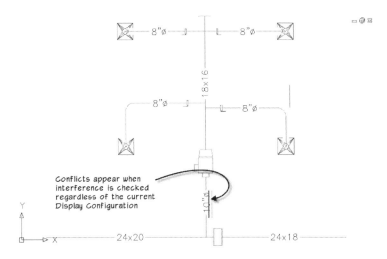

FIGURE 5.25
Objects in a 1-Line Display Configuration still behave like 2-Line objects when it comes to interference detection

2-Line Ductwork—2-Line ductwork is ductwork displayed in its actual size and configuration (see Figure 5.26).

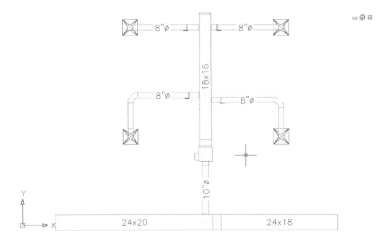

FIGURE 5.26
2-Line ductwork is typically how most ductwork is shown in project documents

All three methods are viable options for placing ductwork in your drawing. The choice is a matter of personal preference. Feel free to stick with your favorite method, or use all three as the mood suits you.

PLACING AND DESIGNING DUCTWORK SYSTEMS WITH AMEP

We have briefly reviewed a few settings available for ductwork in AMEP. We have explored Deign Rules, Design Parameters, Rise and Drop styles, and Display Properties for ductwork. Along the way we have gained a basic understanding of the routing differences and the different types of ductwork in AMEP. You are now ready to begin your design and placing your ductwork systems.

PLACING DUCTWORK

Ductwork can be placed in your drawing by accessing the Duct tool on the ribbon or by clicking one of the tools on the Duct tool palette. To see the Duct tool palette, make sure that the HVAC Workspace is active and that Tool palettes are displayed—refer to Chapter 1 for complete steps. Either method will make the Properties dialog active. The Properties dialog will allow you to configure all necessary parameters for the ductwork that you want to place in your drawing.

The ribbon tool uses overall default settings or simply remembers the settings used for your most recent duct run. Choosing a duct tool from the tool palette will automatically populate some or all pertinent parameters in the Properties dialog. The specifics depend on which tool you pick and how it was configured. Naturally, preconfigured tools can be a big time saver. The default Duct palette is organized by System, Shape and Size. Many common tools have been provided. However, if a common type of duct is used repeatedly in a project, you can create your own tool for it. To do so, simply select a Duct object onscreen, drag it from the drawing and drop it on the palette. This will automatically create a duct tool on the tool palette. When you use this tool, it will behave like any other—populating the Properties dialog with its preconfigured parameters.

One final way you can add ductwork is using the Add Selected command. Select an existing piece of ductwork onscreen, right-click, and choose Add Selected. The experience will be similar to using a tool as all of the parameters of the selected Duct will be transferred to the Properties dialog. Of course you can edit any of these values before you begin to place your

Ducts. So sometimes using tools or Add Selected can be an effective way to begin the command when you need a Duct that is similar to an existing one or a tool you already have.

LOAD A PROJECT

Earlier in the chapter, we worked in stand-alone drawing and template files. Now we will perform the next several tasks in the context of a project. If you are not familiar with projects in AMEP, review Chapter 3.

1. On the Quick Access Toolbar (QAT), click the **Project Browser** icon

2. Click to open the folder list and choose your *C:* drive.

3. Double-click on the *MasterAME 2012* folder.

4. Double-click *MAMEP Commercial* to load the project. (You can also right-click on it and choose **Set Project Current.**) Then click Close in the Project Browser.

> IMPORTANT: If a message appears asking you to repath the project, click Yes. Refer to the "Repathing Projects" heading in the Preface for more information.

> Note: You should only see a repathing message if you installed the dataset to a different location than the one recommended in the installation instructions in the preface.

PLACING DUCTWORK IN A PREDESIGN OR SCHEMATIC DESIGN PHASE

Let's assume for the purposes of this exercise that our project is in the predesign or schematic design phase and we need to show general system configurations to convey our design intent.

5. On the Project Navigator palette, click the Constructs tab.

6. Expand the *Mechanical* folder and then double-click the *03 1-line* file to open it.

You will notice that some of the VAV boxes, supply air diffusers, and 1-Line undefined ductwork have already been placed in this drawing. We will focus our efforts on adding an additional VAV box, supply air diffusers, and more 1-Line undefined ductwork to one of the offices in this drawing. We are going to work in the lower left corner office and our goal will be something that is similar to what is depicted in Figure 5.27.

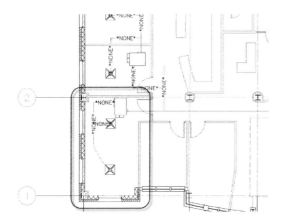

FIGURE 5.27

The end result desired at the completion of this exercise

7. Zoom into the corner office in the lower left corner of the plan.

8. Select the HVAC Workspace and make sure that the Tool palettes are displayed, and the HVAC tool palette group is active.

Refer to the "Choosing your Workspace" topic in the Quick Start chapter if you are not sure how to load a Workspace; refer to the "Understanding Tool Palettes" topic in Chapter 1 for information on how to load and work with tool palettes.

9. On the Tool palettes, click the Equipment tab.

10. In the VAV Box grouping, click the Series Fan Powered tool.

The "Add Multi-view Parts" dialog will appear.

11. From the Part Size Name drop-down menu choose 5 Inch Series Fan Powered VAV box.

12. From the Elevation drop-down menu choose Supply Duct (see Figure 5.28).

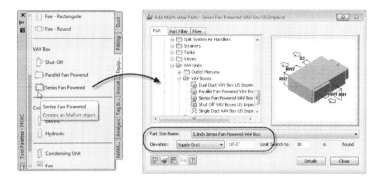

FIGURE 5.28

Cycle through multiple insertion points for the object by typing B ENTER (for basepoint) or simply press the CTRL key to cycle the basepoint. The base points are at the insertion point of the part, as well as the center of each connector.

> **Note:** In this situation, since we already have a termination point of the ductwork in the room and that ductwork is already set at the Supply Duct elevation, we really do not need to set the elevation of the VAV box, as it will inherit the elevation of the ductwork that we are connecting to. However, setting the elevation on all items is a very good habit to get into, and will help to keep you out of trouble when you are making ductwork connections.

You should immediately notice that the insertion point of the VAV box you are inserting is at the corner of the box itself. It will be more useful in this situation to switch to a better point. At the Command Line you can use the Basepoint option to cycle through all the insertion points of this object until the one you need is activated.

13. At the Command Line, type **b** and then press ENTER.

14. Repeat until the insertion point becomes the inlet of the VAV Box (see Figure 5.29).

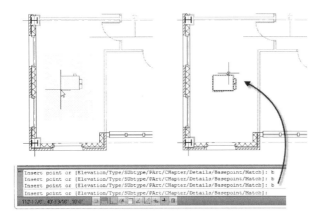

Insert point or [Elevation/Type/SUbtype/PArt/CHapter/Details/Basepoint/Match]: b
Insert point or [Elevation/Type/SUbtype/PArt/CHapter/Details/Basepoint/Match]: b
Insert point or [Elevation/Type/SUbtype/PArt/CHapter/Details/Basepoint/Match]: b
Insert point or [Elevation/Type/SUbtype/PArt/CHapter/Details/Basepoint/Match]:

FIGURE 5.29

MvParts allow you to cycle through multiple insertion points for the object

15. Hover over the end point of the 1-Line undefined duct coming into the Office (near the door).

16. A Duct End Connector snap will appear. When it does, click to place the box and then move the mouse to rotate it.

   The proper orientation is shown in Figure 5.30. If the Compass is not showing as a circle, type **P** at the command line to change the UCS plane so the Compass appears as a circle. This allows you to rotate the VAV box.

> **Note:** If the Duct End Connect Snap is not visible, your Object Snaps might not be turned on. Refer to Chapter 2 to learn more about AMEP's Object Snaps.

> **TIP:** Shift F3 toggles MEP snaps on or off.

17. Press ENTER to complete the command.

> **TIP:** When an object is initially brought into a drawing the object might appear in the correct orientation. However, when connecting to other objects in the drawing the object may flip or rotate unexpectedly. To reorient your part to the correct plane, type **P** (Plane) at the command prompt until the object appears in the desired orientation.

Let's place our supply air diffusers. Since we already have diffusers in the adjacent office, we can use the Add Selected method suggested earlier to place news ones with the same parameters.

18. Select one of the diffusers in the adjacent office, right-click and choose **Add Selected**.

    The "Add Multi-view Parts" dialog box appears with all the existing diffuser information already populated except the flow.

19. Click the Flow tab and in the Flow (Each Terminal) field, type: **150**.

20. Place the diffuser in the office and then press ENTER to accept the default rotation. Add a second one lined up with the first (see the left side of Figure 5.30).

> TIP: Once an object is placed in your drawing and populated with information, it is much faster to use the Add Selected tool or Copy the object to other locations in your drawing rather than choosing a tool from the Tool palette or Ribbon and populating the same information over and over again each time you need it.

Now let's begin placing our 1-Line undefined ductwork.

21. Select the VAV box we just placed and then click the Add Duct (+) grip on the discharge side of the VAV box to begin placing our ductwork.

22. In the Properties dialog change the Routing Preference to Supply – Low Pressure and change the Shape to Undefined.

23. Route your ductwork main to the left and then down, terminating mid-wall between the two windows on the west side of the office space (see the right side of Figure 5.30).

24. Press ENTER to end the command.

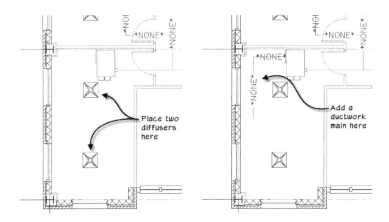

FIGURE 5.30

The start of a 1-Line undefined duct routing.

DUCTWORK PROPERTIES

We just placed 1-Line undefined ductwork and manipulated a few settings in the Properties dialog. Before we move forward let's quickly take a look at the other options that are available to us to control the Properties of our ductwork.

Once you select a duct tool the Properties dialog becomes active. This dialog can be broken down into two distinct sections, Basic and Advanced. The Basic section covers the System type, Dimensions, Sizing characteristics, the Elevation of the ductwork, any offsetting justification applied to the ductwork, and its Routing preferences (see Figure 5.31).

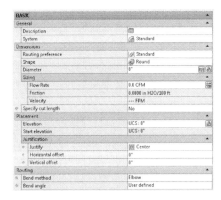

FIGURE 5.31

The Basic Properties of ductwork

TIP: Items in the Properties dialog that have asterisk (*) in front of the box are items that are only available during add commands, they do not appear during edit operations.

Let's breakdown the individual sections of this portion of the Properties for a deeper understanding on how the ductwork you are placing can be controlled further.

Description—A description of the ductwork you are placing can be entered by selecting the dialog icon in the Description box and typing descriptive information in the Description dialog.

System—Select the type of ductwork System that you intent to add to your drawing through this drop down. The System list comes from the Duct System Definitions in the Style Manager. Additional Systems can be added through the Style Manager.

Routing Preference—There are seven (7) out-of-the-box Routing Preferences for your ductwork through this drop down. Those Routing Preferences are Generic Banded, Generic Clipped, Generic Flanged, Generic Slip Joint, Generic Vanstone, Standard, and Undefined. Additional Routing Preferences can be added through the Style Manager.

Shape—There are four options to choose from for the shape of your ductwork. They are Oval, Rectangular, Round and Undefined.

Flow Rate—The number indicated in this box indicates the total CFM that is connected to the ductwork. This information is a sum total of all the flows from the connected objects to this ductwork.

Friction—The number displayed indicates the friction loss of the ductwork based on the flow and the dimensions of the ductwork.

Velocity—The number displayed indicates the velocity of the air moving through the ductwork based on the flow and the dimensions of the ductwork.

Width, Height or Diameter—The dimensional data of the ductwork you are placing. Ductwork sizes can be calculated based on the flow, size of the ductwork, and the parameters provided in the Design Parameters for the individual system identified in the Duct System

Definition through the Style Manager. In addition, sizes can be locked preventing their size to change no matter what flow through the ductwork. Once a size is locked the calculate button next to the size button becomes inactive for that piece of ductwork.

Note: As mentioned above, the ductwork will either be sized based on Friction or Velocity as identified in the Design Parameters of the System. However, if you use the Duct System Size Calculator to size your ductwork, it can be sized on either a maximum Friction value and/or a maximum Velocity value. See the "Duct System Size Calculator" topic below for additional information. In regards to Oval and Rectangular ductwork, the Width or Height Calculate Size button allows you to calculate either the width or the height of the ductwork based on the capacity in the ductwork. Selecting the Width Calculate Size will hold constant the dimension in the Height box and vice versa. The Diameter Calculate Size will calculate the size of the Round ductwork based on the capacity in the ductwork.

Specify cut length—The cut length of the ductwork can be toggled to specify a particular cut length for your ductwork. If you have enabled the "Specify cut length" setting and placed some ductwork in your drawing, AMEP provides a tool that allows you to modify this setting should that become necessary. To access the tool, simply select the piece, or pieces, of ductwork after they have been placed in your drawing; on the Duct tab of the ribbon, on the Modify panel, click the Duct Length button. The "Duct Length" dialog box will appear (see Figure 5.32). In the Duct Length modify tool you can break ductwork further, merge pieces, or even complete runs of ductwork together again.

FIGURE 5.32

Lengths of ductwork can be modified, once established in the drawing, through the Duct Length modification tool

Elevation—Select the elevation where you intend to place your ductwork. Elevation information is configured in the " Options" dialog box as we saw above. If you open this list, you should see any Elevations we added above such as the "Ceiling" Elevation. In addition, an elevation height can be locked as you are routing your ductwork.

Justification and Offset—This allows you to set the insertion point of your ductwork (center, left center, right center, etc.) as you are placing it. You can even identify an offset point of a routing location (see Figure 5.33).

A = Horizontal Offset

B = Vertical Offset

X = Offset Point

FIGURE 5.33
Routing locations and offsets of can be applied to ductwork

Bend method—This allows you to set different bend metods as you are placing your ductwork. The default bend method is "Elbow" and all bends in your ductwork will be drawn as elbows unless you specify "Offset" or "Transition – Offset" as your bend method.

Bend angle—This allows you to set a particular elbow angle that you want to use as you place ductwork. This elbow angle can also be locked so that it will be the only elbow angle selected in the Layout process. Locking the elbow angle is a handy trick when you only want to consider one type of elbow during your routing process. This also helps eliminate the total number of options available for an unconstrained or constrained duct route. See the "Ductwork Behavior" topic above for additional infromation on unconstrained or constrained duct routings.

The Properties palette is a modeless window. This means that it stays onscreen as you draw. If you close it, the command ends and you will not draw any ducts. Get in the habit of moving the palette to a comfortable position onscreen but out of the way of the drawing area. In most cases, moving the cursor from the palette to the drawing area automatically activates the drawing cursor. Values that you type in (such as Flow) require you to press return or tab to re-activate the cursor. Remember that duct properties can be modified "on the fly" through the Properties dialog.

There is one more section of the Properties dialog that we should touch on before we continue with our tutorial on placing ductwork in our project file. That section is the Advanced Properties (see Figure 5.34). Let's breakdown the individual sections of this portion of the Properties for a deeper understanding on how the ductwork you are placing can be controlled further.

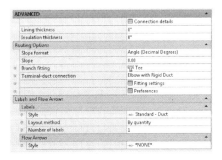

FIGURE 5.34
The Advanced Properties of ductwork

Lining/Insulation thickness—You can apply lining and/or insulation to your ductwork, either as you are placing it in your drawing or after you have placed it in your drawing. It is typically best to handle the application of the lining or insulation on a case-by-case basis by selecting the duct or duct systems after they have been placed and then apply the lining or insulation.

WARNING:
It should be noted that when applying ductwork lining to a duct, this is merely a graphical representation of the lining in the ductwork. The overall size of the ductwork does not increase to allow for the internal lining, nor do the dimensions of the ductwork change based on the application of the lining to the ductwork. In essence, when lining is applied to the ductwork, AMEP will treat the overall size of the ductwork as the inside clear dimensions of the ductwork.

Note: When adding insulation to the ductwork, the Top and Bottom Elevation will be affected by the thickness of the insulation and the tooltip will display the elevation based on the height of the duct plus the thickness of the elevation.

Slope format/Slope—Selecting this drop down allows you to set a slope type (decimal, percentage, or Rise/Run format) for your ductwork and the desired slope of your ductwork.

Branch Fitting—This allows us to set the ductwork branch fitting type to either a Tee or a Takeoff. The Tee or Takeoff used will be based on the Part of the Type that was specified in the "Fitting settings" dialog as you are placing your ductwork.

Terminal-duct connection— In the "Terminal-duct connection" area, there are three options: Flexible, Elbow with Rigid Duct, or Extended Duct. This setting governs the connection from the branch ductwork to the terminal unit (diffuser, register, or grille).

Fitting settings—Click the dialog launcher for the "Fitting settings" dialog. This dialog allows you to set the Part for each Type of fitting based on its connection type (Generic Flange, Standard, etc.) as set in the Routing Preference of the Properties dialog. Use the drop-down of the Part to select the Part that you wish to place for each Type. An image of the Part is displayed once the Type has been selected (see Figure 5.35). The changes in the Fitting Settings window are temporary, and will revert to the settings based on the Routing Preference after the current Duct command has terminated.

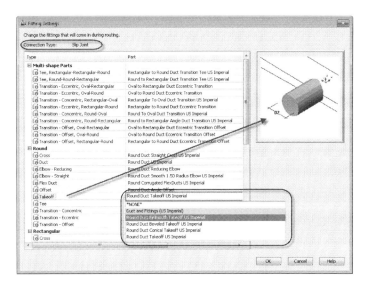

FIGURE 5.35

Different Parts can be set to be placed as you are routing your ductwork

Preferences—Click the dialog launcher to set several user-definable preferences for your ductwork. Some of the settings include: slope format, the creation of risers at elevation changes, providing labels to your ductwork as you place it, and defining elbow layout options. There is some overlap between the Duct Preference and the Properties dialog. Duct Preferences can also be accessed from the ribbon (see Figure 5.36)

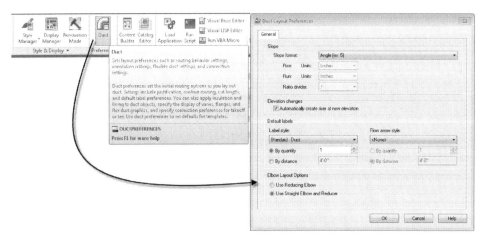

FIGURE 5.36
Duct Layout Preferences can also be accessed from the Preferences panel in the Manage tab of the ribbon

Labels and Flow Arrows—Before you place your ductwork you have options to apply labels and flow arrow to your ductwork as it is being placed. The "Layout method" of the label can be set to Quanity or By Distance. If the labels or flow arrows do not meet your project needs or match your office standards, AMEP allows you to create your own.

If you would like to try out some of what we have learned so far, before we continue, feel free to experiment and add some ductwork runs yourself. You can accept the default sizes or modify the Properties selection to choose your System, Elevation, Shape, etc. Click in the drawing window to begin adding your ductwork. To start a new run without exiting the command, click the New Run button in the lower left hand corner of the Properties dialog.

Once you have finished experimenting let's continue with our exercise.

1. Select the ductwork main.

2. Hold down the CTRL key and click the plus (+) grip to start placing flexible ductwork (see left side Figure 5.37).

3. Hover over the supply air diffuser closet to the lower (south) wall until you see the Duct End Connector and then click to connect to the supply air diffuser (see right side Figure 5.37).

4. Press ENTER to terminate the command.

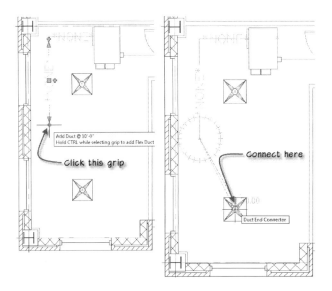

FIGURE 5.37

Using the grips on objects can be a quick and effective means for placing your ductwork systems

5. Place your cursor over the ductwork you just placed and note from the AMEP tool tip that it is flexible ductwork (see Figure 5.38).

Hover over the flexible ductwork that we just placed and note the Elevation in the tool tip which was automatically determined by the selection of the two connection points.

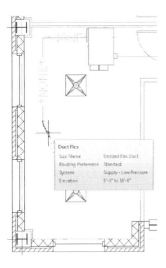

FIGURE 5.38
Ductwork Options established before you begin placing ductwork in your drawing can allow you to place ductwork systems more quickly and effectively

Let's connect the final diffuser in the office to our 1-Line undefined ductwork main. In the previous step we used a grip from the supply air ductwork main to connect to our diffuser to the ductwork main. Let's repeat this process with a few minor modifications.

6. Pick the 1-Line undefined ductwork main running in the north-south direction.

7. Click the Add Duct (+) grip to place a short piece of ductwork routed towards the unconnected diffuser. Stop your ductwork run at the halfway point between the main and the diffuser. Press ENTER to terminate the command.

8. Select the Flex Duct tool from the tool palette and for the start point pick the end of the ductwork branch we just placed.

9. Hover over the supply air diffuser until you see the ductwork connection snap symbol, and then click to connect to the supply air diffuser.

10. Press ENTER to terminate the command (see Figure 5.39).

Notice that we did not modify any duct settings in the Properties dialog and yet an undefined 1-Line flexible duct was added to the project. This is due in part that AMEP objects can take on the characteristics of the objects they are connecting to.

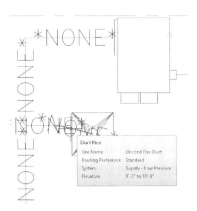

FIGURE 5.39

You can also use ductwork grips to place ductwork and make connections to your supply air diffusers

VERIFY CONNECTIONS

When we placed the two diffusers in the office, we associated 150 CFM to each of the diffusers for a total of 300 CFM in the Office and the main supply duct. Since objects in AMEP know how to communicate with each other, there should be a total of 300 CFM in the 1-Line undefined main duct that is connected to the VAV box. Lets' see how successful you were connecting your diffusers to the main duct.

11. Pick the main duct connected to the VAV box.

    If you were successful with your diffuser connections to the main, the Flow Rate in the Properties dialog will match what is shown in Figure 5.40.

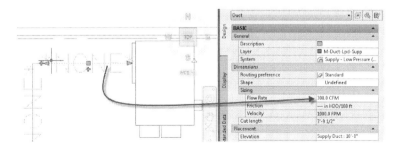

FIGURE 5.40

AMEP objects know their association to one another and can share information between system objects

12. Press ESC to clear the selection set.

That completes our work in the predesign or schematic design phase. Now we are proceeding into the design development phase of the project and we need to provide further definition to our systems. Let's look at how we can use the 1-Line undefined ductwork to our advantage.

REFINING YOUR DUCTWORK DESIGN (FOR DESIGN DEVELOPMENT)

Since we were in the predesign and schematic design phase of our project, the preliminary CFM values that we associated with the supply air diffusers were based on preliminary load analysis or on rule of thumb guidelines. Let's assume that these preliminary CFM values bases are actually close enough for us to use in the design development phase of our project. From here we want to convert all of our 1-Line Undefined ductwork (single line duct) to double line ductwork for our Construction Documents. AMEP has a tool that will allow you to do that.

13. Select the duct main connected to the VAV box.

14. On the Duct tab of the ribbon, on the Calculations panel, click the Calculate Duct Sizes button.

DUCT SYSTEM SIZE CALCULATOR

The Calculate Duct System Size tool is an extremely useful tool to help you define your 1-Line duct, or even redefine your 2-Line ductwork. The Duct System Size Calculator allows you to input project specific requirements, or even override design parameters established in the Design Parameters for a particular Duct System Definition. Following is a description of the components of the Duct System Size Calculator (see Figure 5.41).

FIGURE 5.41

*The Duct System Size Calculator can resize 1- or 2-Line ductwork based
on project conditions and override Design Parameters previously established*

Box 1—This box allows you to select the system, or items, that you would like AMEP to automatically size for you.

- **Calculate Complete System**—Selecting this radio button allows you to have AMEP begin automatically sizing all the ductwork and ductwork fittings for the selected system. This requires the ductwork to have a flow value (CFM) associated to it.

> **Note:** If you differentiate between your systems (i.e. Supply – Low Pressure and Supply – Medium Pressure) AMEP will calculate the complete system based on the ductwork style. Conversely, if you place all your ductwork on one system (i.e. Supply) and there are varying friction losses or velocities in that system (i.e., low pressure and medium pressure supply air ductwork), AMEP will size the complete system based on the values placed in the Duct System Size Calculator for that particular style. This can obviously have adverse effects on the sizing of your ductwork.

- **Calculate Selected Objects**—Performs sizing on a selection of objects rather than the entire drawing. Selecting this radio button grays out everything except the Select Object icon. Click this icon to make a selection in the drawing.

> Note: Use a Crossing or Window to select all the objects for which you wish to calculate duct sizes. You must have a logical start object or the Duct System Size Calculator may ignore the entire calculation process and return you to the dialog box. This means that you should select a logical system; you cannot select a few components here and there in your drawing.

- **Select Object**—Works in conjunction with either the Calculate Complete System or Calculated Selected Objects radio button as noted.

Box 2—This box allows you to identify the shape(s) that you would like to use for all the 1-Line undefined ductwork that you selected in Box 1.

- **All**—Selecting this radio button will allow you to use the drop-down menu in this box to identify the primary shape, Round, Rectangular or Oval, of all the 1-Line undefined ductwork that you selected through Box 1.

- **Individual**—Selecting this radio button will allow you to use the drop-down menus in the Trunk and Runout boxes to select the shape of all the 1-Line undefined ductwork that you selected through Box 1. In the Trunk drop-down box you can choose from Round, Rectangular, Oval, or Inherit from Fan. The Inherit from Fan option will use the dimensional data associated with the fan opening/connection and begin sizing the ductwork from that data. In the Runout drop-down box you can choose from Round, Rectangular, Oval, Inherit from Trunk, or Inherit from Diffuser. The Inherit from Trunk option will use the dimensional data associated with the Trunk duct and begin sizing the ductwork from that data. The Inherit from Diffuser option will use the dimensional data associated with the diffuser connection size, and use that ductwork size all the way back to the Trunk duct. If the Use shape and size from air terminal for runouts box is selected, this option will not be available to you.

- **Use shape and size from air terminal for runouts**—By selecting this box, AMEP will use the dimensional data associated with the diffuser connection size, and use that ductwork size all the way back to the Trunk duct. If this box is selected, the Runout: option in Individual: will not be available to you.

Box 3—This box allows you to override your Design Parameters associated with the Duct System Definition (see the "Duct System Definitions:" topic above). In addition, you can further define your ductwork by specifying allowable heights and limit your ductwork sizes to particular dimensions found in the AMEP duct catalog.

- **Override design parameters from system definitions**—By selecting this box you can override both the Velocity and Friction component(s) specified in the Design Parameters associated with the Duct System Definition. For example, let's say that you established a Design Parameter for your project's low pressure supply air ductwork system of 0.085" w.g per 100' for the friction loss and 1,000 fpm for the velocity. From here you place your ductwork system in the library of a school building. Perhaps the majority of your classrooms might be able to be designed using the established Design Parameters for the low pressure ductwork system, but an acoustical consultant might have other ideas for your ductwork system in the library. Their ductwork design recommendations might be a maximum of 0.05" w.g per 100' for the friction loss or a maximum velocity of 500 fpm. When you are utilizing the Duct System Size Calculator to automatically size your ductwork, you can override your initial Design Parameters here without having to adjust your Design Parameters in the Duct System Definitions through the Style Manager every time a "one off" adjustment is needed. The override to the Design Parameters here can be overridden based on both Friction and Velocity. Remember that a Duct sized via the Design Parameters will be sized either on Velocity or Friction. But the Duct Calculator can be an and/or condition when sizing the ductwork on Friction or on Velocity.

- **Round max size**—By providing a value in this box you can set a maximum size for the diameter of your round ductwork and for anything greater than that size you can tell AMEP to use either Rectangular or Oval ductwork. You can use Round max size in conjuction with Rectangular/Oval Max Height.

- **Rectangular/Oval Max Height**—By providing a value in this box, you can set a maximum height for your rectangular or oval ductwork. You can use Rectangular/Oval Max Height in conjunction with Round max size.

- **Use catalog sizes**—Use the drop-down menu in this box to tell AMEP to use ALL the ductwork sizes found in the Duct catalog, use only the ductwork with 1-inch increments (eliminates the selection of 1/2" increment ductwork), or use Even sizes only (eliminates the selection of 1/2" and 1" increment ductwork).

- **Apply sizing to parts with defined shapes**—By selecting this box, AMEP will apply the information in the Duct System Size Calculator to ductwork that has a defined shape to it and resize this ductwork accordingly if needed.

Box 4—By selecting the Start button in this box, AMEP will begin to caluculate your ductwork sizes based on the information that you provided in the Duct System Size Calculator dialog box and automatically resize the objects that you selected through Box 1.

Box 5—After AMEP calculates and resizes the ductwork you selected, the Duct System Size Calculator will indicate how successful it was calculating, resizingand converting the selected ductwork objects. Selecting the View Event Log button will open the Event Log dialog box (see Figure 5.42).

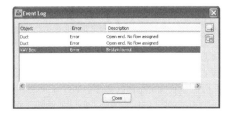

FIGURE 5.42
The Event Log dialog box sums the success of the Duct System Size Calculator in an effort to calculate, resize, and convert selected ductwork objects

You can choose one of the errors that you received in the Event Log dialog box, and then you can use either the Highlight selected objects button of the Highlight or zoom selected objects button to zoom to the ductwork location where the error occurred. Once you are done reviewing your errors, if any, you can close the Event Log dialog box and then close your Duct System Size Calculator dialog box.

> **TIP:** In an effort to help reduce any unforeseen errors, you will need to make sure that there are no open ends in your ductwork system. This means that you should terminate your ductwork with an air terminal or an endcap. You can find endcaps in the duct fitting catalog or using the Endcap tool palette tool on the Fittings palette.

1. In the "Duct System Size Calculator" dialog box configure the following settings (see Figure 5.43):

⇨ In Box 1, make sure that "Calculate complete system" is selected.

⇨ In Box 2, change the shape to Rectangular and check the "Use shape and size from air terminals on runouts" checkbox.

⇨ In Box 3, check the "Round max size" box and type 12" in the size field, check the "Rectangular/Oval Max Height" box, type 12" in the size field and choose Even sizes only.

FIGURE 5.43

Using the Duct System Size Calculator to convert 1-Line undefined ductwork can be a huge time saver

2. In Box 4, click the Start button.

> **Note:** During the conversion process, AMEP might prompt you to make a part selection from multiple choices that are available in one particular location. Here you can make a good engineering judgment to choose the part that you believe would best fit the situation, with the understanding that you can edit that part at anytime in the future.

Converting this particular system associated with this VAV box, you should be prompted twice to tell AMEP what parts to use in two particular situations. In this situation we will choose the two parts shown in Figure 5.44 and Figure 5.45. AMEP will also use the built-in zoom functionality to take you to the actual location where a decision for a part choice is needed so that you can make a more informed choice on the part based on the actual condition.

3. In the "Multiple Parts Found" dialog, make the selection indicated in Figure 5.44 and then click OK.

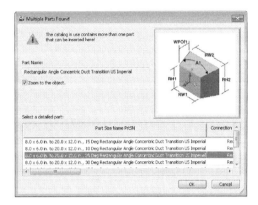

FIGURE 5.44

When multiple choices exist for a part selection, AMEP will prompt you
to choose which part you would like to use in a particular condition

4. In the second and third instances of the "Multiple Parts Found" dialog, make the selection indicated in Figure 5.45 and then click OK.

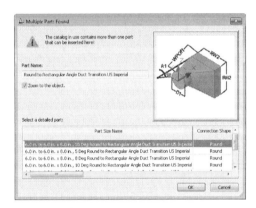

FIGURE 5.45

AMEP will even zoom into the area in question to give you a clearer perspective on the actual condition

5. Before clicking the Close button in the "Duct System Size Calculator" dialog box notice the Success Rate in Box 5 of the Duct System Size Calculator. Click the Close button.

Our 1-Line undefined ductwork for this particular room was easily converted to 2-Line ductwork through the Duct System Size Calculator tool (see Figure 5.46).

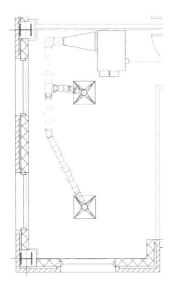

FIGURE 5.46
1-Line undefined ductwork can be easily converted to 2-Line ductwork through the Calculate Duct System Size tool

Since change is inevitable in our field of work, let's assume that right after we converted the 1-Line undefined ductwork to 2-Line ductwork we learned that this Office is now a Conference room.

In this part of the exercise, since we already have a duct layout for this room, we will use the duct modification tools to modify the existing components for the new usage of the room. Let's assume that we calculate the following for our new room:

- 450 CFM required for the room; 225 CFM for each diffuser and;
- An 8" inlet to each diffuser with an 8" branch duct from each diffuser to the main duct and;
- A main size of 12"x10" to accommodate the 8" branches from the diffusers.

In lieu of using the Calculate Duct System Size tool to automatically calculate the ductwork size based on the changes above, let's simply modify the components to match the information.

Note: When making ductwork modifications to a system, it is best to start at the main duct, proceed to your air device, and then work your way through the system branch in an effort to avoid any error messages that you might receive from branch ductwork connections being larger than the main. If your system is connected correctly, once you change the size of your air device AMEP will automatically change the branch ductwork associated with the air device. This can be a huge time saver when implementing changes.

6. Pick the 8" x 6" main duct running in the north-south (vertical) direction.

7. On the Modify panel of the Duct tab select Modify Run.

8. In the Modify Run dialog check the Width and height selection buttons and type **12"** for the Width and **10"** for the Height and click OK.

TIP: You can also use the drop-down lists to choose your ductwork size.

9. Select the diffuser closest to the VAV box.

10. On the Equipment tab, on the Modify panel, click the Modify Equipment button.

11. Click the Flow tab. In the Flow (Each Terminal) field, type: **225** CFM.

12. Click the Part tab. From the Part Size Name drop-down list, choose: 24 x 24 Inch Square Plaque Ceiling Diffuser – 8 Inch neck.

Due to the modification of ductwork sizes, AMEP notifies you that since some parts need to change the ductwork system can no longer remain connected, and AMEP asks you how you would like to proceed (see Figure 5.47).

FIGURE 5.47

Ductwork modifications might cause you system no longer to remain connected; and. AMEP gives you choices on how to tell AMEP to handle the connections

The Modify to maintain connection to next part option allows you to keep the adjacent ductwork sizes that are the same size of the duct you picked, and AMEP will make new connections to the adjacent ducts through new transitions.

The Modify to next junction/transition allows you to automatically resize adjacent ductwork that are the same size of the duct you picked.

In our particular case, we will want AMEP to automatically change all the 6" ductwork to 8".

13. Click OK in the "Multi-view Part Modify" dialog box.

14. Choose Modify to next junction/transition in the "Maintain Connection" dialog box and then click OK.

> **Note:** The OK button is not located in its normal location. This ensures that you pay attention to the selection that you are about to make as it could have a dramatic effect on the system you are modifying.

15. Repeat this process for the other diffuser.

We used the ductwork modification tools to modify our existing ductwork layout here. The advantage to this approach is that we merely modified sizes of existing components in our system and did not have to redraw any new ductwork components. Again this can be a huge time saver. Once completed, your modifications should look something similar to Figure 5.48.

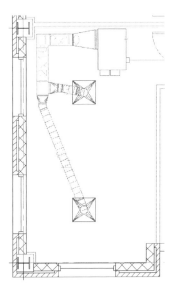

FIGURE 5.48

Modifing existing ductwork helps you avoid a time-consuming process of placing new ductwork objects in your drawing

FURTHER DEFINING YOUR DUCTWORK SYSTEMS

Form here we can modify other existing ductwork objects to suit our system layout needs. Perhaps we wanted to change the transition from the VAV box to a 30°-angled transition in lieu of a 15°-angled transition or perhaps changing the mitered elbow to a radius elbow. Such modifications can quickly be achieved using the Duct Fitting Modify tool. The process is similar to modifying the other components. For the purposes of this exercise, let's assume we are satisfied with our supply air ductwork layout and focus now on the return air system for this room.

16. On the Tool palettes click the Equipment tab and then click the Grille tool.

17. In the "Add Multi-view Parts" dialog box, on the Part tab, choose Return Air Grilles without Trim (US Imperial) and then from the Part Size Name list, choose 24" x 24" inch Return Air Grille.

18. For the Elevation, choose Ceiling.

19. Pick a point between the supply air diffusers near the interior wall to place your register. Press ENTER to accept the default rotation (see Figure 5.49).

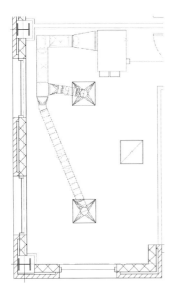

FIGURE 5.49
Add a return air grille

One of the last items that we need to add is the return air ductwork at the VAV box. Let's do that now.

20. On the Tool palettes, click the Duct tab and then choose the Rectangular duct tool.

21. On the Properties Palette, set the System to Return.

22. Hover the cursor over the back of the VAV box until the ductwork connector appears for the return air ductwork connection.

23. Pick the connection location, drag the duct towards the Conference Room door, and terminate the duct halfway between the VAV box and the door opening (see Figure 5.50).

24. Press ENTER to complete the command.

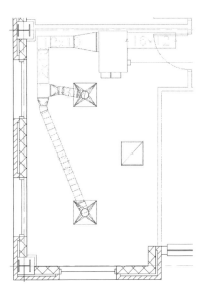

FIGURE 5.50
Add return air ductwork

You will note that we did not provide the size of the ductwork. The size of the ductwork was automatically determined by the ductwork connection size at the VAV box. This can also be a real time saver when placing ductwork in your project. Since there is a fan in the VAV box let's provide some internal insulation for minimum acoustical purposes on our return air ductwork.

25. Select the return air ductwork that you just placed and in the Properties dialog set the thickness of the lining to 2".

If you recall from above, we indicated that we were going to convert our 1-Line undefined ductwork to 2-Line ductwork for the design development portion of our project. For the purposes of this exercise let's assume that we have placed enough components in our Conference Room to show the design intent. We are now ready to move into the construction document phase of our project and fine-tune our placed system components.

REFINING YOUR DUCTWORK SYSTEMS FOR CDS

Let's assume for this portion of the exercise that the Architect has finalized all of their ceiling grid locations and we are now ready to locate our air terminal to correspond to the architect's ceiling grid layout.

26. On the Home tab, on the Layers panel, click the Layer Properties icon.

27. In the Layer Properties Manager, thaw layer 03 PartitionslA-Clng-Grid.

28. Close the Layer Properties Manager.

29. On the Drawing Status Bar, change the Display Configuration to Reflected.

Notice that not only has the Architect placed the ceiling grid in this room, but also it has attempted to provide a preliminary location to the air terminals in this room (see Figure 5.51). For the purposes of this exercise we will use the locations the Architect has shown for the air terminals.

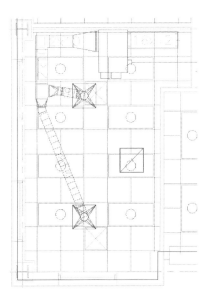

FIGURE 5.51
Architectural ceiling grids can be used to place mechanical ceiling components for coordination purposes

If you recall, as we were placing our air terminals we were placing them on a predefined ceiling elevation. That elevation was preset at 8'-6", as during the schematic and design development phase of our project the Architect had not set the elevations of the ceiling grids. Since the Architect has now provided ceiling grids for the rooms at a set elevation let's relocate our air-terminal units to the Architect's ceiling grid.

30. Select the lower supply air diffuser.

31. Use the diffuser location grip to relocate the diffuser to the intersection of the ceiling grid tiles (see Figure 5.52).

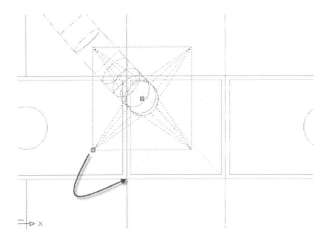

FIGURE 5.52
Use the location grip to move objects to precise locations

32. After moving the diffuser, note on the Properties Palette that the elevation is now 8'-0".

With AMEP, objects understand their relationship to each other. Since the Architect placed their ceiling grid at 8'-0" and we relocated our diffuser to the intersection of the ceiling tiles, the elevation of the diffuser takes the inherent properties of the object to which it is associated. In addition to this you should note that the system remained connected even though you moved one component of the system. This is a huge time saver for project coordination.

33. Relocate the other supply air diffuser and locate the return air grille to the location indicated by the architectural reflected ceiling plan at the intersection of the ceiling grids (see Figure 5.53).

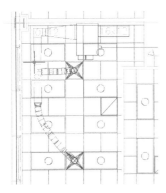

FIGURE 5.53
Using object grips to better associate objects allows you to better coordinate your project

ADDING REFINEMENTS

Now, let's begin to annotate the systems in the Conference Room.

34. Return to the Home tab, open the Layer Properties Manager, and freeze 03 Partitions|A-Clng-Grid layer again.

35. Set the Display Configuration back to MEP Basic 2-line.

36. On the Tool palette, click the Tag & Schedule tab.

37. In the Tag grouping, click the Air Terminal 1 tag tool.

38. At the "Select object to tag" prompt, select one of the diffusers and then pick a location onscreen for your diffuser tag.

39. In the "Edit Property Set Data" dialog box, click OK.

   The "Select object to tag" prompt repeats.

40. Select the other diffuser, pick a location for the tag, and then click OK in the "Edit Property Set Data" dialog box.

41. Press ENTER to complete the command.

Notice that the tags report the information we associated with the diffuser in the steps above. This is a generic tag. If the tags do not match your company standard, they can be created or modified as required. We can use other tags to automatically tag other objects in our drawing, like the VAV box or the return air grille as well. Let's move on to annotating our ductwork.

Let's clean up our drawing a bit. Currently we have several notes that AMEP automatically added as we were placing our ductwork. Based on our final ductwork layout, the quantity of notes we have here is excessive. For example, let's delete the 8" tag that is associated with the supply air flex ducts since the diffuser tag already contains this information. Let's also delete the tag associated with the return air ductwork at the VAV box, since manufacturer openings will vary and we will let a detail address the correct size of the ductwork. Finally, let's delete the duct label in the north-south duct main. To delete such labels, simply select them and then press the DELETE key.

42. Select the excess tags and then press the DELETE key.

We are now left with one ductwork label associated with the main that is running in the east-west direction. Let's see how we can quickly modify this label to meet our annotation needs for our construction documents.

43. Select the duct label on the main running in the east-west direction.

44. Click the location (square) grip and attempt to move the label.

Notice that the label remains constrained along the length of the ductwork.

45. Press the ESC key to cancel. Hold down the CTRL key, and click the Location grip again.

The label is now free to move to any location without being constrained to the ductwork anymore. (To give you the most flexibility in relocating the duct label, make sure your Ortho snap is off.)

46. Place the label in the room above the Conference Room above the VAV box (see Figure 5.54).

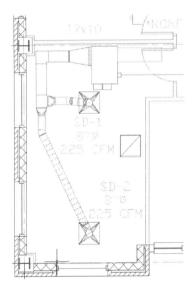

FIGURE 5.54
Once added to a drawing, labels can be quickly modified, removed or relocated to meet your annotation needs

47. On the Tool palette, click the Annotation tab.

48. Click the Text (Straight Leader) tool.

49. Pick the short run of ductwork as the starting point and, following the prompts, create an arrow associated to the note that was relocated.

50. Press ENTER twice and then in the "Text Formatting" window, click OK to terminate the text command (see Figure 5.55).

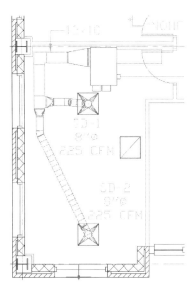

FIGURE 5.55
Add an arrow to the label to associate it visually with the correct object

The tool we chose is actually a text leader; the last few prompts were to add an Mtext object. Since we skipped these prompts, we got only the leader. Although relocated from an object, labels will still maintain their link to the object (i.e., when the object changes, the label will change too).

Finally, now that we have cleaned up the annotation in our drawing we can see a few things that are missing from our ductwork system. Let's add a couple of volume (balancing) dampers to our drawing.

ADD A DAMPER

As we have seen in some other situations, when you add a new component to an object or in a system, it is able to read the size and other properties from the existing objects, saving you time and effort during placement. Further, this cuts down on errors since you can be confident that the sizes and other data match. In this situation, since we have already established the size of our branch ductwork, when we place the damper in the ductwork it will automatically inherit the size of our branch ductwork.

51. On the Tool palette, click the Equipment tab, and then beneath the Damper grouping, click the Balancing – Round tool.

52. Hover near the duct fitting for the branch ductwork until you see the Duct End Connector symbol, and then click (see the left side Figure 5.56).

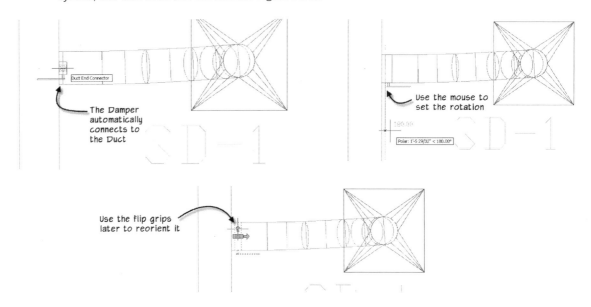

FIGURE 5.56

When placing objects, the object can inherit the information or data for the object to which it is being associated

53. Move your mouse to determine the rotation of the damper and then click (see the right side of Figure 5.56).

While placing the damper, you can use the base point toggle (B at the Command prompt) and/or the plane toggle (P at the Command prompt) to further define the configuration of your damper.

54. Press ENTER to complete the command.

TIP: After placing an object you can use the "flip grips" to flip components of the object if you are not satisfied with their location (see the bottom of Figure 5.56). In the example above we could use the flip grips to flip the handle to the other side of the branch ductwork if we desired.

From here you can add other diffusers to your drawing as your project requires. The same concept above for placing the round balancing damper will apply if you wanted to place the rectangular balancing damper in the main branch duct on the other diffuser.

Note: After you place dampers in your ductwork system and you modify the size of your ductwork, the damper size will not be modified according to the new ductwork size. AMEP will want to provide transitions to the damper. In this situation, it is recommended that you delete the damper, modify the ductwork size, and then place a new damper, or use the Modify Equipment tool to modify the damper size.

ADDING 2-LINE DUCTWORK

Let's add the return air ductwork main for our plenum return for the air handling unit system. What we want to accomplish in the next few steps is depicted in Figure 5.57.

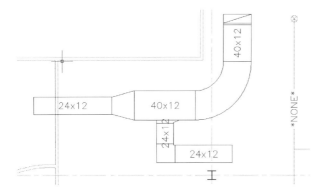

FIGURE 5.57
2-Line ductwork can be added to your drawing as easily as 1-Line undefined ductwork

55. Zoom and pan over to the right in the open area to the right of the reception space and south of the elevator lobby.

We will begin placing the return air ductwork system by placing the 24 x 12 ductwork working our way to the return air riser.

56. On the Tool palette, click the Duct tab and then click the Return duct tool.

57. In the Properties dialog, configure the following settings

- Change Shape to **Rectangular**
- Set the Width to **24"**.
- Set the Height to **12"**.
- Change Elevation to **Supply Duct**.

58. Using Figure 5.57 as a guide, click in about the middle of the reception space, move to the right and then click again in the unoccupied tenant space.

59. In the Properties dialog change the Width to **40"**.

Since you are changing ductwork sizes, AMEP will prompt you for a fitting type.

60. Select the 30 degree fitting type shown in Figure 5.58.

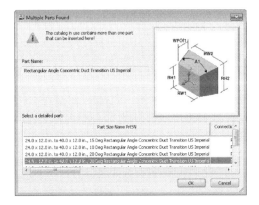

FIGURE 5.58

AMEP automatically recognizes when parts need to be added to systems based on the sizes of the components being connected

61. Continue horizontally and click to the left of the 1-line supply ductwork.

In order to place the return riser, let's use the tracking tools within AMEP to line up our ductwork risers in a row. Here we will use the tracking tool with the supply air ductwork riser that has already been placed in our drawing.

62. With the Duct command still active hover over the center of the supply air ductwork to acquire the point.

63. Move the mouse back to the left (a dotted line will appear on your screen). When you have snapped back to a 90° angle, click to place the vertical ductwork run (see Figure 5.59).

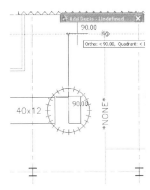

FIGURE 5.59
Using AMEP's tracking tools can help you place objects in relationship to others

64. In the Properties dialog change the Elevation to **12'** and then press ENTER to terminate the duct command.

A mitered elbow was used by default in our return air system. This is indicated in our Fitting settings. See the "Fitting settings" topic above for additional information. Let's change the mitered elbow in this system to a radius elbow.

65. Select the mitered elbow that is visible in plan.

66. In the Properties dialog select the image of the part in the Type box (see right side of Figure 5.60).

67. Select the Subtype: drop down and select Smooth Radius (see left side of Figure 5.60).

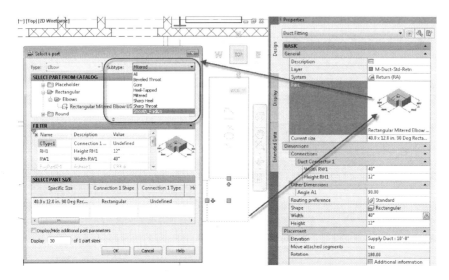

FIGURE 5.60

AMEP Parts can be modified after the Part has already been placed

68. Select the Rectangular Duct Smooth Radius 1.5W Elbow US Imperial and type **12"** for the height and **40"** for the width and click OK (see Figure 5.61).

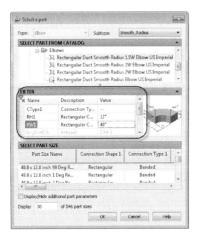

FIGURE 5.61

Part size values can be entered for the Parts

Now let's add the remaining 24" x 12" branch ductwork. Let's use the grip of the 40" x 12" main ductwork to add our branch ductwork.

69. Select the main Duct.

70. Click the Add Duct (+) grip to begin adding our branch duct.

71. In the Properties dialog change the width to 24" and the height to 12" and draw an "L" shaped return branch Duct in the approximate length and direction as shown in left side of Figure 5.62.

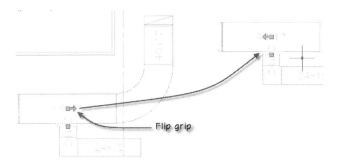

FIGURE 5.62
Complete the layout and flip the fitting

72. Use the flip grip of the ductwork fitting to correctly position the ductwork fitting in the direction of air flow (see the right side of Figure 5.62).

73. Close and save the file.

Our return air system is now complete and we have quickly learned how to place 2-Line ductwork in our project drawing.

DISPLAY THEMES

A Display Theme is an AEC object that can change the way other AEC objects display. This occurs independently of the current Display Configuration. A Display Theme queries the drawing for certain properties, when the values of these properties meet the conditions outlined within the Display Theme Style, the display of the affected objects is modified. The modified display remains in effect as long as the Display Theme is active. While you can insert as many Display Theme objects into the drawing as you wish, only one can be active at any given time. Previously active Display Themes are automatically disabled when a new one is inserted. You can disable a Display Theme at any time.

USING DISPLAY THEMES

Like most objects in AMEP, Display Themes are style-based objects. They key to Display Themes is the Property Set Data attached to the objects, and Display Themes use rules based upon those properties to modify the display. Display Themes do not apply to AutoCAD entities like lines, arcs and polylines. A few styles have been provided as out-of-the-box samples. Let's take a brief look at them now before building our own style below. Let's look at a few Display Themes for HVAC systems.

1. On the Project Navigator, on the Constructs tab, expand the *Mechanical* folder and then double-click on *03 2-line Complete* to open it.

2. On the Tool palette, click the Analysis tab.

3. Beneath the Theme grouping, click the by Velocity theme tool.

4. Pan in the drawing and click to place the Theme in a clear area next to the plan and then press ENTER.

Notice that all the ductwork in the drawing changed colors based on the definitions in the Display Theme (see Figure 5.63).

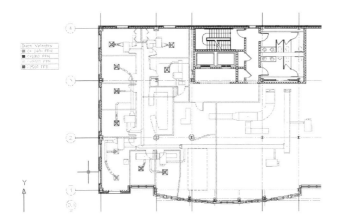

FIGURE 5.63
Display Themes provide a way to quickly analyze your drawings visually

5. On the palette, click the by Friction Loss theme, place it beneath the first one, and then press ENTER.

Notice that the ductwork systems change color to reflect the parameters established in the second Display Theme. Also note that the previous Display Theme has a slash through it, thus indicating that this Display Theme is no longer active (see Figure 5.64).

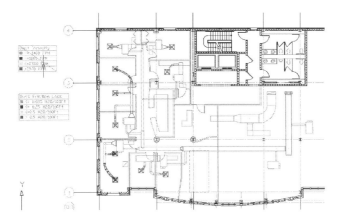

FIGURE 5.64

Multiple Display Themes can be placed in a drawing: however, only one Display Theme can be active at a time

To make other Display Themes active, select the Display Theme you wish to activate and on the Display Theme tab, click the Apply Display Theme button. To deactivate one, either activate another one or select it and click the Disable Display Theme button on the ribbon.

SUMMARY

Throughout this chapter we reviewed many key aspects of ductwork in AMEP. We configured ductwork settings and then applied those concepts to the ductwork placed in our sample commercial office building. We also discussed 1-Line undefined ductwork and how we can easily convert this to 2-Line ductwork, and we looked at how to create and modify 2-Line ductwork. We ended our discussion with Display Themes, giving you a small glimpse at the potential that powerful tool possesses. You should now have a broad understanding of what is possible with the ductwork tools in AMEP and how to begin adding ductwork to your own projects. In this chapter we learned:

✓ Ductwork settings for Display Control for ductwork, establishing Elevations, and configuring which Catalogs will be used to place your HVAC objects.

✓ Ductwork Properties control Routing offsets, lining, insulation, annotations of Ducts, and the specific Parts used for ductwork connections.

✓ Place 1-Line ductwork for your predesign or schematic design phase or your project.

✓ As the design progresses, you can convert 1-Line ductwork to 2-Line ductwork with the Duct System Size Calculator.

✓ Further modify ductwork systems manually as required.

✓ Place 2-Line ductwork directly in your drawings in later phases like construction documents and make modifications as required.

✓ Display Themes can represent information about your systems graphically onscreen to help you understand those systems better.

Chapter 6

Piping Systems

INTRODUCTION

With AutoCAD MEP you can easily design and document Pipe Systems for both Pressure and Gravity applications. We will cover the fundamentals on how piping works in AutoCAD MEP, describe the settings that control piping, and discuss the preferred workflow approaches when laying out piping in both Pressure and Gravity systems.

In this chapter we will look at how Piping works in AutoCAD MEP. We will explore what you need to know about creating a new system definition for your piping system, defining the fittings that will be used in the layouts, how pipe systems are displayed, and how the routing tools work when you are laying out your piping system. Routing tools incorporated into the PipeAdd command support sloped piping and true male x female connections. We will go through the best practices on how to lay out both pressure and gravity piping systems and the differences between them.

OBJECTIVES

In this chapter you will create new system definitions and lay out a fire protection system. We will create a sanitary piping system using the Sloped Piping abilities of AutoCAD MEP. By creating these systems, you will understand the fundamentals of routing, systems and routing preferences. Display configurations and how to control them will be explained and put into practice, and you will learn when to apply display overrides to create the desired look in construction documents. In this chapter you will:

- Learn the fundamentals of Piping in AutoCAD MEP
- Learn about Settings and Controls
- Determine what fittings are used and when
- Explore System Definitions
- Learn about Pressure Piping System design tools

- Learn about Gravity Piping System design tools

FUNDAMENTALS OF 3D PIPING

Let's begin with an exploration of the fundamentals of 3D piping by reviewing what makes piping work. 3D piping has certain characteristics, features, and settings that allow AutoCAD MEP to be an efficient 3D piping application. Like other 3D layout features such as duct and cable tray, piping uses system definitions to determine which settings the pipe run will inherit. These include layers, display settings, system abbreviations, etc.

Piping requires content such as elbows, pipes, tees, etc. to assemble a pipe run. These fittings are brought into the drawing from the pipe catalogs (see Chapter 11 for more information on creating fittings). Unlike duct and cable tray, the pipe feature leverages a style-based approach for storing the fitting preferences that will be used while creating a pipe run. This style is called a "Pipe Part Routing Preference," more commonly known as simply a "Routing Preference." A Routing Preference is a collection of fittings stored inside a style that assigns different fittings based on the size of the pipe. These fittings are automatically loaded into the drawing file and added to the pipe run. Multiple styles can be created based on your needs, see the "Creating a Routing Preference" topic below for more information.

Creating pipe runs leverages the auto routing functionality built into AutoCAD MEP for routed 3D objects to automatically add fittings and required connections within the run. Piping, like the other 3D layout objects, leverages the AecbCompass to restrict your cursor to the predefined angles stored within the AecbCompass dialog. This functionality allows for gravity (sloped) pipe runs to be designed using the fundamentals of real world piping by incorporating Angle of Deflection inside female pipe connections. See the "Auto Routing" topic below for more information.

The ability to display piping is unique to all the 3D layout objects in AutoCAD MEP. Piping gives you the ability to show pipe runs with three different displays within the plan display representation based on diameter for each system definition. What this means is that you can display a single system as Graphical one line, Single Line and as a two line pipe solely based on the diameter specified in each system definition. See the "Understanding System Definitions" topic below for more information.

As noted, System Definitions, Routing Preferences, and Display are all explained in more detail in the tutorials later in the chapter. However, the fundamental feature that leverages all of these controls and settings is the "Auto Routing" feature, which the next topic details. Routing preferences, the pipe fittings in the catalog and the AecbCompass directly impact the results of auto routing. Some items have been briefly explained, but you must understand their importance to auto routing before understanding how to create or configure the settings.

AUTO ROUTING

AutoCAD MEP creates piping layouts using auto routing algorithms for the two different types of layouts; unconstrained and constrained. The unconstrained solutions are based on the angles you select when picking the points inside the drawing. AutoCAD MEP provides the AecbCompass during layout to limit the angles. Constrained layout solutions are determined by the fittings themselves and associated angles they support that are listed inside the current Routing Preference style to determine layout options and make connections. These two types of auto routing behaviors—unconstrained and constrained—react in different ways. Unconstrained applies when you are drawing pipe and you are not attempting to connect to another pipe or fitting. The unconstrained auto routing will adhere to the displayed angle on the AecbCompass (whether the routing preference supports the angle or not) as shown in Figure 6.1, while the constrained solutions use the angles from the content specified in the current routing preference. In the figure, the pipe is drawn to 45° since the third point picked was at 45°.

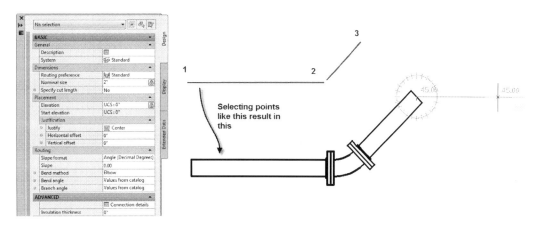

FIGURE 6.1

Unconstrained Pipe Routing (Auto Layout)

TIP: AecbCompass can be best described as Orthomode for AutoCAD MEP.

With Orthomode on in AutoCAD, you are restricted to 90°, but with AecbCompass, you are restricted only to the angles you specify in the dialog. To access the AecbCompass, go to the View tab of the ribbon, on the MEP View panel, click the Compass button (see Figure 6.2).

FIGURE 6.2

Accessing AecbCompass

In the "Compass Settings" dialog you can specify the diameter of the compass, the color, the snap angles, and the angle in which the tick marks appear (see Figure 6.3).

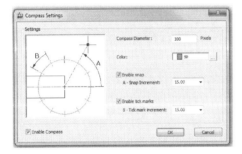

FIGURE 6.3

The Compass Settings Dialog allows you to configure your preferences for the Compass

The Snap Increment can be set to any angle, but you should set this angle to the most common angle available. For example, Figure 6.3 shows a manually typed in value of 11.25. To input such a value, simply type over the existing value. The drop-down list shows all angles that have been typed in. To delete one, you simply select the angle and then press the DELETE key (see Figure 6.4).

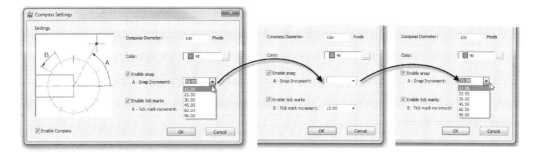

FIGURE 6.4

Adding or deleting Compass Snap Increments in the drop-down list

You can change the tick marks in the same manner as the snap increments. These settings are stored per user on the computer and are not stored in the drawings. We will cover the different ways to use the compass in the "Equipment and Piping Layout" and "Gravity Piping Fundamentals" topics later on in the chapter.

When the Compass is turned on, it will display the snap angle increments before selection, and will display the available elbow/tee/lateral angles at each fitting during constrained layouts. In addition to the angle numbers, the Compass will also display the slope angle being proposed within each solution along the pipe segment. Figure 6.6 below shows the slope value on the pipe segment. It is displayed at one-half the text size of the angle. This allows you to determine the preferred slope value when laying out gravity piping. To turn this text on or off type PIPESLOPEDISPLAY at the command line. This command is only available via the command line. While you can turn it off, it is extremely useful when laying out gravity systems in constrained routing. It is therefore recommended that you leave it on.

Auto routing constrained layout applies when you are drawing pipe and you try to connect to another pipe, fitting, or a Multi-View Part (commonly referred to as an Mvpart) (see Figure 6.5). During the constrained layout, you will get to preview the different connection options based on the available angles within the fittings stored inside the routing preferences (this will ignore the Compass angles specified). The command line will indicate how many possible solutions are available based on the available angles and fittings from your Routing Preference. You can change the preview of the possible solutions before accepting the one you prefer via the command line (see Figure 6.6) or the right click menu.

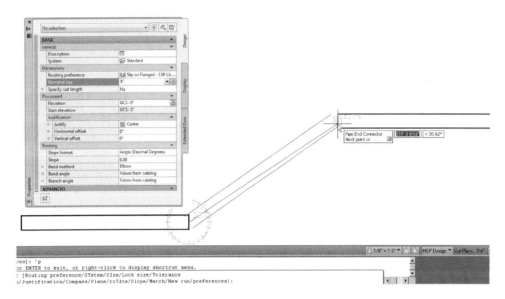

FIGURE 6.5

Constrained Pipe Routing (Auto Layout) Selecting connection point

The Figures show the Slip On Flange 150 lb. and Threaded Routing Preference with 4" size selected. As you can see in Figure 6.6, there are three possible solutions, two with 90-degree elbows at different locations and 1 with a 45-degree elbow. The elbow in the Routing Preference supports both 45- and 90-degree angles; therefore, both are used when calculating possible solutions.

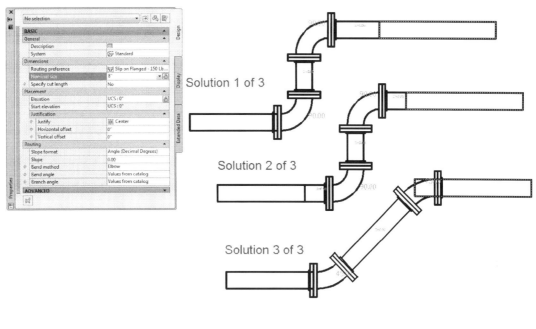

Solution 1 of 3

Solution 2 of 3

Solution 3 of 3

FIGURE 6.6
Three Possible Solutions with Slip On Flange Routing Preference

The Auto Layout feature will create piping solutions based on parts stored in the Routing Preference and will add all required fittings and even change elevation as needed. This works by creating a plane based on the initial pipe's location and another Plane at the connecting pipe's location. The software will then calculate possible routes based on the available elbow angles stored within the Routing Preference or current elbow. When auto layout generates a layout, it can only generate options using up to four fittings; one on each end and two in the middle of the proposed layout. In addition, couplings that support deflection will also be considered when determining valid layouts. We will be covering this more in detail later in the chapter in the "Gravity Piping Fundamentals" topic.

All pipe connections are aware of their gender (Male or Female) based on the Connector Engagement Length (CEL) value specified inside the pipe catalog. This value specifies whether the connection is inserted into another connection (male) or will allow another object to be inserted (female). This applies to all pipe connections on segments, fittings, and Multi-view

Parts (MvParts). Please refer to Chapter 11 for more information and to learn how this value is defined. In addition to defining the gender p on the CEL value specified in the pipe content, another parameter called AoD (Angle of Deflection) is available to specify the allowable angle of deflection within female pipe connections. The parameter and the value determine whether a female pipe connection can deflect. (AoD is also known as Fitting Tolerance.)

The Angle of Deflection (AoD) is *only* allowed when the Connector Engagement Length (CEL) is greater than zero. This approach is based on how installed pipe runs are actually deflected during installation; specifically, the female connection always controls whether joint deflection is allowed. In Figure 6.7, a Bell x Spigot elbow is connected on the bell end to a pipe.

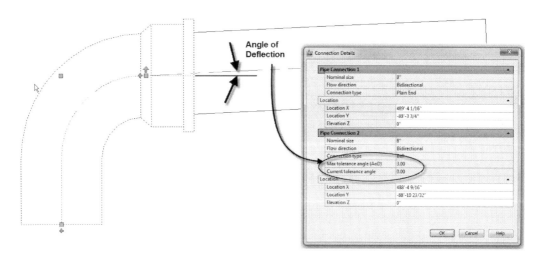

FIGURE 6.7

Angle of Deflection defined on a Bell Connection Type

The connection is not at zero degrees as indicated by each object's center line. The connection is a valid connection since the bell connection (shown in the "Connection Details" dialog) has a specified Max tolerance angle (AoD) of 3.00° and the connection is currently using a tolerance angle of 2.64°. The Max tolerance angle is referring to the specified AoD parameter value inside the pipe catalog content fitting information. (See Chapter 10 for more information on accessing content.) To access the connection details for a pipe object, select the object, go to the Properties palette and beneath the Advanced grouping, click the Connection details worksheet icon (see Figure 6.8).

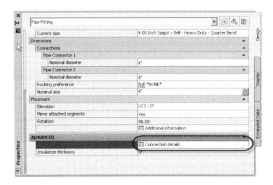

FIGURE 6.8
Accessing Connection details

Within the "Connection Details" dialog you can see the connection type for each connector, the Max tolerance angle (AoD) for each connection, and location coordinates for each connection as well as flow direction. This is an easier way to determine if a connection allows deflection and the maximum value. See the "Gravity Piping Fundamentals" topic later on in the chapter for more information on AoD and controlling fitting tolerances.

PIPE CONNECTIONS

AutoCAD MEP can apply several different behaviors depending on what connection type is defined in the Routing Preference. The basic behaviors are as follows and shown in Figure 6.9:

- Male to Male Connection Types: Butt Welded and Fusion

- Male to Female Connection Types: Glued, Mechanical Joint, Socket Weld, and Threaded

- Male to Female to Male (Joint Required) Connection Types: Grooved and No Hub

- Flanged Connections are treated as a unique connection type in the software as in the real world; two Flange Faces are needed to make a connection.

Connection Genders, Types and Behaviors

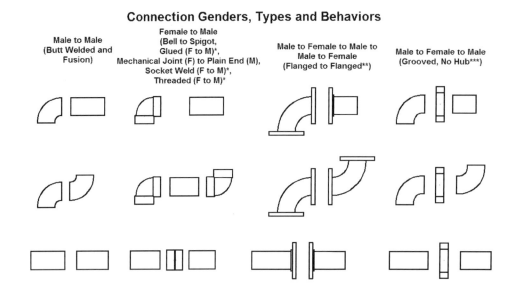

FIGURE 6.9

Pipe Connection Behaviors

*Existing connection types are defined as both female and male to support existing drawings.

**Flanged to Flanged indicates both the face of flange and the connection type on the opposite connection such as slip on or threaded.

***Grooved and No Hub always require a female connection between the male connectors.

> Note: Refer to the "Gravity Piping Fundamentals" topic below to learn how to route Male x Female Piping.

The different connection types have unique rules associated with them. AutoCAD MEP requires that all Pipe Connectors are defined as either Male or Female in the Content file and applies rules based on this information. Please refer to Chapter 11 for more information on how to build fittings.

All connection types requiring a joint will be added automatically based on the joint type specified in the Routing Preference. Automatic joints are not necessary for Butt Welded and Flanged connections; however these may be added as needed. When a fitting to fitting connection is attempted, AutoCAD MEP will read the connection type, determine if it is male or female, and then determine if an additional object is required. For example, when you attempt to draw a fitting to fitting connection with the Glued Routing Preference, the connection on the elbow is determined to be female and a pipe segment will automatically be added making a female to male to female connection. When a Grooved connection is made, the female joint will be added between the 2 male connections making a male to female to male connection as shown in Figure 6.9.

Flanged connections are allowed to have the face of a flange connect to another face of flange and will add additional flanges (Joint Connection) as necessary to complete the connection. When a fitting to pipe connection is made, the pipe will have a flange added to the end of it to match up with the flanged connector built within the elbow. Two flanges will be added when adding a joint connection in a pipe segment. When two flanged fittings meet, no additional flanges will be added between the two fittings.

The Butt Welded Connection is defined in the software as a special connection type that is allowed to connect to itself without an additional joint object. Therefore, if the fittings specified in the routing preference have the connection type defined as Butt Welded and the joint type is defined as another connection type, such as grooved, the fitting to fitting connection or fitting to segment will be allowed to connect to one another without adding the specified joint object.

UNDERSTANDING A ROUTING PREFERENCE

Piping systems are created by accessing the Fittings stored inside the Pipe Catalogs supplied with AutoCAD MEP and specified in Pipe Part Routing Preferences (also referred to as simply "Routing Preference"). A Pipe Part Routing Preference is a style that includes a collection of fitting types. Like other styles, they are accessed through Style Manager (see Figure 6.10).

FIGURE 6.10
Style Manager Button on the Manage ribbon tab

The list of catalog fittings in the Routing Preference allows AutoCAD MEP to assemble a pipe system as it would be assembled during construction. We will begin by explaining what a Pipe Part Routing Preference is and how it is utilized when adding pipe.

You can open the Style Manager from the Manage ribbon tab on the Style & Display panel. On the left panel expand *Piping Objects* and then expand *Pipe Part Routing Preference*. All of the Routing Preference styles will appear both in the expanded list on the left side and on the right panel (see Figure 6.11).

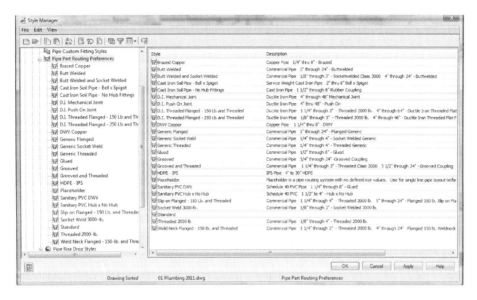

FIGURE 6.11
View available Pipe Part Routing Preferences in the Style Manager

A Pipe Part Routing Preference is defined based on Size Ranges. Size Ranges allow you to specify different fittings based on the lower and upper limit within a size range. Each Pipe Part Routing Preference can have multiple size ranges.

Within each size range you can specify the type of Joints, Crosses, Elbows, Pipes, Takeoffs, Tees, Wyes (Laterals) and both Concentric and Eccentric Reducers. The parts listed are stored in the Part Catalogs that are supplied with AutoCAD MEP. We will cover more about content below.

In Style Manager, you can view the settings of any Routing Preference definition. On the Preferences tab, one or more size ranges are listed with labels like "Size Range 1" or "Size Range 2." Each size range has a description. This description appears in the Size drop-down list on the Properties palette when adding pipe. The "Size Upper Limit" setting appears next and is used two ways. The drop-down list contains the available common sizes for the selected parts. When selecting parts that change the upper limit or the available sizes, a warning dialog will appear stating that the Size Range will be changed. The Size Upper Limit can also be used to set the preferred size that you want to use as the upper limit, which in turn sets the lower limit in the next size range.

The type column specifies the types of fittings that are allowed to be stored in a Routing Preference and the Part Size Range Column specifies the minimum and maximum sizes available for the selected fitting.

INSTALL THE DATASET FILES AND OPEN A PROJECT

In this exercise, we use the Project Browser to load a project from which to perform the steps that follow.

1. If you have not already done so, install the book's dataset files.

   Refer to "Book Dataset Files" in the Preface for information on installing the sample files included with this book.

2. Launch AutoCAD MEP 2012 and then on the Quick Access Toolbar (QAT), click the Project Browser icon.

3. In the "Project Browser" dialog, be sure that the Project Folder icon is depressed, open the folder list and choose your *C:* drive.

4. Double-click on the *MasterAME 2012* folder.

5. Double-click *MAMEP Commercial* to load the project. (You can also right-click on it and choose **Set Project Current**.).

IMPORTANT: If a message appears asking you to repath the project, click Yes. Refer to the "Repathing Projects" heading in the Preface for more information.

> Note: You should only see a repathing message if you installed the dataset to a different location than the one recommended in the installation instructions in the preface.

6. Click the Close button in the Project Browser.

When you close the Project Browser, the Project Navigator palette should appear automatically. If you already had this palette open, it will now change to reflect the contents of the newly loaded project. If for some reason, the Project Navigator did not appear, click the icon on the QAT.

SET UP THE PIPING WORKSPACE

We will begin with creating a Routing Preference and a System Definition for the Fire Protection system. Make sure that the Piping Workspace is active and that the Tool Palettes are displayed (see Figure 6.12).

Refer to the "Choosing your Workspace" topic in the Quick Start chapter if you are not sure how to load a Workspace and refer to the "Understanding Tool Palettes" topic in Chapter 1 for information on how to load and work with tool palettes.

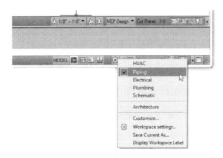

FIGURE 6.12
Enabling the Piping Workspace

LOAD A CATALOG

Provided with the files installed from the book's dataset files is a catalog containing some of the piping items used in the following tutorials. Before we begin the lessons, let's take a moment to load the required catalog.

7. From the Application menu, choose Options.

8. In the "Options" dialog, click the MEP Catalogs tab.

9. In the Catalogs area, select the *Pipe* folder.

10. Click the Add button.

11. Browse to the *C:\MasterAME 2012\MAMEP Pipe* folder, select the *MAMEP Steel Pipes.apc* and then click Open.

    The *MAMEP Steel Pipes* catalog contains the updated Grooved Flanged Adapter required for the lessons below.

12. Use the Move Up button on the right to make the new catalog the first in the list and then click OK to exit the "Options" dialog (see Figure 6.13).

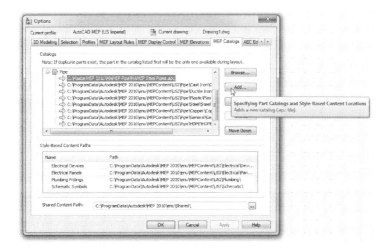

FIGURE 6.13
Add a custom catalog to your list and move it to the top

After loading a new catalog, you must regenerate the catalogs to update the change.

13. On the Manage Tab, expand the MEP Content panel and click the Regenerate Catalog button (see Figure 6.14).

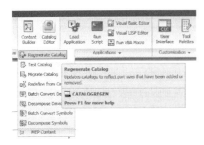

FIGURE 6.14

Regenerating the AutoCAD MEP Catalog

14. On the command line select Pipe (type P), click OK to confirm any messages that appear, and then press ENTER at the command line to finish.

CREATE A ROUTING PREFERENCE

In this exercise, we will be creating a new Routing Preference for the sprinkler system piping.

1. On the Project Navigator palette, click the Constructs tab.

2. Under Constructs, expand the *Fire Protection* folder and then double-click to open the *01 Fire Protection* drawing.

3. On the Manage tab, on the Style & Display panel, click the Style Manager button.

4. Expand the *Piping Objects* folder.

5. Select the Pipe Part Routing Preferences item and on the toolbar at the top, click the New Style icon (see Figure 6.15).

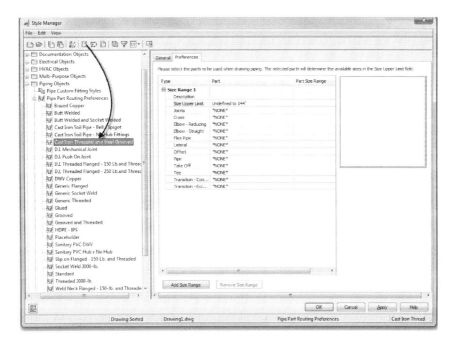

FIGURE 6.15

Creating a new Routing Preference

6. Type "Cast Iron Threaded and Steel Grooved" for the name in the left panel.

The General Tab allows you to edit the name or add a description. We will not make any such edits at this time. On the preferences tab we will begin selecting the fittings to use for this routing preference.

7. On the right side, click the Preferences tab.

   Size Range 1 will appear automatically.

8. For the Description, type **Cast Iron Threaded**.

For now we will accept the default for Upper Size Limit and make no changes to it. Next to each condition *None* is currently selected. Click on *None* to open a pop-up list of available choices.

9. Next to Joints click on *None* to open a pop-up menu.

10. Scroll down to the Cast Iron Pipe (US Imperial) catalog and then select the **Threaded – Class 3000 - Forged Coupling** item (see Figure 6.16).

A message will appear warning you that your selection will change the Size Upper Limit.

11. In the warning dialog, click Yes.

FIGURE 6.16
Selecting a Fitting

CAD MANAGER NOTE: AutoCAD MEP Catalogs are separated by material type. The order of the catalogs can be modified in the "Options" dialog. From the Application menu, choose Options and then click on the MEP Catalogs tab. The order of the catalogs as listed determines the order of the catalogs in any of the commands that access the catalogs.

12. Repeat the process to assign the following parts from the *Cast Iron (US Imperial)* Catalog (see Figure 6.17):

⇨ Cross—Threaded—Class 3000—Forged Cross.

⇨ Elbow—Reducing—No change (Leave this at *None*)

> **Note:** Reducing Elbows are used in place of a straight elbow and reducer when changing size on a bend

⇨ Straight Elbow—Threaded—Class 3000—Forged Elbow.

⇨ Flex Pipe—No change (Leave this at *NONE*).

> **Note:** You are not required to specify all parts in a routing preference. If the part is needed while adding pipe, a dialog will appear asking that you select the appropriate part.

⇨ Lateral—No change (Leave this at *NONE*).

⇨ Offset – No change (Leave this at **NONE***).

> **Note:** You can specify a Pipe offset in place of an Elbow during PipeAdd on the Properties Palette.

For the Pipe setting, we'll access a different Catalog.

⇨ Pipe—Select **Commercial Pipe** in from whatever catalog it is listed. See note.

> **Note:** The Commercial Pipe can be used for all materials and types. This pipe segment has all available sizes and allows for more sizes to be listed in the Routing Preference. In previous releases, the Commercial Pipe segment replaced most material-based segments since they all shared common dimensions. In current releases of AutoCAD MEP, the pipe catalog is now broken out to multiple catalogs based on material, so each of these catalogs has a copy of this pipe segment to support drawings created in previous releases. AutoCAD MEP will display the first instance of duplicate content (pipes and fittings) based on the order of the catalogs.

The remaining settings will come from the *Cast Iron (US Imperial)* Catalog again.

⇨ Takeoff—Threaded—Class 3000—Outlet.

⇨ Tee—Threaded—Class 3000—Forged Tee.

⇨ Transition—Concentric—Threaded—Class 3000—Reducer.

⇨ Transition—Eccentric—No change (Leave this at *NONE*).

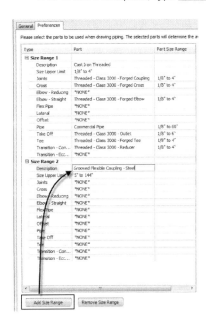

FIGURE 6.17

The completed settings for Size Range 1 of the Cast Iron Threaded style

Size Range 1 is now complete. We are going to add an additional size range for the Grooved portion of the Routing Preference.

13. Click the Add Size Range Button at the bottom of the dialog.

14. For the Description, type **Grooved Flexible Coupling – Steel** (see Figure 6.18).

FIGURE 6.18

15. Assign the following parts from the *Steel Pipe (US Imperial)* Catalog (see Figure 6.19):

⇨ Joints—Grooved—Flexible Coupling 0.75-12 Inch.

⇨ Cross—Grooved—Cross.

⇨ Elbow—Reducing—Grooved—Reducing Elbow.

⇨ Elbow—Straight—Grooved—Elbow.

⇨ Flex Pipe—No change (Leave this at *NONE*).

⇨ Lateral—Grooved – 45 Deg Lateral.

⇨ Offset—No change (Leave this at *NONE*).

⇨ Pipe—Steel Pipe.

⇨ Takeoff—Butt Welded—Outlet.

⇨ Tee—Grooved—Tee.

⇨ Transition—Concentric—Grooved—Concentric Reducer.

⇨ Transition—Eccentric—Grooved—Eccentric Reducer.

FIGURE 6.19

Completed settings for Size Range 2 – Grooved

16. Click OK to exit Style Manager.

We have now created the Routing Preference for the Fire Protection system. Let's test it out to make sure it is working properly.

17. On the Home tab of the ribbon, on the Build panel, click the Pipe button.

18. On the Properties palette set the Routing Preference to: **Cast Iron Threaded and Steel Grooved**.

19. Set the Nominal size to 2" and then draw a small pipe run on screen.

Make a few 90-degree turns to trigger the insertion of fittings (see Figure 6.20).

20. Press ENTER to complete the command.

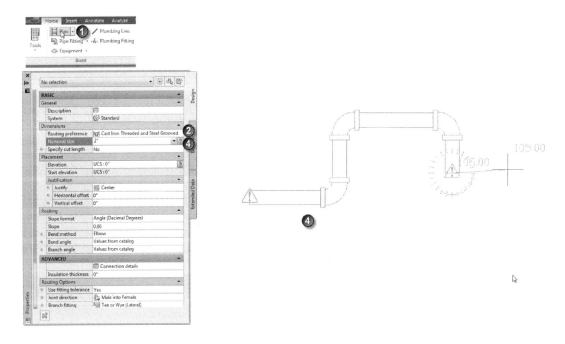

FIGURE 6.20

Draw a short pipe run to check the Routing Preference

21. Start the PipeAdd command again, change the Nominal Size to 6" and draw another run to check the Grooved Size Range.

22. Press ENTER to complete the command.

23. Erase the pipes and fittings.

It is possible to undo here, but be careful you do not go too far and undo the creation of the Routing Preference style.

UNDERSTANDING SYSTEM DEFINITIONS

Now that the Routing Preference is defined for the Fire Protection system, we need to create a System Definition for the sprinkler system. System Definitions assign the Layer Keys, Display System settings, Abbreviations, System Grouping, Rise Drop styles and determine if the piping and equipment assigned to the system will display as single line, graphical single line, or two line. In addition, System Definitions determine whether items assigned to this system will follow the display system drawing default or will be assigned a system level override.

System definitions are used to separate objects based on their use such as a fire protection system versus a chilled water system. This allows you the flexibility to isolate the systems as well as assign uniquely different characteristics to the objects designated to this system such as layer, abbreviation, and display properties.

CREATING A SYSTEM DEFINITION

1. On the Manage tab, on the Style & Display panel, click the Style Manager button.

2. Expand the *Piping Objects* folder.

3. Select the Pipe System Definitions item and, on the toolbar at the top, click the New Style icon (see Figure 6.15).

4. Type **Sprinkler System** for the name.

   With the new style selected in the left pane, you will be able to edit it on the right. As with all style, the General Tab allows you to edit the name and/or add a description (see Figure 6.21).

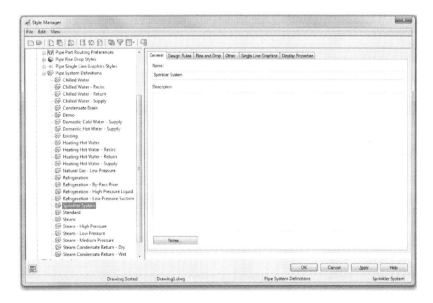

FIGURE 6.21

Creating a new Pipe System Definition

CAD MANAGER NOTE: Routing Preferences, System Definitions and other kinds of styles can be stored in template files (DWT). This means that each new drawing created from the template will have those items already available. You can also create tool palettes to import systems, routing preferences and preferred settings as needed on a per drawing basis. For more information please refer to the online help.

CONFIGURE THE STYLE SETTINGS

On the Design Rules tab you assign an Abbreviation, System group and the associated Layer Key to the system.

5. Click the Design Rules tab.

6. In Abbreviation field, type **SPKR**.

7. From the System Group drop-down choose **Non-Potable Water**.

8. From the Layer Key drop-down choose **F-SY-PIPE-WET_PIPE_SUPPLY** (see Figure 6.22).

The abbreviation is used with some labels and is also appended to the end of the System Name within the Properties Palette.

The System Group allows all System Definitions that have the same group to connect to one another. For example, all the Chilled Water systems and our new Sprinkler System have the same Non-Potable Water System Group, which allows these systems to connect to the same pipe main coming into the building, enabling AutoCAD MEP to determine which System Definitions are allowed to cross connect. The AutoCAD MEP Default systems are set up with basic System Groups.

If no System Group is specified, then the system will be allowed to connect to any system regardless of the assigned system group.

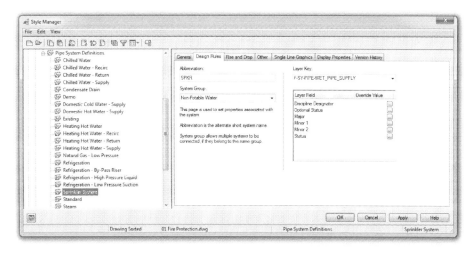

FIGURE 6.22
Design Rules allow us to configure the Abbreviation, System Group and Layer settings

> Note: The System Group is not a fixed list. To add a new system group to a drawing, simply type the new System Group Name into the drop-down list field. To have a custom system group available in all new Drawings, add the System Group to your company's drawing template file DWT. When no system group is assigned, the system will be allowed to connect to any System. This is beneficial for systems like Drain.

A Layer Key automatically creates and assigns layers for objects based on a standard list. For more information on Layer Key Styles refer to the online help.

The Rise and Drop Tab stores which Rise and Drop style will be used for this System.

9. On the Rise and Drop tab, choose the **Pipe Break – Patterned Rise Only** style (see Figure 6.23).

FIGURE 6.23
Setting the Pipe Rise and Drop Style

Note: For more information on how Rise Drop Styles work please refer to Appendix A.

10. AutoCAD MEP allows pipe systems to display in graphical single line, to-scale single line, or as two line display, and is controlled by the Single Line Graphics tab.

11. Click on the Single Line Graphics tab.

12. Select the two checkboxes to enable single line and graphical 1 line display.

⇨ For the single line displays, set the "For pipe size less than or equal to" to: **4"**.

⇨ For the graphical 1 line, set the "For pipe size less than or equal to" to: **3"**.

Set the "Inline/anchored MvPart plot length" to **1/8"** (see Figure 6.24).

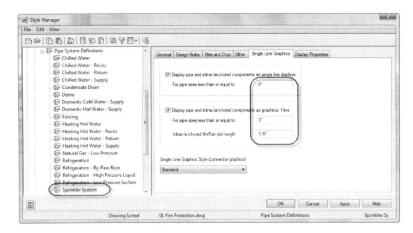

FIGURE 6.24

Configure the settings for Single Line Graphics

This will display pipe as single line display when the size is equal to or under 4". This allows piping with diameters above 4" to display as two line. The intent of 1 line display is to show the piping, inline, and anchored components symbolically, while still representing some of the key spatial factors of the piping and components (see Figure .6.25).

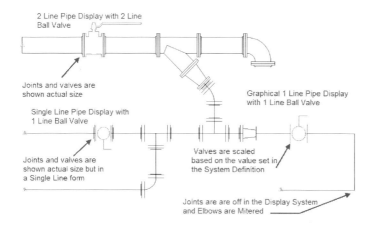

FIGURE .6.25

Graphical 1 Line, Single Line Piping and 2 Line Piping Display characteristics

When Single Line Graphics are being used, (whether Single Line or Graphical 1 Line), the display of inline equipment such as Valves will use the symbolic version of the equipment.

Pipe Single Line graphics styles are controlled by the Pipe Single Line Graphics Styles. To change the display of the symbol used for the connection type go to Style manager, Piping Objects, Pipe Single Line Graphics Styles (see Figure .6.26).

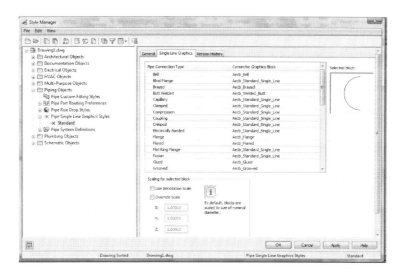

FIGURE .6.26

Pipe Single Line Graphics Styles

Pipe Single Line Graphics Styles use AutoCAD blocks to represent the connection graphic on pipe connectors. You can create your own AutoCAD Blocks to represent connection types.

When creating blocks for Pipe Single Line Graphics, the block size should be based on 1 unit high by up to 1 unit wide (see Figure .6.27).

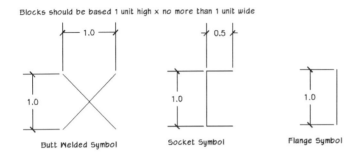

FIGURE .6.27
Pipe Single Line Graphics Style Block Rules

Pipe Single Line Graphics Styles allow you to create multiple styles and assign them on a per system basis. This allows you to determine what the single line connection graphic is based on its use. For example, existing or piping to be demolished can have no connection graphics display to simplify the display of your drawings. The Single Line Graphics Style for a system is defined on the Single Line Graphics style for that system (See Figure .6.26).

DISPLAY PROPERTIES

The basics of the Display System have been covered in detail already in previous chapters. When you configure a style, such as the System Definition in consideration here, you have the option to assign style-level display properties to it. Settings on the Display Properties tab are used to control the appearance of the style (in this case the System Definition). Before making any edits to the display properties of the System Definition, make sure you are comfortable with the display system hierarchy and definitions. Refer to the "Overview and Key Display System Features" heading in Chapter 2 and the "Display Properties and Definitions" topic in Chapter 12 for definitions of the key Display Control terms. You will also find detailed tutorials on working with the Display System in Chapter 12.

CONFIGURE DISPLAY PROPERTIES

1. Click the Display Properties tab.

Notice that there are several Display Representations listed. In this exercise we will focus on the Plan Display Representation. The other Display Representations are discussed as appropriate in other chapters. For the Fire Protection Sprinkler System Definition that we are building here, we want to turn on the Centerline display component for this system. Therefore we will need to add a System Definition override.

2. Next to plan, in the Style Override column, check the box to add the override (see Figure 6.28).

FIGURE 6.28
Applying a Style Override to Plan

The "Display Properties" Dialog will appear. The titlebar will read: "(Pipe System Definition Override – Sprinkler System) – Plan". This indicates that a System Definition level override is now applied.

3. Select the Center Line Display Component and then click the Light Bulb icon next to it to turn on its visibility (see Figure 6.29).

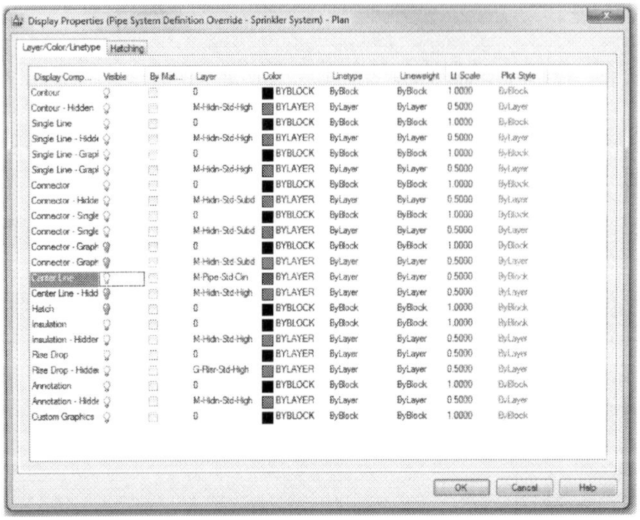

FIGURE 6.29

Turn on the Center Line component

4. Click OK to dismiss the "Display Properties" dialog.

5. Click OK again to dismiss the Style Manager and complete the System Definition setup.

The setup is now complete and we are ready to begin laying out our equipment and piping.

6. Save the drawing.

EQUIPMENT AND PIPING LAYOUT

In general, layouts begin with determining the equipment location to allow others involved with the design to review access requirements, structural issues and additional information critical for making an informed decision.

The Equipment command on the ribbon allows you to add equipment (Multi-view parts) from the AutoCAD MEP catalog. The catalog contains equipment that is modeled three-dimensionally at actual size and has specific information such as dimensions, required or optional connections and 1 line symbols.

ADDING EQUIPMENT

When you click the Equipment button on the ribbon, it calls the MvPartAdd command. The MvPartAdd command opens the Multi-view Parts (MvParts) dialog which allows you to find specific equipment through the catalog's folder structure (see Figure 6.30).

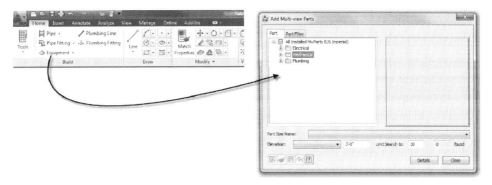

FIGURE 6.30

Adding Equipment starts the MvParts command

By expanding folders you can find parts listed in broad categories. Once you have selected a piece of equipment, click the Part Filter tab to filter the list of parts based on specific criteria. MvParts can be defined as either block-based or parametric-based content, and is identified by different icons within the dialog. Refer to Chapter 9 for more information.

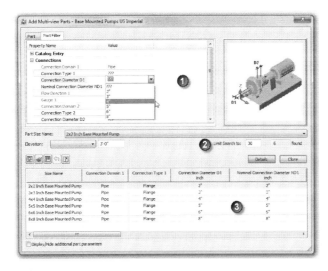

FIGURE 6.31

Filtering Equipment via MvPartAdd

On the Part Filter Tab the top left panel includes many Filter Controls (see number 1 in Figure 6.31). Once you begin assigning filters, the parts found quantity begins to reduce (see number 2 in Figure 6.31). Some parts have the capability to create thousands of options, so using the filter allows you to focus your selection. You can filter the selection set using the Catalog Entries such as:

- Connections (Details such as Diameter)
- Display
- Part Size Name

In addition, the Details list (see number 3 in Figure 6.31) also allows you to select a part based on the parameters from the part catalog. The information here shows the visible parameters in the content builder part parameters. If you want to display the non-visible parameters, you can select the "Display/Hide additional part parameters" check box at the bottom of the details drop list. To access and change the visibility of a parameter you will need to edit the part in the content builder and edit the part's parameters by accessing the parameter configuration dialog. To find out how to do this, please refer to Chapters 9 and 10.

Multi view Parts are style-based objects that display an AutoCAD block for each view. This allows the Multi-View Parts to display differently depending on which way it is being viewed. To access the list of AutoCAD blocks being used for each view, select an instance of the Multi-view part in your drawing, go to the Ribbon and select Edit Style then select the View tab (see Figure 6.32).

Each numbered symbol at the top of the dialog refers to a unique view inside AutoCAD MEP. To view the block name associated with the view, simply select the view in the top section and the settings section updates (see Figure 6.32).

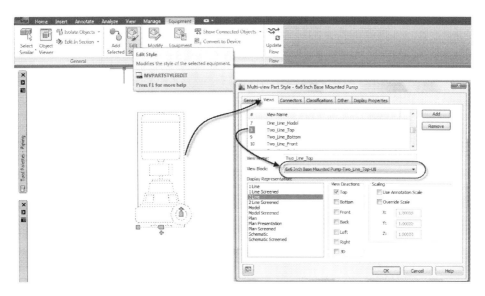

FIGURE 6.32
Accessing Multi-View Part Styles and View Blocks

> **Note:** Editing the blocks or changing the blocks will only affect this drawing. All blocks are created from the catalog. Refer to Chapter 10 for more information on creating equipment.

CREATE A FIRE PROTECTION SYSTEM

Now that we have completed configuring the necessary settings, (we created a Routing Preference, a System Definition and explored the basics of adding equipment) we are ready to begin the layout of a fire protection system. For this exercise, we will work on the first floor of our commercial project.

ADDING THE FIRE PUMP

The first step in laying out your system is to lay out a pump, the associated suction and discharge, and the required valves.

Continue in the *01 Fire Protection* drawing (in the *Fire Protection* folder of Project Navigator).

1. Zoom in on the room in the upper right corner of the first floor plan.

2. On the Home tab, on the Build panel, click the Equipment drop-down and then select the **Pump Control** (see Figure 6.33).

This filters the "Add Multi-view Parts" dialog to show only Pump Equipment.

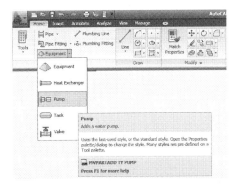

FIGURE 6.33

Adding Equipment using with one of the tools automatically filters the Add Multi-view Parts command

In the "Add Multi-view Parts" dialog, Base Mounted Pumps US Imperial is selected. This is the pump we are going to insert, but before we do, we want to change the size and elevation.

3. From the "Part Size Name" list, choose **4x4 Inch Base Mounted Pump** (see Figure 6.34).

4. To allow a slab to be added below the pump, change the Elevation to **6".**

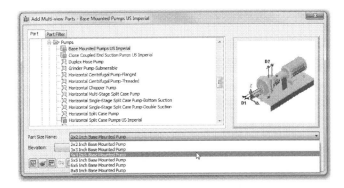

FIGURE 6.34

Selecting the Fire Protection Pump

Do not click Close in the "Add Multi-view Parts" dialog.

5. In the drawing window, click to place the pump on the right side of the Fire Protection room with the motor pointing towards the exterior wall approximately in the position shown in Figure 6.35.

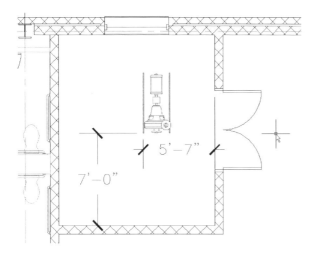

FIGURE 6.35
Adding Fire Protection Pump to Room

6. Click Close to dismiss the "Add Multi-view Parts" dialog.

FIRE PUMP SUPPLY PIPING

For this exercise we are going to assume the supply pipe, for the Fire Protection system is coming from the upper left-hand corner of the room and the discharge will be in the lower left-hand corner of the room. We will not be adding the entire water supply in this exercise, just the specific piping in and around the pump; we will lay out the sprinkler heads and associated piping in a later exercise.

7. Select the pump you just placed.

8. On the Equipment Tab of the Ribbon, click the Equipment Properties button.

9. In the "Multi-view Part Properties" dialog, click the Systems tab, set the Inlet and Outlet Systems to **Sprinkler System,** and then click OK (see Figure 6.36).

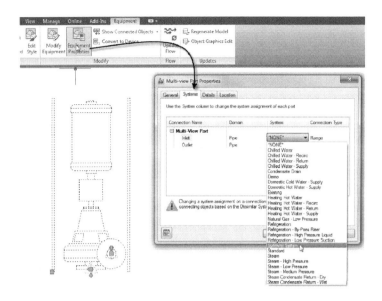

FIGURE 6.36

Configure the Connector Systems for the Multi-view Part

Now that we have the pump placed and its connectors configured, let's copy it to the left to create a primary and backup pump system.

10. With the pump still selected, on the Home tab, on the Modify panel, click the copy button.

11. Click any basepoint, move the cursor to the left (make sure ortho is on with the F8 key), type **4'-0",** and then press ENTER. Press ENTER again to complete the command (see Figure 6.37).

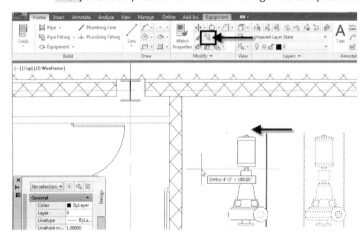

FIGURE 6.37

Copy the pump to the left

12. Select the pump on the right and then click the Add Pipe grip (+) on the suction side of the pump (the bottom Add Pipe grip).

The "plus" grip starts the "Add" command associated with the connection type as defined in the part catalog. In this instance the connection type is defined as a Pipe connection and as the connection type Flanged, therefore the PipeAdd command is started and will require a flanged joint. Recall above in the "Create a Routing Preference" topic that we assigned joint types for various situations. In this case, however, we are going to temporarily override the Joint type in the Routing Preference with a Grooved – Class 150 – Flanged Adaptor 2-12 Inch from the Steel Pipe Catalog. We can do this using the Fitting Setting override capability.

13. On the Properties palette, in the Advanced grouping, click on the Fitting Settings worksheet icon (see Figure 6.38).

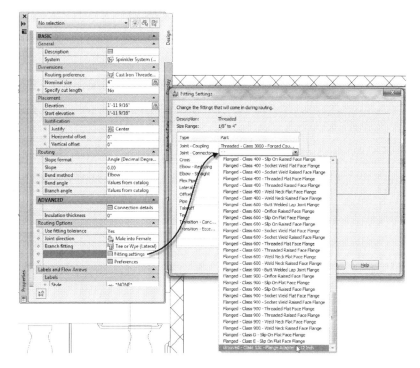

FIGURE 6.38

Selecting the Joint Connector Override

14. In the "Fitting Settings" worksheet that appears, from the Joint Connector drop-down list, under Steel Pipe (US Imperial), choose **Grooved – Class 150 – Flanged Adaptor 2-12 Inch**.

Note: If the Grooved Class 150 – Flanged Adapter parts are not listed, you will need to edit your catalog and change the Autolayout setting to True. Start Catalog Editor from the Manage Tab, MEP Content Panel. Browse to the *C:\ProgramData\Autodesk\MEP 2012\enu\MEPContent\US\Pipe\Steel* folder and select the *Steel.apc* file. Next expand the Grooved Chapter, then scroll down to the Grooved Class 150 – Flange Adapter 2 – 12 Inch, select the Constants, find the entry AlayF and change it to TRUE. Save the catalog and run a CATALOGREGEN in AutoCAD MEP.

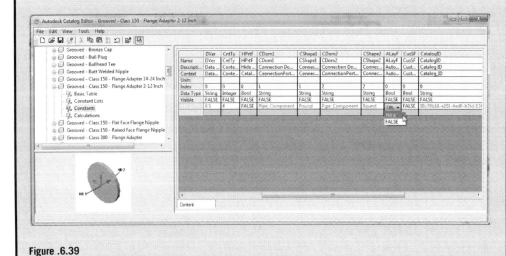

Figure .6.39

Editing the Grooved Flange Adapter Part to be used during Pipeadd

This will be used for the Pipe connection at the pump.

15. Move your cursor straight down so the compass reads 0.00° angle, type in **20** and then press ENTER.

16. On the Properties palette, set Justify to **Top Center** (see Figure 6.40).

17. Change the pipe Nominal size to **6"**

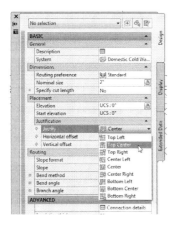

FIGURE 6.40

Setting Justification during Pipe Layout

The "Choose a Part" Dialog will appear because the size change goes across Range 1 and 2; therefore you need to verify the part you want to use as shown in Figure 6.41.

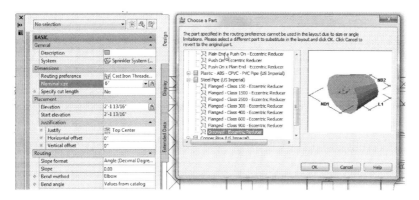

FIGURE 6.41

You must verify your part choice when the size change spans more than one range

18. In the "Choose a Part" dialog, select the Grooved – Flexible Coupling 0.75" to 12" and then click OK.

19. Continue to drag your cursor in the same direction, type **30"** (2'-6") and then press ENTER.

   This will add an Eccentric Reducer and pipe.

20. Next, pick a point towards the left pump so an elbow is added (see Figure 6.42).

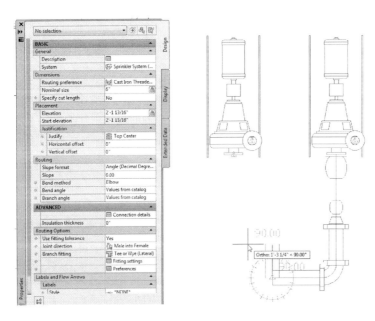

FIGURE 6.42

Suction Piping for Fire Protection Pump

21. Press ENTER to exit the PipeAdd command.

22. Select the suction piping coming from the Pump up to the elbow at the bottom of the screen. (See Figure 6.43 to see the copied pipe.)

23. Copy the selection using the AutoCAD MEP snaps: For the base point, snap to the Pipe End Connector (PCON) snap point on the Discharge connection on the pump on the right, and then snap to the PCON on the discharge of the pump to the left.

    This copies the 4" and 6" piping with fittings to the left pump.

24. Select the 6" pipe, click the Plus (+) Grip. For the next point, select the end of the horizontal pipe using the PCON snap and cycle through the solutions by selecting Next from the right click menu. When the solution using a 90-degree elbow is shown, right click and select Accept (see Figure 6.43).

25. Press ESC to complete the command.

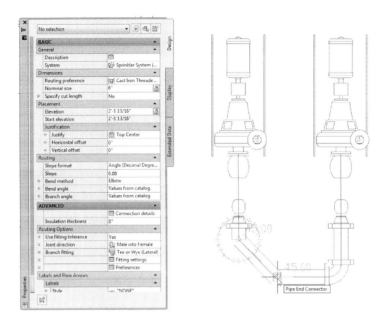

FIGURE 6.43

Connecting the Second Pump to create the header piping

26. Select the left 6" elbow. Click the left plus (+) grip to change the elbow into a tee (see Figure 6.44).

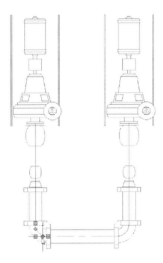

FIGURE 6.44

Upgrading an Elbow to a Tee

Once the Tee is placed you can set the Bend angle on the Properties palette to 90°. This means that only 90-degree elbows will be used in routing. Therefore, if you pick a point in the upper left corner of the room, a 90° bend and a pipe along the wall will be created instead of the "Choose a Part" dialog appearing and prompting for an odd angle.

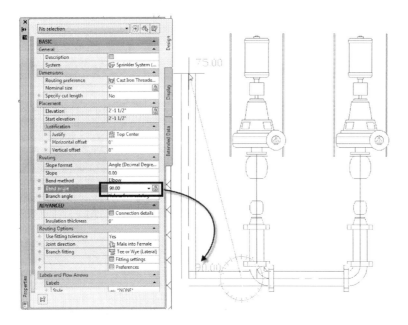

FIGURE 6.45

Locking the Bend Angle to control Auto Layout

27. In the palette, set the bend angle to 90 (see Figure 6.45).

28. Make sure ortho is off (F8), and then click a point near the corner of the room as shown in (see Figure 6.45).

29. Change the elevation to **-4'-0"** to draw the pipe down below the slab, and pick a point outside the exterior wall to complete the suction piping.

30. Press ESC to complete the pipe command.

Later in the chapter we will be adding valves to these pipes, but for now this completes our work on the suction piping. Next we are going to lay out the discharge piping for the two pumps and combine them into a single discharge header.

FIRE PUMP DISCHARGE PIPING

1. Select the right pump and then click the discharge plus (+) grip.

2. On the Properties palette, change the Justification to **Center**. Change the size to **6"**,

   The "Choose a Part" dialog will **appear** to select a Flanged Reducer.

3. Under Steel Pipe (US Imperial), select the Flanged – Class 150 – Reducer, and then click OK.

4. On the Properties palette, set the Elevation to **6'-0"**.

5. Select the Grooved Flanged Adapter 2 – 12 Inch in Choose a Part from the MAMEP Pipe catalog.

6. Move the mouse down and to the right away from the pump at 45 degrees (towards the Suction Pipe) approximately 24" in length (by the elbow on the suction pipe), where the elbow is shown in Figure 6.46.

7. Pick the next point straight down by the lower wall, and then another in the lower left corner of the room (see Figure 6.46).

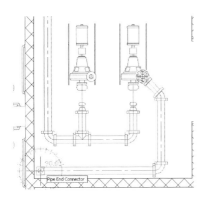

FIGURE 6.46
Laying out the Discharge Header

8. Change the Elevation to **12'-0"** and then press ENTER to exit the command.

Changing the elevation has the effect of adding a vertical run of pipe at the current point. Later we can view the model in section or 3D to see this more clearly,

9. Use a window crossing selection to select the riser from the pump discharge and all the piping up to and including the 45-degree elbow. Zoom in closely to avoid selecting components on the suction side of the pump.

10. Using AutoCAD copy, snap the basepoint to the lower left corner of the pump using the insertion point osnap (INS) to correctly select the basepoint, and then copy the selection to the left pump, snapping to the same relative point, again, using the INS snap.

11. Select the 45-degree elbow and then click the Plus (+) grip to start PipeAdd.

12. Click a point on the horizontal pipe along the bottom wall (see Figure 6.47) to connect the discharge piping.

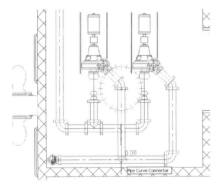

FIGURE 6.47

Connecting to the Discharge Header

13. Select the solution that routes with no additional elbows by using the Next/Accept options on the command line and then exit the command.

14. Press ESC to end the pipe command.

> Note: Depending on the accuracy of the point you click relative to your desired angle, Auto Layout solutions may change the order using the coordinate of your selection point relative to your starting point as the driving factor on which solution appears first.

That completes the work we will do on the discharge piping for now. We will be adding the rest of the sprinkler system piping later on in the chapter.

ADDING VALVES TO THE FIRE PUMP SUCTION AND DISCHARGE PIPING

Next we will be adding the suction and discharge valves to the piping.

1. On the Home tab, on the Build panel, click the Equipment drop-down button and choose the Valve tool.

2. Expand the *Valves* folder and then select the Gate Valves US Imperial item.

> Note: You do not need to set the size—the system will select the correct size and connection type for you.

Remember, do not close the "Add Multi-view Parts" dialog; simply click in the drawing to shift focus there.

3. At the command line, type PCUR and then press ENTER.

4. Click in the middle of the right pump's 6" suction pipe.

5. Type P and then press ENTER.

This changes the plane of the compass to align to your view.

6. Press ENTER again.

The "Choose a Part" dialog will again appear asking for the appropriate flange to use.

7. Expand the *Steel Pipe* Catalog. Scroll down and select the Grooved – Class 150 – Flanged Adaptor 2-12 Inch and then click OK (see Figure 6.48).

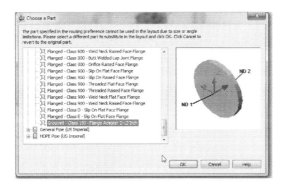

FIGURE 6.48

Choose a Part to select a Flange for the Gate Valve

Note: If the valve comes in with the stem facing down, simply select the valve and then select the Flip Grip that appears next to the Valve to rotate 180 degrees.

8. Pick on the 6" pipe leading to the left pump's suction end and repeat the process to add another valve (see Figure 6.49).

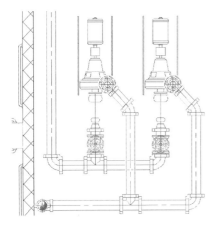

FIGURE 6.49

Placing the Suction shut off valves

9. Repeat the process to add two more valves to the discharge piping. Place one right before the elbow (right pump) and the other right before the Tee (left pump) as shown in Figure 6.50.

10. After placing the valve, select it, hold down the CTRL key, and click the flip grip to rotate the valve 90 degrees instead of 180 (see Figure 6.50).

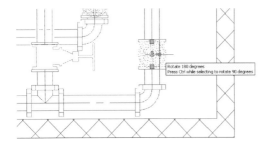

FIGURE 6.50

Use the CTRL key to flip a valve 90° (the valve on the left is shown already flipped)

Our next task is to add a check valve to the elbow coming out of the pump. We will use the Plane option again to orientate the valve to the correct plane.

11. On the Home tab, use the Equipment drop-down to add another Valve.

12. In the "Add Multi-view Parts" dialog, select the *Check Valves US Imperial* item.

13. Click on the discharge pipe before the gate valve in the locations shown in Figure 6.51. Use the mouse to indicate the rotation. (Remember you can always flip it with the grip later if necessary).

14. In the "Choose a Part" dialog, select the Grooved Flanged adapter again.

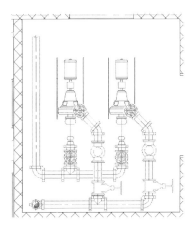

FIGURE 6.51
Completed Layout of Fire Protection Room

EDITING PIPE LAYOUTS

A major advantage of working with AEC objects is that you can manipulate them at any point in your project cycle instead of the more traditional erase and redraw methodology common in most firms today. Pipes, fittings, and other objects in your drawing know what they are and have built-in behaviors. They are able to maintain logical relationships to one another as manipulations are made. In this topic we will look at some of those built-in behaviors as we fine-tune our fire protection pump layout.

REMOVE UNNECESSARY PIPES

To fine-tune the layout, we are going to remove the unnecessary pipes between the valves using the "Associated Movement" ability in AutoCAD MEP, otherwise known as "StickyMove."

StickyMove allows you to edit AutoCAD MEP objects with the grips, while the system automatically adds and/or deletes objects as needed to keep the system connected. This powerful behavior is used on all 3D objects in AutoCAD MEP. To use StickyMove, you simply need to select an object, then select the rectangular move grip and drag it to the appropriate connector. AutoCAD MEP will take care of the rest.

1. Select the Gate Valve on the Suction Piping for the Left Pump.

2. Click the rectangular grip on the side opposite the pump.

3. Drag the rectangular grip to the connector of the tee and snap using the PCON snap (see Figure 6.52).

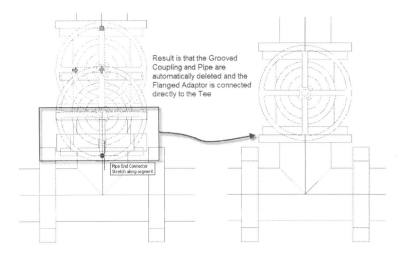

FIGURE 6.52

Using Grips to edit layout

4. Repeat this process to move all the valves to the associated fittings (see Figure 6.53).

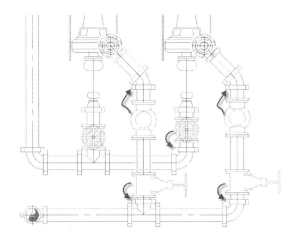

FIGURE 6.53

Move valves to connect directly to fittings

Next we'll break the pipe between the Check valve and Gate valve to raise the header to 8'-0".

5. On the Home tab, click on the Modify panel title.

   The panel will expand to reveal several additional standard AutoCAD drafting tools.

6. Click the Break tool (see Figure 6.54).

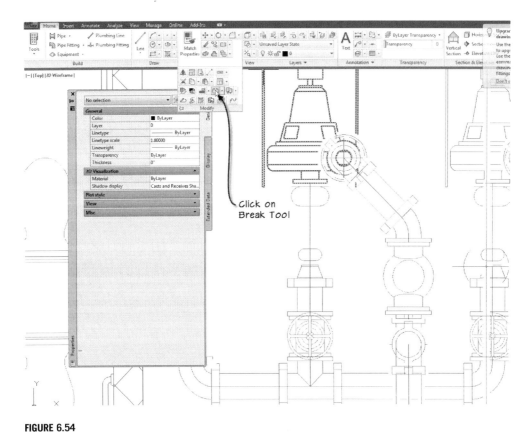

FIGURE 6.54
Move valves to connect directly to fittings

If you prefer, you can type BREAK at the command line (BR is the shortcut).

7. At the "Select object" prompt, select the pipe between the Check valve and the Gate valve for the right pump.

8. Move the mouse slightly and then click the second point of the break so that the pipe is in two parts (see Figure 6.55).

9. Repeat for the left pump piping.

10. Select the broken pipe connected to the Gate valve. On the Properties palette change the elevation to 8'-0".

All the horizontal piping and valves connected to this pipe will move to 8'-0". To verify this, simply hover your cursor over the objects and the Tooltip will indicate the new elevation (see Figure 6.55).

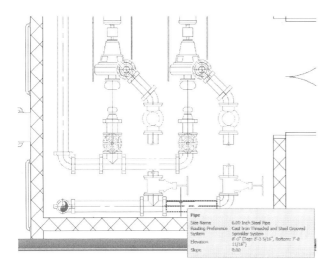

FIGURE 6.55

Discharge piping moved to 8'-0"

Now that the discharge piping has changed elevation we can connect the two pipes and add the associated elbows.

11. Select the pipe connected to the Gate valve, click the plus (+) grip, and then click the PCON (PIPE END CONNECTOR) snap on the pipe connected to the Check valve.

12. Multiple solutions will appear; please select number one as indicated in Figure 6.56.

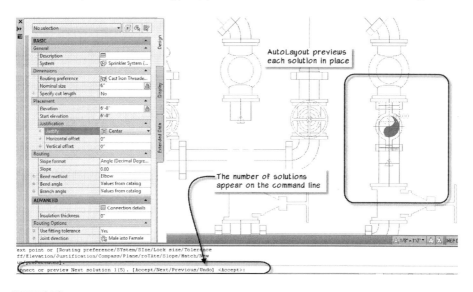

FIGURE 6.56

Auto layout finding multiple solutions

13. Repeat the process for the other pump.

We have completed the Suction and Discharge of the pump run. We have learned how to create Routing Preferences, System Definitions, add equipment (pumps and valves), create pipe layouts, and edit the layouts using the Associated Movement (StickyMove) functionality. As you can see, AutoCAD MEP gives you many benefits, including the ability to create complex 3D layouts from a simple plan view and to make edits quickly and easily with the StickyMove functionality. Let's take a look at what we have accomplished in a 3D view.

14. On the View ribbon, Appearance panel, choose **View, NE Isometric** from the View list.

15. On the View ribbon, Visual Styles panel, choose **Conceptual** from the Visual Styles drop-down list.

> Note: the Home tab of the ribbon has a View panel that may be more convenient than switching to the View ribbon (see Chapter 1 for more details on the ribbon). You can also 'tear off' the View panel so it is floating and always available.

16. Click and drag the ViewCube in the upper right corner of the screen to orbit the view to a steeper angle. (You can also hold down the SHIFT key and drag with the mouse wheel) see Figure 6.57.

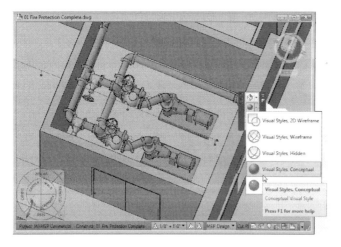

FIGURE 6.57
The Fire Protection Pump System in 3D Conceptual Visual Style, shown with the View panel 'torn off' and floating.

17. Using the Visual Styles button, return display to the 2D Wireframe visual style.

18. Using the View button, return to Top view.

19. Close the *01 Fire Protection.dwg* file, and when prompted, save the file.

GRAVITY PIPING FUNDAMENTALS

Gravity piping in AutoCAD MEP provides the ability for male connections to deflect when connecting to their female connection counterparts. This ability is allowed by reading specific values stored inside the Pipe catalog's fittings and pipe connections on Multi View Parts. These values are Angle of Deflection (AoD) and the Connector Engagement Length (CEL). In this topic you will lay out a gravity piping system for the restrooms and gain additional understanding on how the male x female piping in AutoCAD MEP works.

To begin, let's first review how fittings determine whether they are male or female. When the Connector Engagement Length (CEL) value stored within each pipe connection equals zero the connection is defined as male. When the value is greater than zero, then the pipe connection is female. Once the connection is determined to be female, the software will take into account the Angle of Deflection (AoD) value that is stored within each pipe connection to allow the connection to deflect up to the maximum angle. This fundamental connection method matches how female pipe connections are made during construction. To find out more about Content Builder and how to create and edit content, refer to Chapters 9 and 10.

Furthermore, the actual connection location on female connections is located inside the fitting at the true location based on the defined CEL value. This is also used as the deflection point of the connected object, such as male fittings or pipe segments. In addition, since all pipe connections are defined as either male or female, the software also knows whether the fittings defined in the Routing Preference can connect directly together or need another object (such as a pipe between two female joints or a coupling between two male joints). This makes your layouts more accurate and reflective of the real world. Several options appear on the PipeAdd Properties palette during the PipeAdd command, and this fully supports sloped piping, allowing for better layouts, more precise slope control, and the conveyance of more information during layout (see Figure 6.58).

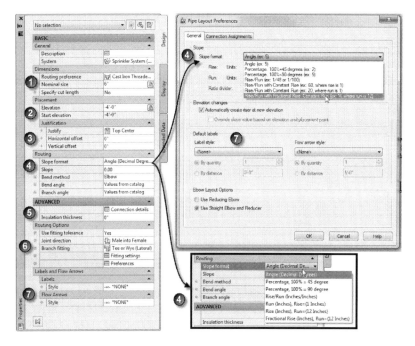

FIGURE 6.58

PipeAdd Property Palette

2. **Dimensions**—The Routing preference allows you to select the Routing Preference that will be used during Pipe Layout. The Nominal Size is where the size is specified and Specify Cut Length allows you to set a specific cut length after the pipe run is complete. Cut lengths are applied to Pipe Segments only.

3. **Placement**—The Start Elevation control reports the elevation of the first point of the pipe. This allows the Elevation control to update based on the slope value and simultaneously inform you of the original elevation from which you began the pipe run.

4. **Justification**—The Justify control allows you to specify the justification of the pipe objects during placement. The Horizontal and Vertical offsets allow you to specify a value to draw the pipe away from the current cursor position. This is very useful for piping layouts along a wall for example. You can specify a Horizontal offset distance to have the pipe drawn that distance away from the wall.

5. **Routing**—In this grouping, Slope format and Slope are always active. The Slope format is a Drop List control that is configured within the pipe preferences and informs you of the current pipe routing format. Pipe Preferences support several different formats such as Percentage, Fractional, Constant Rise or Run values and a multitude of measurement units The Slope format is a drop list to allow you to select the preferred format without editing the Pipe Preferences. (We will look at Pipe preferences in more detail in the "Create a Floor Drain System" topic below.) You can change the current format by selecting the control and selecting the new format from the list available. The Bend Method allows you to select either place an Elbow or a Pipe Offset. The Branch angle control supports angles other than 90° when branching off a pipe with either a Tee or a Wye (Lateral).

6. **Advanced**—This section allows you to access the Connection Details that state the current connection type on the Pipe Segment and the Flow direction. Insulation is where you specify the thickness of the insulation that will be applied on the Pipe and Fittings.

7. **Routing Options**—The Use fitting tolerance control determines if the Angle of Deflection value will be considered during Auto Layout or not. This is the setting that determines if female fittings will deflect or not. The Joint Direction control allows you to specify the orientation of Male x Female fittings during layout, to determine if the male end of the fitting should go into the female connection or have the female object placed first. The Branch fittings supported are: Takeoffs, Tee Only, Tee or Wye (Lateral), Wye Only and Wye (Lateral) or Tee. Using the Tee or Wye / Wye or Tee methods allows Auto Layout to determine solutions using both fittings and will place the first fitting when both fittings have the same angle values. This allows for the most comprehensive set of solutions to be found with constrained layout. We will be using these controls in the exercises below.

8. **Routing Options**—The Labels and Flow arrow section is where you can specify if the pipe should have Labels and / or flow arrows added to the pipe run during layout. These controls are also in the Pipe Preferences to allow you to set the values in the templates.

Using Auto Layout in sloped pipe layouts with fitting tolerances turned on will split the deflection to each connector equally, thus allowing twice the maximum Angle of Deflection to be utilized during layout. In addition, when the joint specified is female and supports Angle of

Deflection, Auto Layout will add couplings as necessary to make up for minor angle differences instead of adding a custom angled elbow. For example when you draw a pipe at slope zero and then change the slope to -1, the coupling stored in the Routing Preference will be added to account for this slope change (see Figure 6.59).

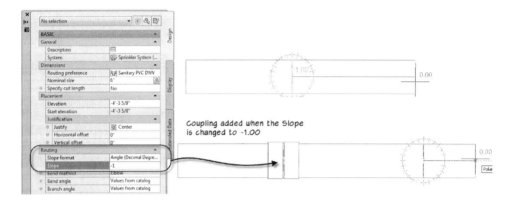

FIGURE 6.59
PipeAdd automatically adds couplings when needed

In the following topics we will explore these settings as we lay out a floor drain system and a sanitary piping system.

FLOOR DRAIN SYSTEM LAYOUT

In the next two exercises we will be laying out a floor drain system and a sanitary system using Piping. We will explore how piping works in sloped conditions and use the associated controls to leverage the slope functionality.

CREATE A FLOOR DRAIN SYSTEM

Before we begin, we will change our Display Configuration from MEP Design to MEP Basic 2 – Line. The Basic 2 Line Display Configuration displays all piping as 2 line regardless of the Single Line Graphics settings in the Pipe System Definition, and turns off Hidden Lines. This allows you to see two line piping, even for small diameter piping that would otherwise display as Single Line or Graphical One Line in the MEP Design display configuration. For more information on Display Configurations, see Chapter 12.

1. On the Project Navigator palette, on the Constructs tab, under the *Constructs* folder, expand the *Plumbing* folder.

2. Double-click to open the *01 Plumbing* construct drawing.

3. Zoom in on the restroom area in the upper right-hand corner.

   Floor drains are already located in both restrooms. Please see the "Plumbing" topic in the Quick Start Chapter for more information on how to layout the floor drains.

4. On the Drawing Status Bar (lower right corner of the drawing window) change the Display Configuration to: MEP **Basic** 2-Line (see Figure 6.60).

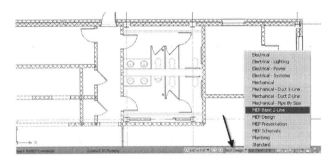

FIGURE 6.60
Switch the Display Configuration to Basic 2-Line

Let's add the piping required for the floor drains for the two restrooms on the first floor and locate the main drainage stack into which the other floors will connect.

5. On the Gravity Pipe Tool palette, scroll down to the By System group.

6. Click the Sanitary tool.

7. On the Properties palette, set the Routing Preference to: **Cast Iron Soil Pipe – Bell x Spigot**.

8. At the command line, type PCON and then press ENTER.

9. Select the floor drain in the upper (Women's) restroom.

10. On the Properties palette, change the Elevation to **-1'-0"**.

CAD MANAGER NOTE: When routing pipe using the Plus (+) Grip from plumbing equipment MvParts, by default Plumbing Line (2D) will be drawn. You can route 3D Pipe by holding the CTRL key when clicking the grip. To permanently change this behavior to instead route 3D Pipe by default, you can edit the *AecbDomainConfig.xml* file located at *C:\ProgramData\Autodesk\MEP 2011\enu\Shared\AecbDomainConfig.xml*. The string that needs to be changed is: <PartTypeEnum globalStr="Appliance" isPlumbingFixture="True" /> You must change the word "True" to "False" for each part type that you want to start the PipeAdd command instead of the Plumbing Line Add command. Please note: You should use Notepad.exe to edit this file so the header information is not corrupted. Remember to create a copy of the original file before you begin editing any XML file.

Before we set the slope, we are going to change the slope format as mentioned above.

11. On the Properties palette, scroll down to the Routing Section, select the Slope Format Drop list..

12. Change the Slope format to: **Fractional Rise (Inches), Run=(12 Inches) (ex. ¼ where run is 12)** (see Figure 6.61).

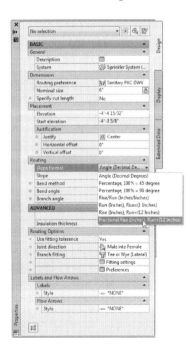

FIGURE 6.61

Setting the Slope Format in Pipe Preferences

When you input a slope value such as -¼ it will be applied over 12" making a ¼" / 1'-0" slope pitching down. A positive number slopes up and a negative number slopes down.

13. On the Properties palette, in the Routing grouping, change the slope to **-12"** to pitch the pipe away from the drain at 45 degrees.

14. Move the mouse horizontally to the left and then click a point in the middle of the Stall Door (see item 1 in Figure 6.62).

15. On the Properties palette, in the Routing grouping, change the slope to **-1/4"** to pitch the pipe away from the drain.

16. Move the mouse horizontally to the left and then click a point just beyond the Stall to add a 45-degree elbow (see item 2 in Figure 6.62).

17. Move the mouse horizontally to the left and then click a point just beyond the Corridor hallway (see item 3 in Figure 6.62).

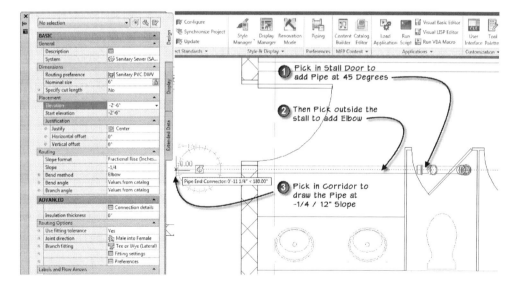

FIGURE 6.62

Laying out the Sanitary Floor Drain Piping

18. Move at 45° (the compass will read either 45 or 135) and then click another point in the middle of the corridor.

19. Click the next point just inside the upper horizontal wall.

20. At the command line, type E, press ENTER and type **-5'-0"**, and then press ENTER.

21. Finally, pick a point outside the wall and then press ENTER to exit the PipeAdd command (see Figure 6.63).

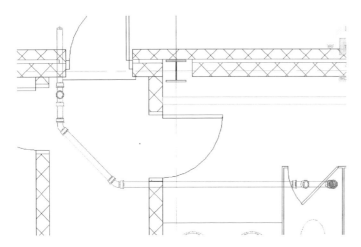

FIGURE 6.63
Completed Floor Drain layout

CONNECT A SECOND DRAIN

Following a similar process, we will connect to the drain in the Men's restroom.

22. Select the 45-degree elbow in the corridor, and then click the plus grip (+) that appears pointing toward the Men's Room (see Figure 6.64). This will upgrade the 45-Degree Elbow into a Lateral.

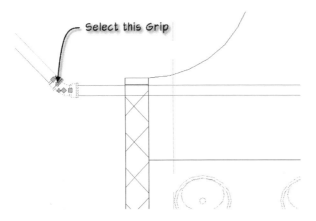

FIGURE 6.64
Upgrading the 45 Degree Elbow into a Lateral

> **Note:** When upgrading Male x Female Elbows into a Lateral or a Tee, the part in the Routing Preference must support the same Male x Female Connection Orientation to be successfully completed.

This starts the PipeAdd command again, updates the Properties palette with all the settings from the fitting you selected, and upgrades the elbow.

Since we will be drawing this pipe toward the drain instead of away from it, verify that the slope changed from negative to positive and you will need to change the Joint Direction from Male into Female to Female out to Male.

23. On the Properties palette, in the Routing grouping, verify the slope is **1/4"** (see Figure 6.65).

24. Beneath Advanced change the Joint Direction to Female out to Male (see Figure 6.65).

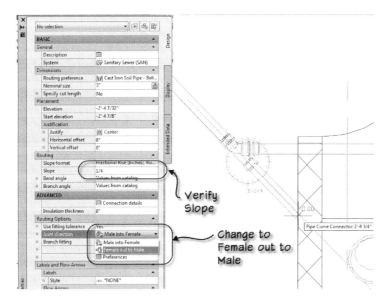

FIGURE 6.65

Upgrading the 45-Degree Elbow into a Lateral

25. Click a point in the Men's restroom under the right most sink (see Figure 6.66).

26. Type PCON at the command line to use the Pipe Connector snap and select the Drain.

The auto layout controls will give different options for you to choose. You can review each option by selecting next at the command line.

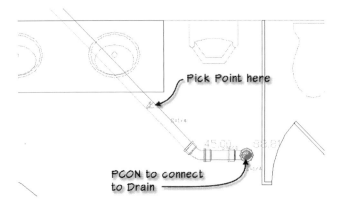

FIGURE 6.66

Connecting to the Drain in the Men's Room.

We have just covered how to route Male x Female piping in AutoCAD MEP. The Joint Direction control allows you to route with the flow using Male into Female and opposite the flow using Female out to Male. You can also upgrade Elbows into Tees when upgrading a 90-Degree elbow or a Lateral when upgrading a 45-degree elbow. In addition, you can add Laterals to pipe segments by using the Plus Grip in the Pipe and pointing away at 45 degrees.

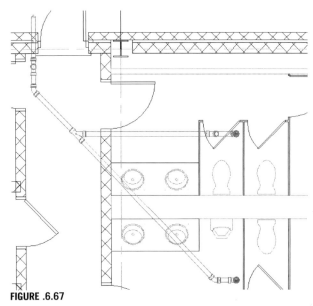

FIGURE .6.67

The Complete Floor Drain Layout

BATHROOM SANITARY PIPING SYSTEM

In the Quick Start Chapter we used Schematic Plumbing Line to lay out the sanitary piping from the Water Closets to the outside of the building. In this exercise we are going to create a similar system using 3D Piping.

> Continue in the area around the restrooms.

To begin, you need to overlay an AutoCAD MEP Water Closet on top of the Architectural Water Closets if you want to maintain connectivity.

> **Note:** An alternative is to supply the Architect with the Water Closet Mvpart that you use and ask that they replace their mvblock with your AutoCAD MEP version. Also, you can convert their mvblock into an Mvpart and send the updated object back to the Architect for inclusion into their drawings. For this exercise, you will simply place the Mvpart on top of the Architectural one.

27. On the Home tab, on the Build panel, click the Equipment button.

28. Scroll to the bottom and expand the *Plumbing* folder and then the *Water Closet* folder.

29. Select the Floor-Mounted Flush Valve Water Closet and place one at each Water Closet location.

> **TIP:** You can use the AutoCAD insertion object snap (INS) to snap to the architectural toilets.

30. Expand the *Urinals* folder, select the Wall-Hung Urinal.

31. Set the Elevation to 1'-0"

32. Toggle on the Replace Z value with current elevation control.

33. Snap to the insert of the architectural urinal, and rotate the urinal into place.

34. Press ESC to finish adding equipment.

35. On the Gravity Pipe Tool palette, in the By System grouping, click the Sanitary tool.

36. Set the Routing Preference to **Glued**.

Start with the right-most water closet in the upper (Women's) restroom.

37. Type PCON at the command line. Click the sanitary connection on the bottom of the water closet.

38. On the Properties palette, change the Elevation to **-8"** to go below the slab.

39. On the Properties palette, change the Slope to **-1/4"**, verify that the Slope format is still set to Fractional Rise.

40. Click a point near the middle of the cavity wall.

41. Pick a point perpendicular to the last (moving to the left) by the end of the accessible stall.

42. On the Properties palette, change the size to: **4"**.

43. Click the next point along the same line approximately at the midpoint between the sinks.

44. On the Properties palette, change the size to: **6"**.

45. Click a point just inside the wall.

46. On the Properties palette, change the elevation to **-6'-0"**.

47. Click the next point in the hallway at 45°, and then click the final point outside the building by the drain pipe (see Figure 6.68).

48. Press ESC to end the pipe run.

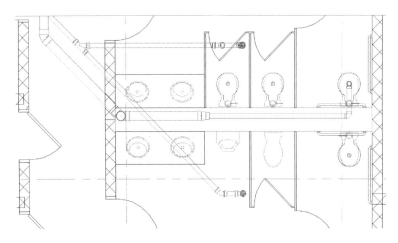

FIGURE 6.68

Complete the main run of Sanitary Piping

49. Select the elbow behind the water closet and then click the bottom plus (+) grip.

50. On the Properties palette, change the slope to ¼".

51. Click a point towards the opposite water closet to apply the slope.

52. Click on the sanitary connection on the bottom of the water closet and then accept the solution that mirrors the layout in the women's restroom.

   The piping command continues and the command line prompts you for a new Start point. Note that even though a tee was placed, we will modify this later to utilize a more appropriate fitting type.

53. Click on the sanitary connection on bottom of the next water closet in the Women's room.

54. On the Properties palette, change the size to: -8".

55. Click a point on the main pipe.

56. Accept the tee solution.

57. Repeat the process for the water closet on the left in the upper (Women's) restroom.

58. Press ENTER to complete the command.

59. Select the Tee between the back-to-back water closets and then click the plus (+) grip.

60. Change the slope to ¼" and then change the Nominal size to 3".

61. Click on the sanitary connection of the water closet (see Figure 6.69).

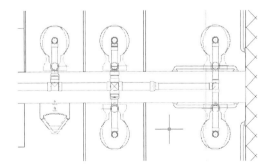

FIGURE 6.69
Complete the Water Closet Layout

We can change the part applied in the automatically generated solutions with a different part where needed. In this case, we are going to swap out the standard cross with a double sanitary cross. We will do this using the Properties palette.

62. Select the Cross. On the Properties palette, click the image next to Part.

The "Select a part" fitting selection dialog will appear and will be filtered by the Subtype Straight.

63. Change the Subtype filter to All (see Figure 6.70).

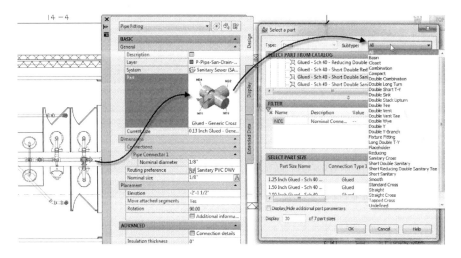

FIGURE 6.70
Modifying an existing Fitting

64. In the *Glued* folder, select the Glued – Sch 40 – Short Double Sanitary Tee and then Click OK

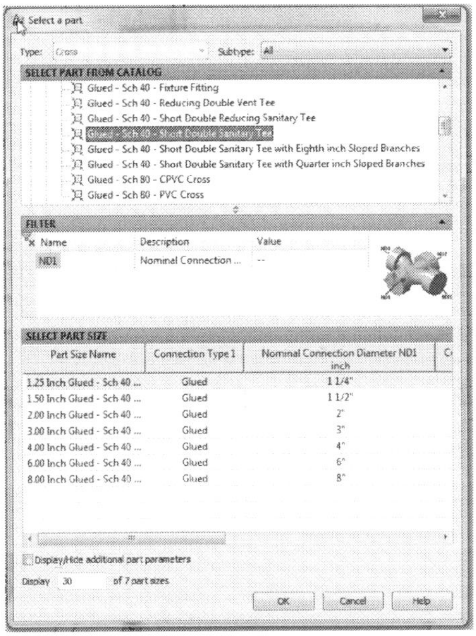

FIGURE 6.71

Selecting the 4" Short Double Sanitary Tee

We now have the correct fitting, but it points the wrong way.

65. With the fitting still selected, simply select the flip grip to orient the fitting in the correct direction (see Figure 6.72).

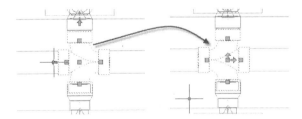

FIGURE 6.72

Flip the Fitting to the proper orientation

Now we want to upgrade the end tee to a cross to vent the pipe.

66. Select the tee at the right end of the run, and then click the plus (+) grip.

67. Draw the new pipe towards the wall on the right.

68. Make sure the Slope Format is still Fractional Inches, and that the Slope is set to 1/4"

69. Change the elevation to: **12'-0"** to go to the next floor and then press ENTER (see Figure 6.73).

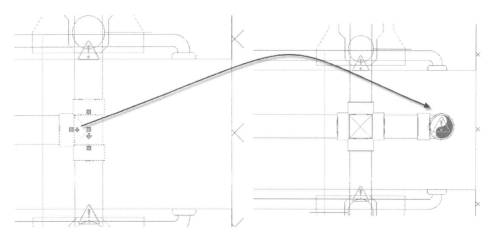

FIGURE 6.73
Upgrade Tee to Cross then change Elevation to span to the floor above

Before we replace the Cross with a Double Sanitary Tee we need to rotate it so the Connectors are in the correct orientation and the cross will be oriented with the flow direction.

70. Select the cross, and then click the midpoint rectangular grip.

71. Press the SPACEBAR twice to invoke the grip rotate command.

72. Type **270** and then press ENTER.

The water closet connections to the cross will become disconnected, due to the different laying length of the two crosses and this is expected; we just need to redraw the pipes from the elbows under the water closets to the cross (see Figure 6.74).

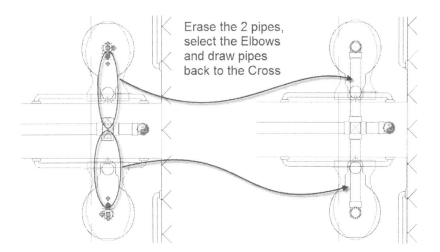

Erase the 2 pipes, select the Elbows and draw pipes back to the Cross

FIGURE 6.74
Reconnecting the Toilets to the Main

We just learned how to lay out Gravity Piping using Slope to pitch the pipe away from the equipment using a negative slope value or towards the equipment using a positive value. We also learned how to set access and change the Slope Format during layout. And we covered how to change fittings using the Properties palette to swap out a Cross with a Double Sanitary Tee (Cross) and to use the flip grips to orientate the fitting in the correct direction. Leveraging the Plus (+) Grips to continue the pipe run is essential to inherit the settings on the Properties palette for the Pipe.

We are done with this exercise. We have learned how to lay out piping with a slope, how to replace existing fittings using the Properties palette, and how to upgrade fittings from elbows to tees and from tees to crosses.

The elbow that turns down at the end of the wall by the corridor can be upgraded to a tee in the vertical to become the stack for the other floors. Feel free to continue to lay out the restroom and extend the stack up to the next floor. The completed water closet and urinal piping layout is provided in the *Plumbing Complete* folder as a file named: *01 Plumbing Complete* with the files you installed from the book's dataset. Feel free to open the file and compare it to your work.

> Note: You can run interference detection to check to see if the sanitary piping and the floor drain piping are clashing. To do this, go to the Analyze tab and click the Interference Detection tool. Select Pipe and then click the Start Analysis icon at the bottom of the palette. For more information on interference detection, see Chapter 14.

We also covered how to use AutoCAD's grip functionality to rotate the fitting to the correct orientation so the pipes pitch in the right direction.

PREPARATIONS FOR SPRLINKLER LAYOUT

Our final task in this chapter will be to complete the sprinkler piping layout for the first floor of the project.

1. On the Project Navigator palette, on the Constructs tab, under the *Constructs* folder, expand the *Fire Protection* folder.

2. Double-click to open the *01 Fire Protection* construct drawing.

Before we begin with the sprinkler piping layout, we need to determine the correct elevation for the sprinkler main and branches. To do this, we will add the second floor framing to this file as an XREF and then create a live section through the building. This will provide a clear understanding of the space and allow us to easily assess the available clearances.

3. On the Project Navigator palette, on the Constructs tab, under the *Constructs* folder, expand the *Structural* folder.

4. Drag *02 Framing* from the *Structural* folder and drop it into the drawing window.

5. On the Home tab, on the Sections & Elevations panel, click the Section Line tool (see Figure 6.75).

FIGURE 6.75

The Section Line tool on the Home tab

6. Click two points to draw a horizontal line through the middle of the building.

A Section line can have more than two points. Therefore, the "Specify next point" prompt continues to repeat after you click the second point.

7. Press ENTER to finish placing section line points.

8. At the "Enter length" prompt, press ENTER again to accept the default.

The default creates a section that is 20'-0" deep. This is how far it "looks" into the building. You can always select the section line and edit the depth with the grips later. Now that you have the section line, you can use it to generate a 2D section or enable a live section. In this exercise, you will enable a live section. For more information on sections, refer to Chapter 13.

9. On the View tab, on the Appearance panel, choose SE Isometric to change the view to a 3D view.

10. Select the section line.

11. On the Building Section Line tab of the ribbon click the Enable Live Section button.

As you can see, this limits the model view to display only what is within the section box. This tool is extremely helpful in isolating sections of buildings and layouts to facilitate working in 3D (see Figure 6.76).

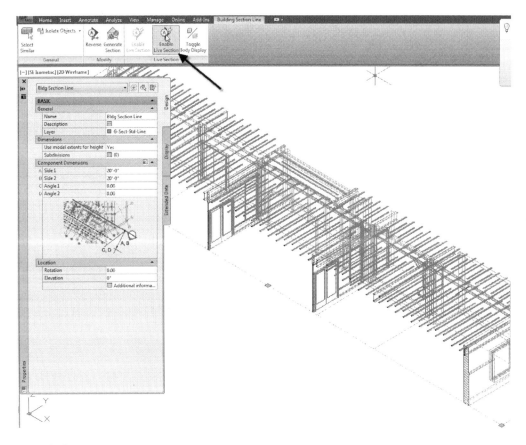

FIGURE 6.76

Enabling a Live Section crops the 3D model to display only the AEC geometry that falls within the section box

12. On the View tab, on the Appearance panel, choose Left view.

Now you need to determine the bottom of steel.

13. On the Analyze tab, click on the Inquiry panel title.

The panel will expand to reveal several additional tools.

14. Click the ID Point tool.

Note: If you prefer you can type ID at the command line instead.

15. Using the Nearest osnap, ID the lowest points on the Steel objects (see Figure 6.77). If you are getting a Z value of 0, make sure the Replace z value with current elevation is toggled off.

The results will appear on the command line. If they have scrolled past, you can press the F2 key to open the text window. The Steel should be at a Z elevation of 9'-11 3/32", and the HVAC at a Z elevation of 9'-0"

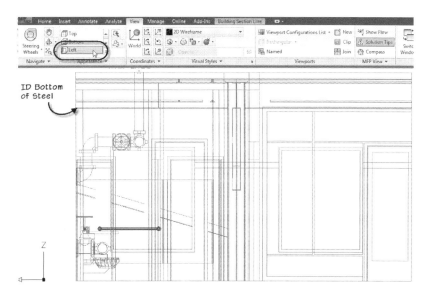

FIGURE 6.77
Identify the Elevation of the bottom of Steel and HVAC

CREATE THE SPRINKLER MAIN

Now that we have the elevations of the steel and the HVAC in the drawing, we can lay out the sprinkler main and the branch locations with this information in mind.

> **Note:** In your own projects, if you do not receive backgrounds that contain 3D geometry, you will not be able to perform this process to determine the clearance heights. In that case, simply ask your Architect, Structural Engineer, and/or Mechanical Engineer to provide this information to you. Even when 3D models are provided, you should communicate with your team to ensure that heights and other clearances are drawn accurately to avoid costly change orders later.

To begin the process of laying out the sprinkler system, it will be helpful to see our work progress in both plan and section. Therefore, we will split the screen into two viewports.

16. On the View, tab on the Viewports panel, click the Viewport Configurations List drop-down button and then choose **Two: Vertical** (see Figure 6.78).

    This will split the screen into two tiled viewports.

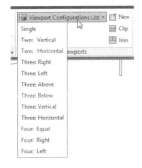

FIGURE 6.78

Setting up multiple tiled viewports

You will now have two viewports of the model on screen. You can make either active by simply clicking in it. The outline will appear bold to indicate the active viewport.

17. Click on the viewport on the left to activate it.

18. On the View tab, on the Appearance panel, select Top view.

Notice that the entire model is restored. Live section does not apply to plan view. The model is still cropped in the right side viewport, however.

19. Click in the right viewport to activate it.

20. On the Appearance panel, scroll the view buttons to locate Front view and then click it (see Figure 6.79).

Notice that the Front view is cropped by the section line. If necessary, you can adjust the section box in the plan view using the grips to make the live section larger or smaller. The results will apply immediately to the front view in the right viewport.

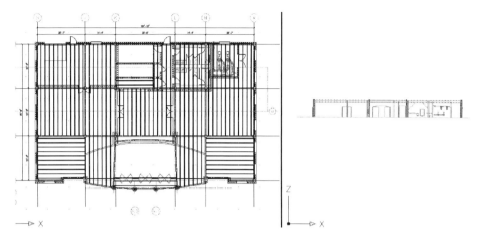

FIGURE 6.79

Set up the tiled viewports to show a plan on the left and front view on the right

21. Zoom in on the Fire Protection room in both viewports.

22. In the Front view, select the Fire Protection Pipe that goes up to elevation 12'-0" and erase it (see Figure 6.80).

23. Select the Coupling off the Elbow going up, and then click the plus (+) grip (see Figure 6.80).

24. On the Properties palette, set the Elevation to **9'-6"** and then click the Lock icon.

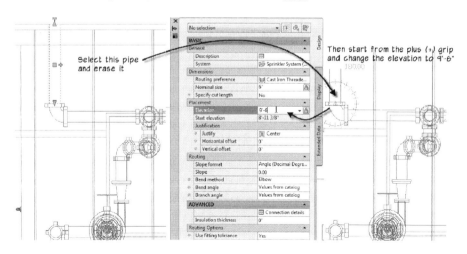

FIGURE 6.80

Erasing and replacing the vertical sprinkler pipe

This locks the specified elevation and constrains all the pipes drawn to this elevation. You can still set a new elevation in the palette, but when you connect to another pipe the specified elevation will remain. Also, when you are drawing close to other objects at different elevations, those objects will be ignored.

25. Click in the Plan view viewport to begin drawing the pipe header.

26. Type P and then press ENTER.

27. This will change the AutoCAD MEP Compass to the plan view. You should now see the compass circle (see Figure 6.81).

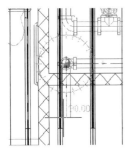

FIGURE 6.81
Setting the Compass to Plan view

28. Click a point just beyond the wall so an elbow is added.

29. On the Properties palette, change the size to 4".

30. Draw the pipe to the middle of the building and click.

31. In the "Choose a Part" dialog that appears, under Steel Pipe (US Imperial), select Grooved - Flexible Coupling 0.75 – 12 Inch, and then click OK.

32. Move horizontally to the left side and click again (see Figure 6.82).

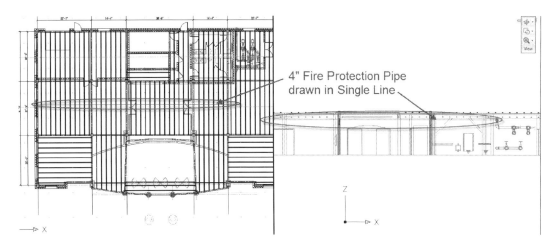

FIGURE 6.82

Lay out the 4" Fire Protection Main down the middle of the building

33. Select the elbow and click the plus grip along the horizontal pipe to upgrade the elbow to a tee.

34. Continue the pipe to the right end of the building.

We have now created the Fire Protection Header.

CREATE THE BRANCH LINES

Now we need to begin to lay out the branches for the sprinkler system, but first we can detach the Structural Framing XREF. This will reduce some of the visual clutter on screen and make it easier to work in plan.

35. On the Project Navigator palette, right-click on the *01 Fire Protection* construct and choose **External References**.

36. In the "External References" dialog, right-click on *02 Framing,* choose **Detach** — (see Figure 6.83).

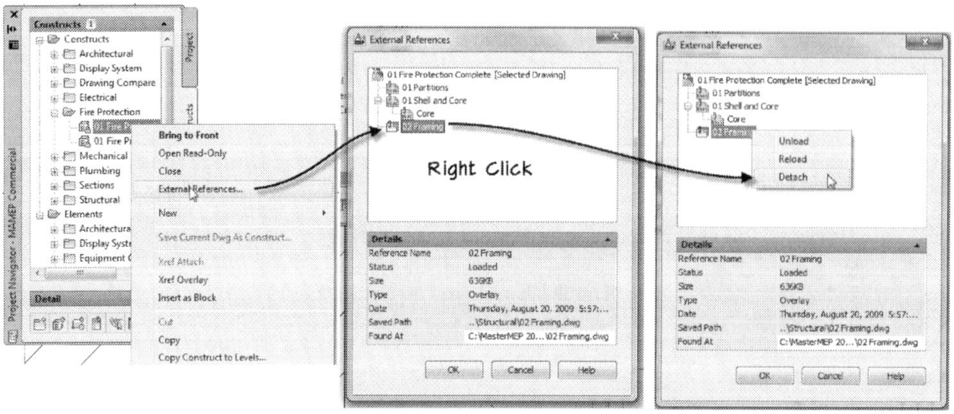

FIGURE 6.83
Detaching an XREF with Project Navigator

37. On the Home ribbon tab, on the Build panel, click Pipe.

38. Click a point on the header approximately 4'-0" away from the wall at the left end of the building (the distance is not that important—we will move the pipe later).

39. On the Properties palette, change the size to 2" and then pick a point toward the north side of the building just inside the wall.

We now want to set the distance between the pipe and the wall to be 4'-0".

40. On the Annotate tab, on the Dimensions panel, click the Dimension, Linear tool.

41. At the "first extension line origin" prompt, click a point on the inside face of the wall.

42. At the "second extension line origin" prompt, click the end of the pipe using the PCON snap.

The dimension will tell you the current distance. Since we want the distance to be 4'-0", you can simply do the math to figure out how far to the left or right the pipe needs to move. Once you have this value, you can move the pipe this amount. Use the AutoCAD move command or the Location grip on the pipe.

43. Select the Pipe and then click the square grip at the midpoint.

44. Move your cursor horizontally in the direction that it needs to move.

45. Type in the distance that will bring you to 4'-0" and then press ENTER (see Figure 6.84).

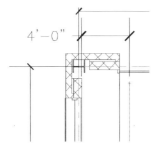

FIGURE 6.84

Use a dimension to assist in locating the Fire Protection Pipe.

To verify your work, you can edit the dimension with grips to the new location. You can delete the dimension when you are finished.

46. Select the tee that branches off the main, select the plus (+) grip, change the size to 2", and then route a pipe to the south wall of the building.

That completes the basic branch setup.

ADDING SPRINKLER HEADS

47. Select the Branch Pipe at the northwest corner of the building, click the middle plus (+) grip, and then on the Properties palette in the Routing Options grouping (under Advanced) change the Branch fitting to: **Takeoff only** (see Figure 6.85).

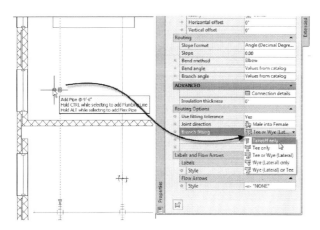

FIGURE 6.85

Changing Branch Fitting to Takeoff only

48. Set Nominal size to ½".

49. Set Elevation to **8'-6"** and then press ENTER to exit the pipe command.

50. Click in the right (Front) viewport to activate it. On the View tab, change to the Left view.

> **TIP:** In the Plan view you can select the Section Box and, using the grips, drag the box smaller to surround only the branch pipe you just drew. This allows for the Left viewport to show just the branch pipe.

51. Zoom in on the pipe you just added.

As you can see, we added a vertical drop of pipe at the spot indicated in plan. We now need to add the sprinklers to the line.

52. On the Home tab, on the Build panel, click the Equipment tool.

53. In the "Add Multi-view Parts" dialog that appears, expand the *Mechanical* folder, then the *Fire Protection* folder, and finally the *Sprinklers* folder.

54. Select the Pendent Sprinkler – US Imperial item.

55. At the bottom of the dialog, select the **0.5 Inch Pendent Sprinkler - Standard Coverage Standard Response** from the "Part Size Name" list.

56. In the Left viewport, place the part at the end of the vertical pipe you just drew. Press ENTER to accept the default rotation (see Figure 6.86).

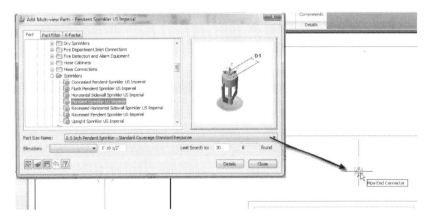

FIGURE 6.86
Place the sprinkler at the end of the ½" Pipe

57. Close the "Add Multi-view Parts" dialog.

58. Click in the left (Plan) viewport to activate it.

Use the stretch command, using a window around the sprinkler, and stretch the sprinkler so it is about 4'-0" north of the main. Now we have a single Sprinkler head on the branch. We are going to use the Array command to add the rest of the sprinklers to this side of the main and then use the mirror command to add them to the opposite side.

59. In the Plan viewport, click above and to the left of the sprinkler and then click down and to the right to create a window selection.

   This will select the sprinkler, pipe and takeoff. On the ribbon, a Multiple Objects tab will appear.

60. On the Home tab, click on the Modify panel title.

   The panel will expand to reveal several additional tools.

61. Click the AEC Array tool (see the top left corner of Figure 6.87) (AEC Array is also available on the right click menu on selecting AEC Modify Tools.) You can also use the AutoCAD Array command to complete this step.

62. At the "Select an edge to array" prompt, click on the main.

   A light blue line will appear on the main before you click to confirm selection (see the bottom left corner of Figure 6.87).

63. Move the mouse out toward the upper outside wall.

Several squares will appear to indicate where the copies will appear. The default distance between arrayed items is 3'-0". To change this distance in the AEC Array command, you can use Dynamic Input. You can turn this on with the toggle on the Application Status bar at the bottom of the screen (see the bottom of Figure 6.87). Alternatively, you can type P at the command line to Pick array distance.

64. Type **8'-0"** in the dynamic dimension on screen (it will appear near the midpoint of the main) and then press enter (see the middle of Figure 6.87).

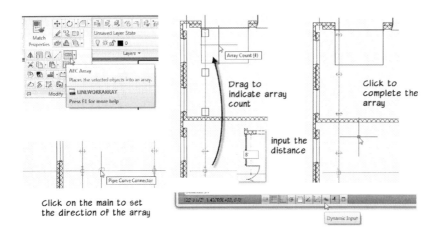

FIGURE 6.87
Array the sprinkler heads using the AEC Array tool

65. The squares will adjust and a tooltip will confirm the new spacing. To accept the array, click the mouse (see the right side of Figure 6.87).

The objects have been arrayed along the pipe and automatically connected.

> **Note:** Array, Mirror, Copy and Move are enhanced AutoCAD MEP commands which allow AutoCAD MEP objects to align automatically to other AutoCAD MEP objects when the objects are on the same UCS plane.

Let's copy the arrayed sprinklers to the lower branch.

66. On the Home tab, on the Modify panel, click the Copy tool (see Figure 6.88).

FIGURE 6.88
The Copy Command on the Ribbon

67. Make a window selection (click from left to right) to surround all of the previously arrayed sprinklers and then press ENTER.

68. Select a basepoint, move your cursor straight down towards the bottom of the screen, type **32'-0",** and then press ENTER.

This will copy the sprinklers to the other header.

Using array again, we can make copies of the entire branch and all its sprinklers to complete the layout.

69. Window Select both branches. (Surround the entire branch to select the branches, the sprinkler heads, and fittings, including the tee at the main.)

70. On the Home tab, click on the Modify panel title to expand it and then click the AEC Array tool.

71. Click on the branch at the edge to array.

72. Type **7'-9"** and then press ENTER in the dynamic dimension

73. Drag out the array across the building and then click to complete it (see Figure 6.89).

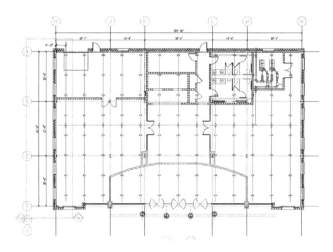

FIGURE 6.89
Using AEC Array to copy the branches and sprinklers across the building

The branches are laid out along the entire ceiling. You can go through and delete unnecessary sprinkler heads and terminate the pipes that go into the Foyer as well as into the Elevator area (see Figure 6.90). Note also, the crosses don't break into the main with this array operation, you may also want to break the main across each of the tees, then use grip edits to attach the main into the tees.

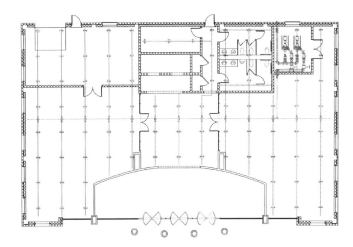

FIGURE 6.90

The completed First Floor Sprinkler System

Make any other fine-tuning adjustments that you like.

74. When you have completed your layout, save and close all project files.

SUMMARY

Throughout this chapter, we have reviewed many key aspects of the 3D piping tools in AutoCAD MEP 2011. We have put these concepts into practice in our sample commercial office building. You should now have a good feel for how 3D piping works as well as an understanding of specific systems such as fire safety layouts and restroom gravity piping. In this chapter we have learned:

✓ The Piping Feature in AutoCAD MEP allows for accurate creation of models in a fast and accurate manner.

✓ Routing Preferences are used to establish the preferred fittings during layout.

✓ System Definitions define how the System is abbreviated in annotation, and how it is displayed in Graphical 1 Line, Single Line, and 2 Line displays.

✓ System Definitions also control overrides to the default display, allowing for Centerlines to be displayed.

✓ Sloped piping controls provide for female fittings to have an allowable angle of deflection to make a 90°-elbow bend at angles other than 90°.

✓ The ability to route Male x Female piping.

✓ The ability to automatically lay out Laterals during Auto Layout improves the overall control of how the pipe system is created.

✓ The ability to upgrade 45-Degree Elbows into Lateral. Utilizing the grip functionality in AutoCAD MEP is essential to creating real world layouts.

✓ The Associated Movement ability, or StickyMove, displays the power of AutoCAD MEP. This functionality allows us to use the grips to move and automatically remove unwanted fittings and pipe segments.

✓ Leveraging AutoCAD commands like Copy and Array helps us to create repetitive layouts quickly and easily while automatically maintaining relationships to other AutoCAD MEP objects.

Chapter 7

Electrical Systems Layout

INTRODUCTION

The key differentiator between AutoCAD MEP (AMEP) and standard AutoCAD, or other electronic drafting methodologies, is the functionality provided by connectors on Devices. Connectors enable the designer to embed power characteristics in electrical devices, circuit the devices to panels, and ultimately report total connected and demand load in panel schedules. In addition to Devices and Panels, AMEP provides Wires for annotating circuits with "tick marks," as well as Conduit and Cable Tray objects for modeling such objects for coordination in three-dimensional space with other trades.

Historically, AutoCAD users have depended on a myriad of flyout buttons to provide easy access to all the various devices we may need. Typically, this requires the user to select an icon to run a particular LISP routine to ensure proper layering, block selection, and placement orientation. With AMEP, Devices and Panels have built-in functionality to provide intuitive placement, and since all electrical objects, including Wires, are style-based, changing a Device from one type to another, or even modifying the ticks on a Wire, is as simple as changing the style of the selected object on the properties palette.

OBJECTIVES

In this chapter, we will explore all the settings on the Properties palette and how they can be used for efficient device placement and editing. In this chapter you will:

- Learn how to place Devices and assign load.
- Learn how to place Panels and create Circuits.
- Learn how to circuit connectors on Devices.
- Learn how to place Wires
- Learn how to convert AutoCAD geometry into AutoCAD MEP objects.

DEVICE PLACEMENT

Placing a Device in AMEP is fairly simple and straightforward. All Devices are easily accessible from the Electrical workspace: on the Home tab of the ribbon, on the Build panel, using the Device tool (see the top of Figure 7.1). Using this method removes a lot of the administrative overhead of configuring toolbar buttons or tool palette tools, as traditionally done in AutoCAD, for each conceivable Device that may need to be used.

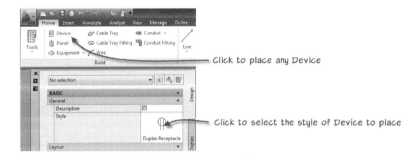

FIGURE 7.1

The Device tool on the Home ribbon, Build panel, and the Style setting on the Properties palette

When you click the Device tool, the Properties palette will open automatically, if it is not open already. As outlined previously, it is recommended that you get accustomed to having the Properties palette readily accessible, as you will use it for many common placement and edit operations.

On the Properties palette, Design tab, clicking the Style preview image will allow you to select a Device Style to insert (see the bottom of Figure 7.1). This will open the "Select a style" dialog that provides access to all available Devices (see Figure 7.2).

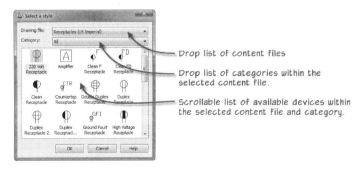

FIGURE 7.2

The "Select a style" dialog

Devices are organized in the individual drawing files. Within each drawing file devices are organized in categories. You can view all styles within a file by selecting the "All" category or you can filter the list of devices by selecting a specific category. In the "Select a Style" dialog you can either double-click on a style, or select a style, and then click OK to place it.

> **CAD MANAGER NOTE:** The list of drawings is automatically populated by the files that reside in the *Electrical Devices* folder. This folder is specified in the "Options" dialog on the MEP Catalogs tab. At the bottom of this tab several Style-Based Content Paths are listed, including Electrical Devices. The list of devices and associated preview icons are automatically generated from the contents of the drawings in this location.

PROPERTIES DURING DEVICE PLACEMENT

The Properties palette is used both during placement of objects and to edit them later. Some properties are only available during actual placement and will not appear when the object is selected later. Such properties are indicated on the Properties palette by an orange star next to placement command (see Figure 7.3). Let's take a look at some of these properties and their functions.

Layout method—Determines if Devices will be placed individually or in multiples.

Setting the Layout method to One by one will allow you to specify the location of each device instance, as in traditional workflows (see Figure 7.3).

FIGURE 7.3
Select the One by one method to place Devices one at a time

The other two Layout method options, Distance around space and Quantity around space, require that you have Spaces objects visible in your drawing. Spaces may be in the drawing itself, or XREFed from another drawing (to learn more about Spaces, refer to Chapter 4). Devices will automatically snap to the Space boundary and orient themselves perpendicular to the inside face of the Space, even if all OSNAPs are disabled.

Setting the Layout method to Distance around space allows you to specify the distance between devices. The number of devices that will fit in the space will automatically update in the Number of devices field. The location of the cursor will determine the placement of the first device, and the devices will be spaced at the specified distance along the Space boundary going counter clockwise (see Figure 7.4). On the right side of the figure you can see that the last device will end up at a distance less than that specified.

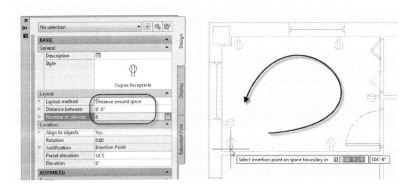

FIGURE 7.4
Device Layout set to Distance around space

The Quantity around space option lets you specify the Number of Devices you wish to place. All Devices will be spaced evenly around the Space. The Distance between fields will automatically update to give an indication of how far apart the Devices will be (see Figure 7.5).

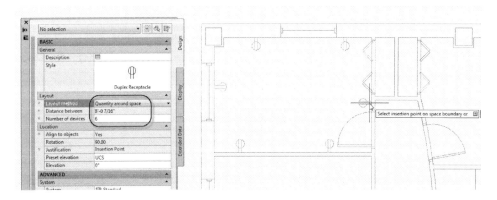

FIGURE 7.5
Device Layout set to Quantity around space

TIP: Start with the Distance around space option to find out how many devices you may need around a Space, then switch to the Quantity around space to get evenly spaced devices.

Regardless of the layout method employed, you can always move Devices, if required, after placement.

Align to objects—When you choose the One by one layout method, you can set the Align to objects property to either yes or no. When set to yes, AMEP will automatically align Devices to geometry near the cursor during placement, such as to align Devices to walls. When set to no, you have the option of specifying a rotation, similar to when placing a standard block.

INSTALL THE DATASET FILES AND CREATE A DRAWING

In this exercise, we use the Project Browser to load a project from which to perform the steps that follow.

1. If you have not already done so, install the book's dataset files.

   Refer to "Book Dataset Files" in the Preface for information on installing the sample files included with this book.

2. Launch AutoCAD MEP 2012 and then on the Quick Access Toolbar (QAT), click the Project Browser icon.

3. In the "Project Browser" dialog, be sure that the Project Folder icon is depressed, open the folder list and choose your *C:* drive.

4. Double-click on the *MasterAME 2012* folder.

5. Double-click *MAMEP Commercial* to load the project. (You can also right-click on it and choose **Set Project Current**.).

IMPORTANT: If a message appears asking you to repath the project, click Yes. Refer to the "Repathing Projects" heading in the Preface for more information.

Note: You should only see a repathing message if you installed the dataset to a different location than the one recommended in the installation instructions in the preface.

6. Click the Close button in the Project Browser.

When you close the Project Browser, the Project Navigator palette should appear automatically. If you already had this palette open, it will now change to reflect the contents of the newly loaded project. If for some reason, the Project Navigator did not appear, click the icon on the QAT.

7. On the Application Status Bar, set the Workspace to Electrical.

PLACE DEVICES USING LAYOUT OPTIONS

The Project Navigator palette should have opened onscreen when you loaded the project (or it remained open from the previous chapter). If it is not open, you can click the icon on the QAT to open it now. For more information on Projects and Project Navigator, refer to Chapter 3.

1. On the Project Navigator palette, click the Constructs tab.

2. Expand the *Electrical* folder, then the *Power* folder, and then double-click to open the *03 Power.dwg* Construct file.

3. On the View tab of the ribbon, on the Appearance panel, click northwest Offices (see Figure 7.6).

FIGURE 7.6
Select to restore a named view

4. On the Home tab, on the Build panel, click the Device button.

   The Properties palette should appear. If it is already open, it will be populated with the settings for the Device you are adding.

5. On the Properties palette (on the Design tab) click the Style preview image (see Figure 7.1 above).

6. In the "Select a style" dialog, select the Receptacles (US Imperial) from the Drawing file list (see Figure 7.2 above).

7. Double-click the Duplex Receptacle.

8. Set the Layout method to Quantity around space, and specify 4 as the number of devices.

9. Toggle off the OSNAPs (F3); look at the application status bar to ensure the Object Snap icon is dimmed.

10. Move the cursor around the inside edge of the northwest corner office. Note that as you hover, the locations of the receptacles preview in their final locations. Pick when there are no receptacles in a doorway (see Figure 7.7).

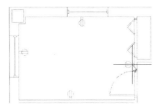

FIGURE 7.7

Device layout preview

11. Place receptacles in each of the three offices beneath this one.

12. Press ENTER to complete the command.

LOCATION OPTIONS

The Location options allow you to specify the alignment/rotation and elevation of the Device as you place them (see Figure 7.8).

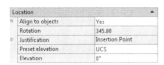

FIGURE 7.8

Location settings on the Properties palette

If there are no Spaces available for your drawing, the Align to objects option is a great alternative. The Align to objects option will automatically set the rotation of the Device so that it is perpendicular to lines, arcs, circles, walls, spaces, splines, and other such geometry. This occurs whether that geometry is in the local file or in an XREF. Similar to an alignment parameter within a Dynamic Block, the Device will orient to the side from which the cursor approaches (see Figure 7.9).

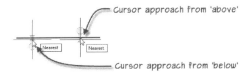

Cursor approach from 'above'

Cursor approach from 'below'

FIGURE 7.9
Device aligns perpendicular to object on the side from which the cursor approaches

If Align to objects option is set to No, you will be prompted for the rotation of the device after you pick the insertion point. Generally, you should use the NEA (nearest) snap to ensure that the device is "on" the geometry.

CAD MANAGER NOTE: The orientation of the block that is used to define the Device is critical for the Align to objects functionality, including layout around Spaces, to work as expected. Since this functionality is newer than some of the content that ships with AMEP, some blocks haven't been updated accordingly (refer to Chapter 9 for more information).

PLACING DEVICES USING LOCATION OPTIONS

Since AMEP allows you to design in 3D, you may want to consider the elevation of Devices as you place them. There are two ways to specify the elevation, by manually typing in a value in the Elevation property (see Figure 7.8 above), or by selecting a Preset elevation from a list. The Preset elevation list is defined in the "Options" dialog. Let's learn how to define elevations that may be used for Device placement.

13. From the Application menu, click Options.

14. On the MEP Elevations tab, click the Defining Systems Elevations icon (at the bottom left corner of the dialog).

15. Set the Name to **Receptacle**, and set the Elevation to: **1'-6"** (you may also type **18"**).

16. Add a second value. Set the Name to **Over Counter** and the Elevation to **3'-6"** (see Figure 7.10). In both cases, adding a description is optional.

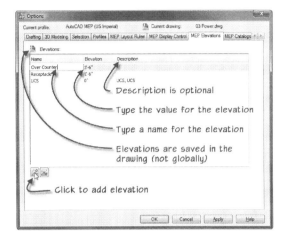

FIGURE 7.10

You can create preset elevations in the "Options" dialog

17. Click OK to close the "Options" dialog.

18. Right-click the Object Snap icon on the Application Status Bar (at the bottom of the application frame) and choose Settings (see Figure 7.11)

FIGURE 7.11

Edit the Object Snap Settings

19. Check the Object Snap On checkbox and, beneath General, make sure only Nearest is selected and then click OK.

20. Select one of the receptacles placed in the previous steps.

21. On the Device tab of the ribbon, click Add Selected button.

22. On the Properties palette, set the Layout method to One by one. Align to objects should be set to Yes.

23. Set the Preset elevation to Over Counter (alternatively, you could type 42 in the Elevation box).

24. Place several receptacles along the walls of the secretarial area (see Figure 7.12).

25. Press ENTER to end the command.

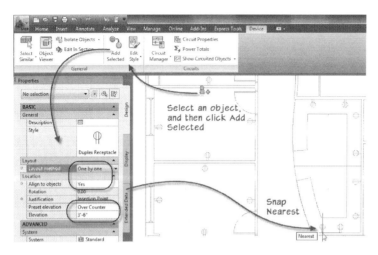

FIGURE 7.12

Placing additional objects using Add Selected and a preset Elevation

> **TIP:** Add Selected is a quick way to create objects matching existing ones you already have in your drawing. If necessary, you can update any properties on the Properties palette prior to placing the next instance. This is more convenient than using the Copy command since copy doesn't give you the opportunity to use the Align to objects functionality or change properties on the fly. You can, however, use the grip edit/copy to utilize the align functionality.

ADVANCED OPTIONS

The Advanced options (see Figure 7.13) described below are options less frequently used during Device placement.

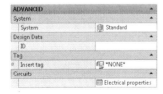

FIGURE 7.13

Advanced Device Placement Properties

System—Each Device is associated with a System. The primary purpose of a System is to specify the Device's layer when placed. However, Devices typically layer according to their style's layer key, and thus, the System's layer setting is usually ignored. This means that the default system named Standard is suitable for placement of Devices in most cases. In cases where there is no layer key associated with the Device style, the System will determine the layer for the placed Device. If neither the Device style nor the System has a layer key assignment, the DEVICE layer key will be used instead. In most cases, layering per the layer key of the Device is preferable since it alleviates your having to remember to set an appropriate System for each device you place. There are exceptions, but this could safely be considered the "rule."

Design Data ID—The ID of a device is used to provide an identifier for the Device instance. For example, you may want to use the ID to identify Devices representing connections to mechanical equipment such as "AHU-1" or "EF-4."

> Note: The default lighting fixture tag uses this ID property; however, in such cases it is better practice to use a property of the lighting fixture style to define the fixture type. Refer to Chapter 15 for more information.

Insert tag—This option lets you automatically place a Tag when placing a Device. The Tag is placed near the insertion point of the Device. The Tag can be moved later if desired, but there are no location placement options available during Device placement.

Electrical Properties—This item opens a worksheet where you can specify the electrical characteristics and circuiting of Devices as you place them (see Figure 7.14).

FIGURE 7.14

Optionally configure the Electrical Properties of a Device as you place it

However, in most cases you will likely define the electrical characteristics in the definition of the Device style to ensure consistency across all instances of a particular style. Furthermore, while this option is provided here as you place a Device, circuiting is typically done at a later stage in the design process.

USE GRIPS TO MODIFY DEVICE PLACEMENT

As mentioned previously, Devices can automatically align to geometry during placement. This same functionality exists when modifying a Device using grips. For example, when an architectural background changes, instead of using the move and rotate commands to reorient Devices to the architecture, grips can be used to simplify the task. You can also find this useful to fine-tune placement of automatically placed Devices. Before we modify the placement of some Devices using the grips, let's add a few more Devices to our plan.

1. Select one of the duplex receptacles in your file.

2. On the Device tab of the ribbon, click the Add Selected button.

3. Use the Quantity around space layout option to place six receptacles around the perimeter of the large office at the southwest corner of the building (see Figure 7.15).

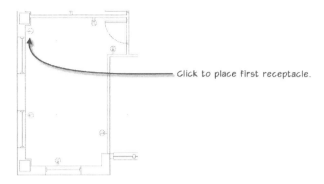

FIGURE 7.15

Device layout in southwest corner office.

4. Use the Distance around space option to place receptacles 70'-0" on center in the corridor. Place the first receptacle at the north end of the corridor (see Figure 7.16).

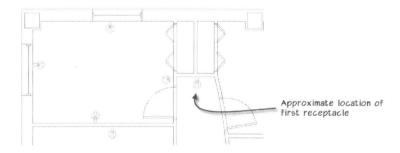

FIGURE 7.16

Device placement in the corridor

5. Press ENTER to complete the command.

When a Device is selected, it has at least two grips: Show all insertion points (grey circle), and Align device (cyan pentagon shaped). Additionally there is a plus (+) grip to Add Wire for each electrical connector. Clicking the Show all insertion points grip reveals additional alignment grips. Clicking on an Align device grip lets you move the Device and align it to geometry. Press and release CTRL to toggle the alignment (see Figure 7.17).

Show insertion points

Align device
Press Ctrl to toggle between:
- Automatic alignment ON
- Automatic alignment OFF

FIGURE 7.17

Device insertion grips

Typically, for receptacles and other wall-mounted devices, the bottom-center grip is frequently coincident with the Device insertion point, and is the most commonly used grip for relocating Devices when the background changes, or to make other such modifications. After you click the alignment grip, the Device will move with the cursor. You can also use typical grip editing command line entry, such as typing C, to copy the Device. Let's use grips to relocate two of our receptacles, one in the corner office room and one in the corridor, to more appropriate locations.

6. Select the receptacle near the strike side of the door in the corner office.

7. Click the Show all insertion points (round grey) grip.

8. Using the bottom-middle alignment grip and the nearest OSNAP, move the Device to the adjacent wall (see Figure 7.18).

9. Press ESC to deselect the Device.

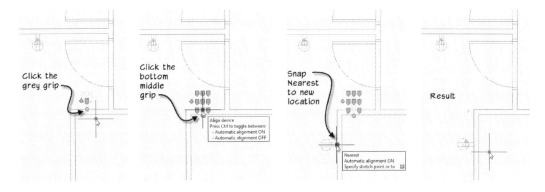

Click the grey grip

Click the bottom middle grip

Align device
Press Ctrl to toggle between:
- Automatic alignment ON
- Automatic alignment OFF

Snap Nearest to new location

Nearest
Automatic alignment ON
Specify stretch point or to

Result

FIGURE 7.18

Use grips to relocate a Device

Notice that as you move your mouse over the new wall, it will reorient to the face of that wall.

10. Repeat the process with the receptacle indicated in the corridor to move it onto the wall nearby (see Figure 7.19).

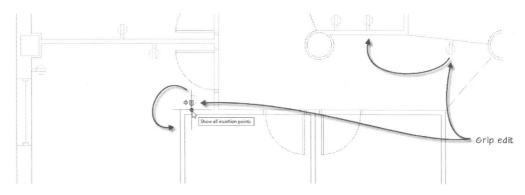

FIGURE 7.19

Move receptacles using the grips

DEFINE THE ELECTRICAL CHARACTERISTICS OF THE DUPLEX RECEPTACLE

In this exercise, you will modify the Duplex Receptacle Device style to constrain it to a 120V/1P circuit, and a standard 180VA load.

1. Select one of the Duplex Receptacles in the drawing.

2. On the Device tab of the ribbon, click the Edit Style button (see Figure 7.20).

FIGURE 7.20

Edit the style of a receptacle

3. On the Connectors tab, set the following values and then click OK.

⇨ Number of Poles: for Value, choose 1; for Prevent Override, choose **Yes**.

⇨ Voltage: for Value, choose 120; for Prevent Override, choose **Yes**.

⇨ Load Phase 1: for Value, type **180**; for Prevent Override, choose **Yes**.

⇨ Load Category: for Value, choose **Receptacles**; for Prevent Override, choose **Yes**.

All Duplex Receptacles in this drawing will now only connect to a 120V/1P circuit and will contribute 180VA. Using separate Device styles for specific conditions helps ensure consistency in standards. For example, using separate Device styles for duplex receptacles that are intended for microwaves, copiers, or refrigerators, for example, allows the designer to specify the value on the style ahead of time, and then simply select the receptacle style when needed, automatically having the corresponding load information update on the Device.

> **Note:** If the Number of Poles or Voltage properties are disabled in the Device style window, this is because there is at least one instance of the Device style already connected to a circuit. This keeps you from inadvertently setting a style to a voltage and pole combination that would invalidate circuited Device connections.

4. Click OK.

5. Save the file.

DEFINE A SPECIAL PURPOSE RECPTACLE

In this exercise, we are going to modify a Device for a special condition. The executive in the corner office likes her coffee. Let's make the receptacle in the corner for a coffeemaker (see Figure 7.21).

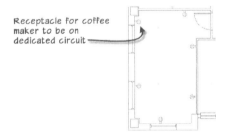

Receptacle for coffee maker to be on dedicated circuit

FIGURE 7.21
Device to be modified for a special purpose receptacle

This means that it will have a load greater than the typical duplex receptacle. To set this receptacle up as for a coffeemaker, we will copy the receptacle style, assign a new load to the connector and then update the Device instance to the new style.

6. On the Manage tab of the ribbon, on the Style & Display panel, click the Style Manager dropdown button and choose Electrical Device Styles (see Figure 7.22).

Alternatively, you can open Style Manager, expand the *Electrical Objects* folder and then select *Device Styles*.

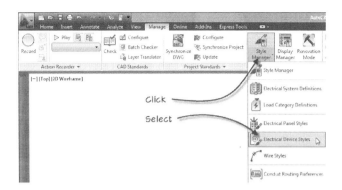

FIGURE 7.22

Open the Style Manager to Electrical Device Styles

7. Select the Duplex Receptacle style, on the toolbar, click Copy and then click Paste (see Figure 7.23).

This will create a new Device Style called Duplex Receptacle (2).

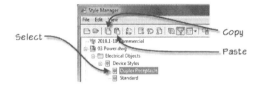

FIGURE 7.23

Copy and paste a style in Style Manager

8. In the tree view for the Device Styles, select Duplex Receptacle (2), on the right side, on the General tab, change the name to **Dedicated Coffee Maker Receptacle**.

9. On the Connectors tab, set the following values, and then click OK.

- Load Phase 1: for Value, type **1000**; for Prevent Override, verify **Yes is specified**.

10. Select the receptacle in the northwest corner of the office (see Figure 7.21).

11. On the Properties palette, click the Style property (if the Properties palette is not open, right-click and select Properties).

12. In the "Select a style" dialog, for the Drawing file choose <Current Drawing> and then double-click on the Dedicated Coffee Maker Receptacle (see Figure 7.24).

13. Press ESC to deselect the Device.

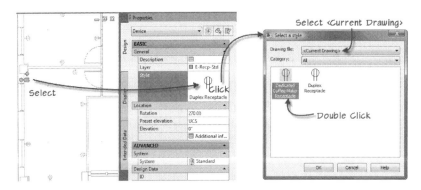

FIGURE 7.24
Change the Style of a Device instance

Using this method to define a style for each type of receptacle or connection helps facilitate project-wide changes. For example, when working on a commercial kitchen, you may receive a layout of a variety of loads, each keyed to a schedule with the voltage and load requirements. Each load type may have multiple instances, and thus having to manually specify the load information at each instance can be cumbersome, especially when the design gets revised. By using separate styles for each condition, you can easily update the load information for the styles in Style Manager, and each instance of each style is updated accordingly.

14. Save the file.

PANEL PLACEMENT

As indicated at the start of this chapter, Panels are used to Circuit devices. The Panels can exist in the same drawing with the Devices, or they can reside in a different drawing. For example, it is common to place all panels for a given floor plan in the "Power" drawing. For a lighting plan, the Electrical Project Database allows lighting fixtures in the lighting drawing to be Circuited to Panels in the power drawing. Placement of panels is very similar to placing Devices (see Figure 7.25); the similar functionality will not be reiterated here.

FIGURE 7.25
Basic and advanced Panel placement options are similar to Devices

Some key differences are that Panels don't have Layout options (i.e., to place multiple panels around the perimeter of a Space). However, the align-to-objects functionality does exist for Panels, ensuring that panels are easily placed at any angle. Another difference is that panel styles don't have a layer key. Panel layers are controlled via the PANEL layer key which uses or creates the E-Panl-Std layer in the default *MEP Object - AIA 256 Color* layer key style file. This layer can be overridden by a System's layer key.

If you accept the default Standard System, there are no layer key overrides, and thus, E-Panl-Std will be the layer used for the panel. The Devc-Power - 208V system has overrides for Minor 1 and Minor 2 of -Pwr and -208V, respectively. Thus, the resulting layer will be E-Panl-Pwr-208V. If the system definition has a layer key associated with it instead of just overrides, the system layer key will be used when placing the Panel. This is similar to Devices. Unless you need to use different layers for your various Panels and Systems, the Standard system with no keys or overrides is likely suitable. You can edit the System styles or define other Systems with layer key overrides to suit your needs if necessary.

Circuits and Design Data—A variety of properties specific to Panels may be specified when a Panel is placed. When placing panels, it is most critical that you pay attention to the Panel type (ANSI vs. ISO), and the Phases, as you cannot change these options after a panel is placed. All the other settings can be modified later.

> **TIP:** If you ever find that you have these settings incorrect on a placed Panel, you can create a new Panel with no circuits and move the circuits from the old panel to the new panel in Circuit Manager. This will ensure that you don't have to re-circuit anything you may have already connected.

When creating panels, you have the option of whether to create the circuits or not. In some cases, you may have no idea what circuits you will need, so you may opt initially not to create circuits. However, if you do create circuits when placing the Panel, it is best that you specify the voltages phase-to-neutral and phase-to-phase; having these set will ensure that any circuits created during panel placement have the proper voltage associated with them.

When placing a Panel, the following properties are available. A brief description of these Panel properties follows (see Figure 7.26).

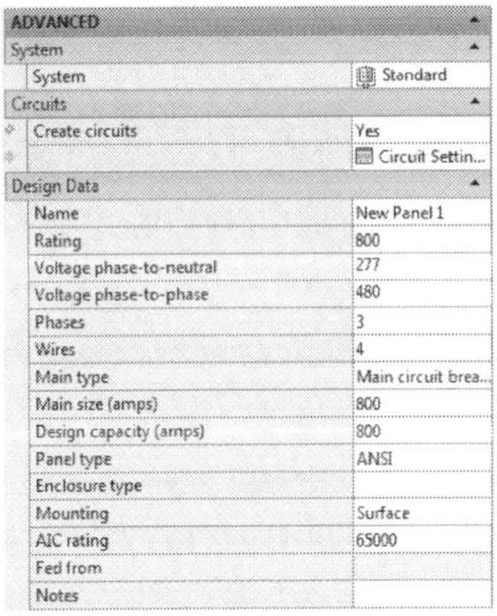

ADVANCED	
System	
System	Standard
Circuits	
Create circuits	Yes
	Circuit Settin...
Design Data	
Name	New Panel 1
Rating	800
Voltage phase-to-neutral	277
Voltage phase-to-phase	480
Phases	3
Wires	4
Main type	Main circuit brea...
Main size (amps)	800
Design capacity (amps)	800
Panel type	ANSI
Enclosure type	
Mounting	Surface
AIC rating	65000
Fed from	
Notes	

FIGURE 7.26
Panel placement options

Name—The name of the panel, such as MDP, LA, PP1, etc.

Rating—The rating of the bus in the panel (amps).

Voltage phase-to-neutral—Sets voltage of 1 pole circuits.

Voltage phase-to-phase—Sets voltage of 2 and 3 pole circuits.

Phases—Number of phases for the panel – 1 or 3.

Wires—The number of wires to the panel (3 or 4 for 3-phase, 3 for 1-phase).

Main type—Main circuit breaker (MCB) or Main lug only (MLO).

Main size (amps)—Only editable if the main type is Main circuit breaker (MCB).

Design capacity (amps)—Used to compute the spare capacity of a panel.

Panel type—ANSI or ISO; effects phase naming. ANSI is typically used in the US for phases named A, B, and C. ISO uses phases named L1, L2, and L3.

Enclosure type—For example, NEMA 1.

Mounting—Surface, Recessed, or Floor; for scheduling only, does not effect the placement/graphics of the panel itself.

AIC Rating—Available interrupting current.

Fed from—Indicates which panel feeds this panel and populates automatically.

Notes—Other notes for scheduling purposes.

When the Create Circuits option is set to Yes, the circuits will be created according to the settings in the "Circuit Settings" dialog. To edit these settings, click the Circuit Settings worksheet icon (see Figure 7.27).

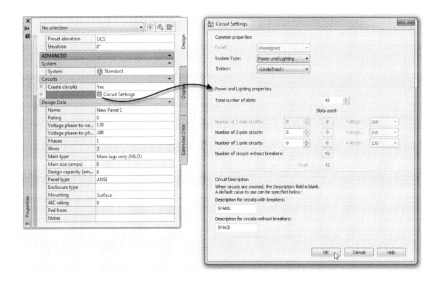

FIGURE 7.27
Click the circuit settings worksheet icon to open the circuit settings

In the "Circuit Settings" dialog, you can set the System Type to Power and Lighting, General or Other. The list of Systems available comes from those System Definitions that are configured as either Power and Lighting or General. In the Total number of slots field, you can input how many breaker spaces exist in the panel. In the Slots used area, you can specify how many of each type of breaker (3-pole, 2-pole, and 1-pole) to create. The total number of breaker poles cannot be more than the number of slots. The voltages are automatically defined based on the panel's phase-to-phase and phase-to-neutral properties.

When creating circuits as you create a panel, any unused slots will become 1-pole SPACES. The spaces have a rating of 0 amps, and will not let you circuit to them. If you later need to use a space, simply modify the rating to something other than 0.

The circuit descriptions specified for SPARE and SPACE circuits are defaults only; you can modify these on individual circuits if desired. For example, you may rename a SPACE to SHUNT-TRIP to indicate the additional space required by such a breaker.

After you have placed a panel, you can later modify the circuits in Circuit Manager.

Circuits are most commonly utilized when circuiting components with load, such as on 480/277v systems and 208/120v systems. To circuit such Devices, you need to use a Power and Lighting system type. General and Other system types are used to zone or associate Devices together, for example, fire alarm or nurse call systems. Device connectors of such systems don't have a load associated with them in AMEP, and as such, the circuit serves the purpose of computing approximate wire length and tag annotations.

PLACE PANELS

In this exercise, we will place Panels for the Commercial Building dataset. In a later exercise we will create Circuits for the Panels.

> Continue in the *03 Power* Construct.

15. On the View tab of the ribbon, on the Appearance panel, select the Electrical Room named view.

16. On the Home tab of the ribbon, on the Build panel, click the Panel button.

17. Click the Style preview image on the Properties palette.

18. Make sure the Drawing file is set to *Panels (US Imperial)*.

19. From the Category list, choose **Surface Door**.

20. Double-click Surface Door 3 (see Figure 7.28).

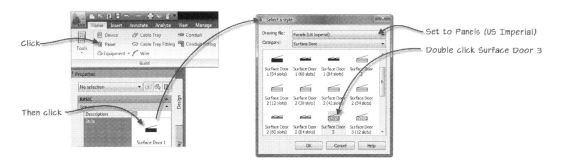

Click→
Then click→

FIGURE 7.28
Use the Properties palette to choose a Panel

21. On the Properties palette, configure the following settings:

- Align to objects: **Yes**
- Elevation: **6'-0"**
- System: **Standard**
- Create circuits: **No**
- Name: **L3**
- Rating: **125**
- Voltage phase-to-neutral: 120
- Voltage phase-to-phase: 208
- Phases: 3
- Wires: 4
- Main type: **Main circuit breaker (MCB)**
- Main size: **125**
- Design capacity: **125**
- Panel type: **ANSI**
- Mounting: **Surface**
- AIC Rating: **10000**

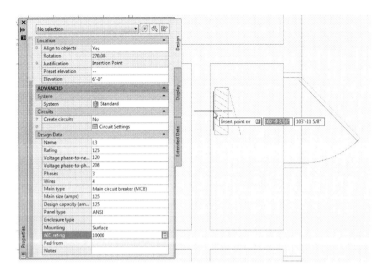

FIGURE 7.29

Configure the properties and insert a Panel

22. Click to place the panel in the electrical room using the nearest snap to place on the wall (see Figure 7.29).

23. Press ENTER to complete the command.

24. Copy the panel 24" plan-south of the first one.

25. Select the new Panel.

As noted above, notice that the Phases and the Panel type can no longer be modified.

26. Make the following changes in the Properties palette (see Figure 7.30):

- Style: Surface Door 1
- Name: **H3**
- Rating: 200
- Voltage phase-to-neutral: 277
- Voltage phase-to-phase: 480
- Main Size: 200
- Design capacity: 200
- AIC Rating: 22000

27. Press ESC to deselect the Panel.

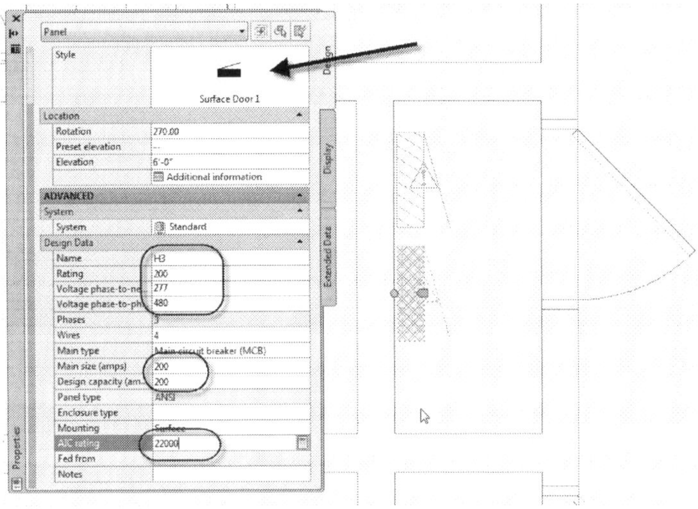

FIGURE 7.30
Copy a Panel and modify the properties

Note: If you have "Create Circuits" set to Yes when placing panels and no Electrical Project Database (EPD) has been specified yet, you will be prompted to create a new EPD file or to open an existing EPD file. You can cancel this prompt and specify the EPD at a later time by opening Electrical Preferences on the Manage tab of the ribbon (or typing the command ElectricalPreferences at the Command Line). Specifying an EPD will be covered in the Electrical Project Database topic later.

CIRCUIT MANAGER

The AMEP Circuit Manager is a project-wide repository of electrical design data. All Panels and Circuits in the project are visible within Circuit Manager. Circuit Manager is used to create, delete, and edit Circuits, and can be used to move Circuits from one Panel to another within the same drawing.

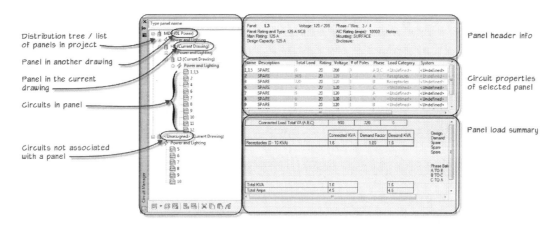

Distribution tree / list of panels in project

Panel in another drawing

Panel in the current drawing

Circuits in panel

Circuits not associated with a panel

Panel header info

Circuit properties of selected panel

Panel load summary

FIGURE 7.31

Main components of Circuit Manager

There are four main components to the Circuit Manager (see Figure 7.31). On the left is the list of Panels in the project in a tree view, similar in structure to an electrical distribution diagram. Panels in the current drawing are indicated with Current Drawing in parentheses next to the Panel name. Panels in other drawings are indicated with the drawing name in parentheses next to the Panel name. Under each Panel is the list of Circuits associated with that Panel. On the right-hand side is information related to the Panel currently selected; on the left side is the distribution tree. The right side is split into three panes. The Panel header pane provides a summary of the Panel properties. These properties are editable on the Properties palette when a Panel is selected in a drawing. The middle pane is a table containing the Circuit properties for the currently selected Panel. This is where you modify circuit properties, such as rating and description. The information is shown "read-only" when a Panel from a non-current drawing is selected. The bottom pane presents a Panel load summary for the selected Panel, including load information from subfed panels. Finally across the bottom of the Circuit Manager is a bank of icons with functions to create circuits, delete circuits, etc.

CREATE CIRCUITS

In this exercise, we will define circuits in the panels created earlier.

1. Select panel named H3 from the previous exercise.

2. On the Panel tab of the ribbon, on the Circuits panel, click the Circuit Manager button.

3. On the left side of Circuit Manager, select H3 (Current Drawing).

4. At the bottom of the Circuit Manager, click the Create Multiple Circuits icon (see Figure 7.32).

FIGURE 7.32

Create Multiple Circuits

5. In the "Create Multiple Circuits" dialog, configure the following settings and then click OK:

   - System Type: **Power and Lighting**
   - System: <Undefined>
   - Number of 3-pole circuits: **4**
   - Number of 2-pole circuits: **4**
   - Number of 1-pole circuits: **22**
   - Description for circuits with breakers: **SPARE**

6. When prompted about the Electrical Project Database, click Cancel (see Figure 7.33). We will specify the EPD in the Electrical Project Database topic below.

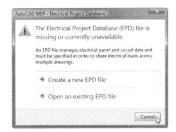

FIGURE 7.33

Ignore the warning about the Electrical Project Database file by clicking Cancel

7. On the left side of Circuit Manager, beneath H3 (Current Drawing), select the Power and Lighting node.

   You should see the list of circuits on the right side of the Circuit Manager. By default, all circuits are created with a 20A rating. The 2 and 3 pole circuits are indicated as 480 volt, and the 1 pole circuits are indicated as 277 volt.

8. Right-click on panel L3 in Circuit Manager and choose Connect To (see Figure 7.34).

9. Select panel H3 (Current Drawing), then click OK.

10. In the "Panel to Panel Connection Method" dialog, select the Circuit breaker in panel option, check the "Panels are connected through a transformer" option, and then click OK.

If prompted to select an EPD, click Cancel.

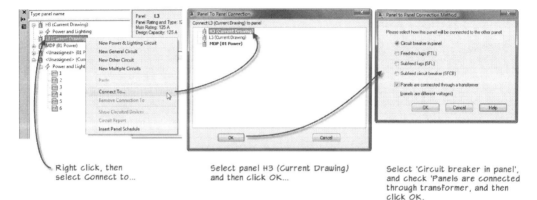

Right click, then
select Connect to...

Select panel H3 (Current Drawing)
and then click OK...

Select 'Circuit breaker in panel',
and check 'Panels are connected
through transformer, and then
click OK.

FIGURE 7.34

Connect the L3 Panel through the H3 Panel

Panel L3 now moves under Panel H3 in the distribution tree.

11. Reselect panel H3 on the left side of Circuit Manager.

12. On right side of Circuit Manager, double-click in the Rating field for circuit 1,3,5 and set the value to **70** (see Figure .7.35).

Subfed panels are nested
under their parent

Circuit description updates to
reflect connection to panel

Double click to set rating

FIGURE .7.35

Edit the Rating of circuits 1, 3, 5 on Panel H3

Note: Circuit manager automatically updates the breaker, subfeeding a Panel with a rating that matches the subfed Panel's rating. In this case, since we are circuiting through a transformer at a different voltage, we need to change this value as follows: 45,000 VA (transformer rating) / (480 * sqrt(3)) = 54.1 A * 1.25 = 67.7A. Round up to the next common breaker size = 70.

13. Select panel L3 in the tree and click the Create Multiple Circuits icon.

14. Configure the following settings and then click OK.

    If prompted to select an EPD, click Cancel.

    - System Type: **Power and Lighting**
    - System: **<Undefined>**
    - Number of 3-pole circuits: **1**
    - Number of 2-pole circuits: **0**
    - Number of 1-pole circuits: **39**
    - Description for circuits with breakers: **SPARE**

    If prompted to select an EPD, click Cancel.

15. On the left side of Circuit Manager, select the Power and Lighting node under Panel L3.

16. On the right side of Circuit Manager, select circuit 31. Hold down the SHIFT key, and then click circuit 42.

    This will select circuits 31 through 42.

17. Double-click in any of the circuits' description fields, type **SPACE**, and then press ENTER.

18. Select circuits 31 through 42 again, double-click in the Rating field for one of the circuits, and set the value to **0**.

When a circuit has a 0 rating, the Circuit cannot be circuited to. This is to help ensure that you don't inadvertently connect to a SPACE in your panel.

19. Close Circuit Manager.

20. Save the drawing.

You can connect any Panel in the current drawing to any Panel in the project, regardless of the drawing in which it exists. You can't specify or modify the connection of a Panel in a drawing that is not current. You can think of it this way: When you "tell" a Panel what it is fed from, you are actually modifying the Panel's properties, and thus, you need to be in the drawing where that Panel exists (just as when you wanted to modify a line in a drawing, you needed to have its drawing current). For example, if you have panel H3 in *03 Power.dwg* and Panel MDP in *01 Power.dwg*, you need to have the *03 Power.dwg* current to connect H3 to MDP; you can't connect H3 to MDP if *01 Power.dwg* is current.

Similarly, you can only make changes to a Circuit's properties, such as its rating or description, when the drawing where the Circuit's Panel exists is current. For example, to modify the breaker size or description of a circuit in MDP, you need to be in *01 Power.dwg*; to modify a circuit in H3, you need to be *in 03 Power.dwg*.

ELECTRICAL PROJECT DATABASE

The Electrical Project Database (EPD) is a special file that manages all the electrical design data from all electrical drawings in the project. Each project will have its own EPD, and in some cases, you may even use a single EPD across multiple projects. For example, multiple remodels within the same building. Typically, however, each EPD is for a single building. If you have a multi-building campus and you want to be able to compute the total connected and demand load at a site main service drop for the entire site, you can certainly choose to use the same file for multiple buildings.

When you first create an EPD file, it is empty. You associate your electrical drawings with the EPD in the drawing's "Electrical Preferences" dialog. The Electrical Preferences are accessible from the Electrical workspace on the Manage ribbon tab (see Figure 7.36).

FIGURE 7.36
Electrical Preferences on the Manage ribbon tab, Preferences panel

To create an EPD, in the "Electrical Preferences" dialog, click the New button. If you already have an EPD defined for the project, you simply select it by browsing for the file using the Open button. In either case, it is recommended that you use the Relative Path option. This provides the greatest flexibility in the event your project needs to move to a new server or folder location (see Figure 7.37).

Note: The "use relative path" option will not be available if the drawing you are working in has not yet been saved; for example, when you create a new drawing from a template. Always make sure to save your drawing before specifying the EPD so the relative path option is available.

FIGURE 7.37
Electrical Project Database tab in Electrical Preferences

Once the EPD is set for a drawing, every time the drawing is opened or saved, data is automatically synchronized with the EPD. When changes are made to the EPD by other users or drawings, a notification icon will appear to alert you to reload the EPD. Clicking on the icon itself will reload the EPD (see Figure 7.38).

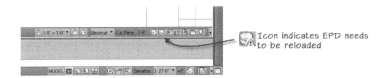

FIGURE 7.38
Electrical Project Database update notification. Note the yellow triangle icon; this indicates the EPD is out of date.

CREATE AN ELECTRICAL PROJECT DATABASE

Continue in the *03 Power* Construct

1. On the Manage tab of the ribbon, on the Preferences panel, click the Electrical button.

2. In the "Electrical Preferences" dialog, click the Electrical Project Database tab.

3. Click the New button.

4. Browse to the *C:\MasterAME 2012\MAMEP Commercial* folder.

5. For the File name, type **MAMEP Commercial**, and then click Save (see Figure 7.39).

FIGURE 7.39
Specify EPD name and save the file

This will return you to the "Electrical Preferences" dialog.

6. Click the Use Relative Path option, and then click OK (see Figure 7.40).

FIGURE 7.40
Choose the Use Relative Path option

7. Save the drawing.

The actual location and name of the EPD is not important. It makes sense to store it with the rest of your project files. In this case, we placed the EPD file in the same folder as the Project Navigator APJ file. If the EPD file is ever erased, you will receive a message when AMEP tries to synchronize to the file such as the one shown in Figure 7.41.

FIGURE 7.41

EPD can't be found when trying to synchronize

When you open a drawing associated with an EPD that no longer exists, you will receive a message such as the one shown in Figure 7.42.

FIGURE 7.42

EPD not found when drawing opened

In either case, you can simply create a new EPD. Just make sure to reopen and resave all the drawings associated with the EPD to ensure all the electrical data is accounted for.

ADD A MAIN PANEL

In this topic, we are going to add a main distribution panel (MDP) for the building service entrance. We will then connect the panels on the third floor to this new panel.

8. On the Project Navigator palette (on the Constructs tab) in the *Electrical\Power* folder, double-click to open the *01 Power* Construct file.

9. On the Manage ribbon tab, on the Preferences panel, click the Electrical button.

10. On the "Electrical Project Database" tab, click the Open button.

11. Select the *MAMEP Commercial.epd* file, and then click Open.

12. Click the Use Relative Path option, and then click OK.

13. On the Home tab, on the Build panel, click the Panel button.

14. In the Properties palette, click the Style image.

    Make sure that the Drawing file is Panels (US Imperial).

15. Double-click Surface Door 1.

16. Set the following in the Properties palette:

    - Align to objects: **Yes**
    - Elevation: **6'**
    - System: **Standard**
    - Create circuits: **Yes**
    - Name: **MDP**
    - Rating: **800**
    - Voltage phase-to-neutral: 277
    - Voltage phase-to-phase: 480
    - Phases: 3
    - Wires: 4
    - Main type: **Main circuit breaker (MCB)**
    - Main size: **800**
    - Design capacity: **800**
    - Panel type: ANSI
    - Mounting: **Surface**
    - AIC Rating: **65000**

17. Click the Circuit Settings worksheet icon (see Figure 7.43).

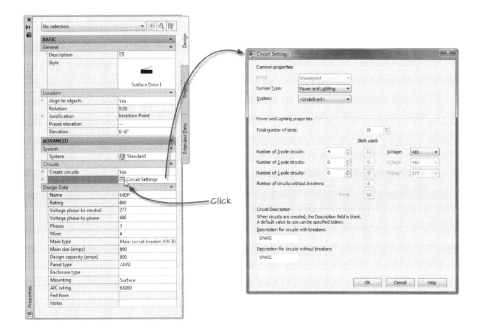

FIGURE 7.43
Configure the Circuit settings while placing the Panel

18. In the "Circuit Settings" dialog, configure the following settings, and then click OK.

- System Type: **Power and Lighting**
- System: **<Undefined>**
- Total number of Slots: **18**
- Number of 3-pole circuits: **4**
- Voltage: **480**
- Number of 2-pole circuits: **0**
- Number of 1-pole circuits: **0**
- Description for circuits with breakers: **SPARE**
- Description for circuits without breakers: **SPACE**

19. Click to place the panel outside the back door/stairwell on the exterior wall (see Figure 7.44).

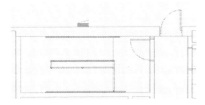

FIGURE 7.44
Panel placed outside of the building stairway

20. Press ENTER to exit the Panel command.

21. Select panel MDP and, on the Panel tab of the ribbon, click the Circuit Manager button.

22. On the right side of Circuit Manager, rename the circuits as follows by double-clicking in the Name column, typing the new value, and then pressing ENTER:

- Rename 1,3,5 to **1**
- Rename 2,4,6 to **2**
- Rename 7,9,11 to **3**
- Rename 8,10,12 to **4**

23. Close Circuit Manager.

24. Close the *01 Power* drawing. When prompted to save, click Yes.

CONNECT PANELS BETWEEN DRAWINGS

The *03 Power* Construct should still be open. If you had closed it above, double-click on the Project Navigator now to reopen it.

25. On the Drawing Status Bar, click the Electrical Power Database icon (see Figure 7.45).

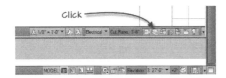

FIGURE 7.45
EPD reload required notification

26. On the Analyze tab, on the Electrical panel, click the Circuit Manager button.

27. Select H3 (Current Drawing) in the distribution tree.

28. Drag H3 and drop it onto MDP (see Figure 7.46).

FIGURE 7.46

Drag and drop Panels to interconnect them

29. In the "Panel to Panel Connection Method" dialog, select the Circuit breaker in panel option, and then click OK.

In this case, since panel MDP and panel H3 are of the same voltage, it is not necessary to connect through a transformer.

The distribution tree indicates that MDP (in drawing *01 Power.dwg*) subfeeds H3, and H3 subfeeds L3. In the Circuiting Devices section below, we will circuit receptacles to panel L3. As a result of the Panel interconnections, all loads from L3 feed into H3 and up to MDP. This can be verified by viewing the circuits on panel MDP. For example, if Panel L3 had a total load of 1620, this will feed through panel H3, and ultimately feed into the Total Load summary on Panel MDP (see Figure 7.47).

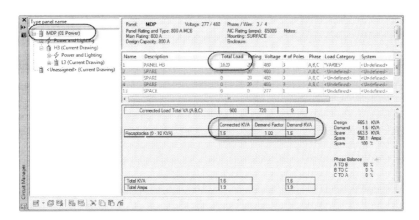

FIGURE 7.47

Panel interconnection relationship in Circuit Manager distribution tree, and total load

30. Close Circuit Manager.

31. Save the drawing.

ELECTRICAL EQUIPMENT

AMEP includes a Multi-View Part (MvPart) catalog filled with various types of electrical equipment such as transformers, generators, transfer switches, and switchboards. These components are all for graphical and physical representation only, they contain no capabilities to interconnect or circuit the components. In an earlier set of exercises, you were able to connect panels of different voltages without actually having a transformer object in the project—the transformer in this case was figurative. Only panels and devices have any circuit connectivity intelligence. If you want to have a larger Panel object to represent a switchboard section or other such distribution gear, you can define a panel style to have such a physical representation. In Chapter 9, we will create a large distribution board to replace the panel style used for the MDP (Main Distribution Panel) in this project.

On the Home tab, on the Build panel is an Equipment drop-down button. The various tools on this drop-down filter the Electrical MvPart catalog for certain classifications of MvParts such as Generators, Junction boxes, Switchboards, etc. Selecting the Equipment button itself opens the catalog with no filtering (see Figure 7.48).

FIGURE 7.48

Insert Multi-View Part Equipment from the Home ribbon, Build panel

In this topic, we will explore the electrical MvPart catalog and insert a transformer into our project.

INSERT ELECTRICAL MVPART TRANSFORMER COMPONENT

Continue in the *03 Power* Construct.

1. On the Home tab, on the Build panel, click the Equipment button.

    The "Add Multi-view Parts" dialog opens.

2. Verify that the Parts tab is current and then browse to (expand): All Installed MvParts (US Imperial) > Electrical > Power Transformers.

3. Select Dry Type Transformer US Imperial.

4. From the Part Size Name list, choose 45 kVa Dry Type Transformer.

5. Set the Elevation to 7' (the transformer will be suspended above the floor) (see Figure 7.49).

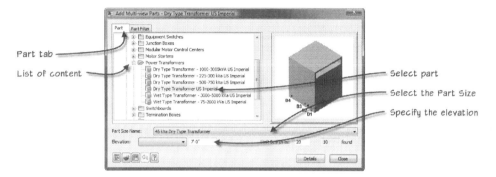

FIGURE 7.49

Multi-View Part selection for 45kVA Dry Type Transformer

Do not close the "Add Multi-view Parts" dialog.

6. Pick in the drawing window to activate it.

7. Pick in the electrical room to place the transformer and then press ENTER to accept 0 rotation (see Figure 7.50).

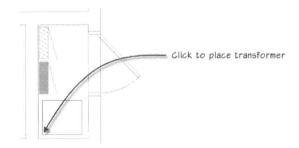

FIGURE 7.50

Placement of transformer in electrical room

8. Click Close to close the "Add Multi-view Part" dialog.

9. Save the drawing.

By default, the "Add Multi-view Parts" dialog remains active when you move your cursor into the drawing window, requiring you to pick in the drawing window before picking to place the Multi-view Part. In Windows XP, there is a pin icon on the window that may be deselected to allow the Add Multi-view Part window to automatically roll up and activate the drawing, resulting in one less click. In Windows Vista and Windows 7, the same may be enabled by clicking the window icon and selecting Enable Rollup (see Figure 7.51).

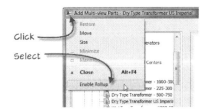

FIGURE 7.51
Enable Rollup for auto-hide of the Multi-View Part window

CIRCUITING DEVICES

A Device may be circuited to a Panel in any drawing associated with the Electrical Project Database. Circuiting is accomplished by assigning a Circuit to the connector of a Device (Devices may have more than one connector). Circuiting may also be accomplished by using Wires. The following exercises will demonstrate these methods.

CIRCUIT DEVICES

This exercise demonstrates how to associate a Circuit with a connector.

Continue in the *03 Power* Construct file.

1. On the View tab, on the Appearance panel, select northwest Offices.

2. Select the four receptacles in the top/left office and the receptacle in the corridor.

3. On the Device tab, click the Circuit Properties button.

4. In the "Electrical Properties" dialog, from the "Show circuits from panel" list, select panel L3.

Since we assigned the connector on the receptacle Device to 120v/1p when we edited the style, only circuits matching these criteria are listed in the Circuit list. The 3-pole circuit (2,4,6), is not shown. Also, circuits 31 through 42, with 0 rating intended as spaces, are not listed.

5. From the Circuit list, select 2.

    After you select a circuit, the load updates to reflect the total load on the circuit. In this case, we have five receptacles @ 180 VA each = 900 VA (see Figure 7.52).

In this case, several of the connector properties are disabled from edit in the Device instance "Electrical Properties" dialog (see the left side of Figure 7.52). Load Phase 2 and 3 are disabled because this is a single pole connection. The Load, Voltage, Number of Poles, and Load Category of the Device style all had Prevent Override set to Yes in an earlier exercise (see the right side of Figure 7.52).

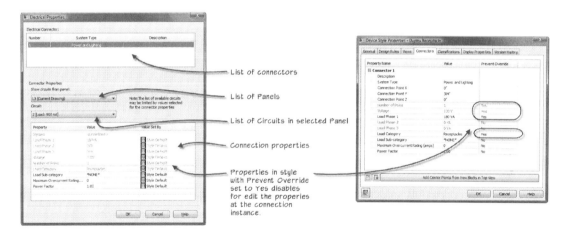

FIGURE 7.52

Relationship between Device instance connector properties and its Style

6. Click OK to close the "Circuit Properties" dialog, and then press the ESC key to deselect the devices.

7. Repeat the process above to circuit the four receptacles in the next office "south" to circuit L3-4.

    The total load on Circuit 4 should update to reflect 720VA.

In the next topic, we will draw wires and show how circuits may be defined as the wires are being drawn.

WIRES

Wires are used to represent circuiting between Devices, as well as for home runs where the wire is shown with an arrowhead. The Wire tool is on the Home tab of the ribbon, on the Build panel (see Figure 7.53).

FIGURE 7.53
Wire tool on the Home ribbon, Build panel

When you click the Wire tool, the Properties palette will appear if it is not already visible. The General properties of the Wire include the Style and Description. The Style defines how the tick marks and homerun are configured. Refer to Chapter 9 for more information on defining wire styles. Additionally, each wire instance can have a description, though this is not commonly used (see Figure 7.54).

FIGURE 7.54
Wire General properties

LOCATION GROUP SETTINGS

The Location settings define how the wire will be drawn (see Figure 7.55).

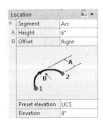

FIGURE 7.55
Wire Location properties

The Segment setting defines the general shape of the wire graphics such as line, arc, etc. After a wire is drawn, the Segment type can't be changed; to do so, one would have to erase and redraw the wire.

Note: Erasing Wires has no detrimental effect on the circuited components. The circuiting information is defined within the Devices.

Depending on the segment type, there are additional geometric parameters that define the wire geometry. For most segment types, there is Height, which defines how far off axis the wire is drawn. The Offset defines on which side of the axis the wire is drawn and can be toggled at the command line. In the case of the Polyline, there is a radius parameter that defines the radius of the arc segment where the polyline changes direction. The Height and Radius values are in model units and not annotation units.

FIGURE 7.56
Wire segment types — line, arc, snake, polyline, chamfer, and spline

The Arc segment type is probably the easiest to use for devices placed on walls, and the Chamfer segment type works well for arrays of lighting fixtures in a grid.

Wires are drawn as 2D geometry only; however, the Elevation property effects AMEP's estimate of the circuit length. As with other Elevation settings in AMEP, the Preset elevations may be used in lieu of typing in an elevation value.

ADVANCED SETTINGS

Wires are placed on the layer defined by the WIRE layer key. If the System has a layer key defined, it will supersede the WIRE key (see Figure 7.57). Additionally, if the System has layer key overrides defined, these will affect the associated WIRE key or the Systems key.

FIGURE 7.57
Wire System setting

Note: The Wire's System is in no way related to the System of a Device, unless you happen to manually set them to the same thing, nor are they related to the System of the Device's connector(s). Systems on Wires are purely for layering control.

Design Data—The Design Data grouping of the Properties palette provides information about the circuit that is being used while drawing wires. The Show circuits from panel property is a drop-down list of the panels available in the project. If you don't want to assign a panel/circuit as you draw wires, set the value to <Unassigned>. When a panel is selected, the list of available circuits is shown in the Circuit list. If you specify a Panel and Circuit, as you snap to the electrical connectors on devices, the selected Circuit will be assigned to the connector. If the connector on the Device is already selected, the panel and circuit will automatically populate into the design data fields. As you are routing wires, you can change the Panel or Circuit to make a multi-circuit wire and home run. After the panel and circuit are set in the design data fields, the next time you start a wire, the circuit will auto-increment to the next empty circuit (see Figure 7.58).

Note: Auto-incrementing is generally in numerical order, and not phase order (i.e., 1, 2, 3, 4, 5, 6, etc, not 1, 3, 5, 2, 4, 6, etc.)

FIGURE 7.58

Wire Design Data properties

Additionally, the Design Data grouping displays what Circuits are represented by the Wire.

Each unwired connector on a device will display a solution tip if Solution Tips are turned on.

TIP: You can toggle the visibility of the Solution Tips on the View tab of the ribbon, on the MEP View panel (see the right side of Figure 7.59).

When a device is selected, the wire grip (+) will start the wire command using the previously configured wire geometry settings (see Figure 7.59). When a Wire is connected to a Device, the solution tip disappears. Whether a Device connector is circuited or not the solution tips will appear if no Wire is connected.

FIGURE 7.59

Device solution tips and the add wire grip

MANUALLY ROUTE WIRES

8. Zoom in on the top left office.

9. Make sure Object Snaps are turned on, and the Electrical Connector (ECON) OSNAP is active.

10. Select the "north" receptacle in the top/left office.

11. Click the plus (+) grip.

On the Properties palette, "Show circuits from panel" should indicate L3, and Circuit should indicate 2 with a load of 900 VA (see Figure 7.60).

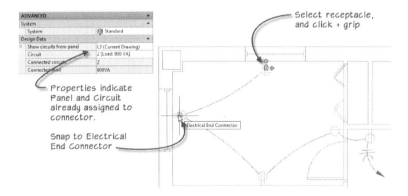

FIGURE 7.60

Manual Wire routing in the northwest corner office

12. On the Properties palette, set the following parameters:

- Style: **1H 1N 1G**
- Segment: **Arc**
- Height: **6"**
- Offset: **Left**
- Elevation: **0**
- System: **Standard**

13. Wire to the "west" receptacle, then the "south" receptacle, then the "east" receptacle.

14. On the Properties palette, change the Offset from Left to Right.

    If you prefer, you can type o at the command line to make this change.

15. Wire to the receptacle in the corridor.

16. Press the SPACEBAR to place the home run.

17. Click to place the home run arrow; make the home run long enough to show the tick marks.

> Note: This Wire style has ticks only on the home run segment.

18. Select the home run to view the available grips (see Figure 7.61).

FIGURE 7.61

Grips on a home run wire segment

There are several grips on the home run. The square grips on the wire modify the location of the segment. The square grip next to the homerun annotation modifies the location of the tick marks along the wire segment. The arrow grips flip the tick marks.

AUTOMATICALLY GENERATE WIRES

An alternative to manually drawing wires is to generate the wires for a specified circuit.

19. On the Home tab, on the Build panel, click the Wire tool.

20. On the Properties palette, set the following properties:

21. Wire Style: **1H 1N 1G**

22. Segment: **Arc**

23. Height **6"**

24. System: **Standard**

25. Show circuits from panel: **L3**

26. Circuit: **4**

27. Right-click in the drawing area, and choose **Generate** from the pop-up menu.

28. Right-click again, and choose **Cancel** to end the Wire command.

29. Select the home run. Use the square grips to edit the home run to a more reasonable configuration.

> TIP: When gripping editing wires, make sure not to snap to anything that is at a non-0 Z value, otherwise the graphics may be foreshortened (see Figure .7.62). To avoid this when editing wires, turn on "Replace Z value with current elevation" and set the elevation to 0.

 Wire graphics are foreshortened because it is not flat.

FIGURE .7.62
Foreshortened wire graphics

MULTI-CIRCUIT HOME RUN

In this exercise, we will modify the wiring to create a multi-circuit home run.

30. Erase the first home run drawn above (see item 1 in Figure 7.63).

31. Select the indicated Device on the corridor wall of the lower office, and then click the plus (+) grip (see item 2 in Figure 7.63).

32. Snap to the Electrical Curve Connector (ECON) on the Device in the corridor (see item 3 in Figure 7.63).

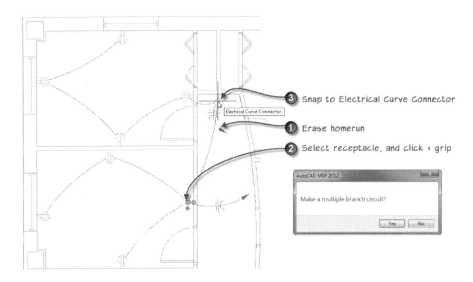

③ Snap to Electrical Curve Connector

① Erase homerun

② Select receptacle, and click + grip

FIGURE 7.63
Create a multi-circuit home run

33. When prompted to make the multiple branch circuit click Yes.

34. Right-click and choose Cancel to end the Wire command.

MODIFY WIRE STYLE

In the previous exercise, the result of the multiple branch circuit is not quite as desired. Note that the tick marks still indicate 1H 1N 1G due to the associated Wire Style. Additionally, you may like to see two arrowheads instead of one, since there are now two circuits represented by the home run.

35. On the Manage tab, on the Style & Display panel, click the Style Manager drop-down button, and then choose Wire Styles (see Figure 7.64).

FIGURE 7.64
Wire styles on the Manage ribbon, Style & Display panel

36. Select 1H 1N 1G style. Click the Copy icon, and then click the Paste icon.

    This will create a new style named 1H 1N 1G (2).

37. Select 1H 1N 1G (2), and on the General tab, rename to 2H 1N 1G.

38. On the Specifications tab, under Hot, specify Number of: 2 (see Figure 7.65).

FIGURE 7.65

Specify 2 for the number of Hot conductors

39. On the Annotation tab, in the Home Run Arrow settings check the Display one arrow for each circuit checkbox (see Figure 7.66).

FIGURE 7.66

Show one arrow for each circuit

40. Click OK to close Style Manager.

41. Select the home run, and on the Properties palette, change the Style to: 2H 1N 1G (see Figure 7.67).

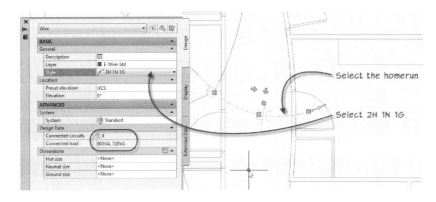

FIGURE 7.67
Select 2H 1N 1G from the list of available styles on the Properties palette

After the new style is selected, the tick marks and home run arrows update to reflect the new style. When multiple Circuits are connected through the Wire, the Connected circuits in the Properties will display them separated with a comma. The Connected load in the Properties also displays the load on the Circuits represented by the Wire. As with the Circuits, loads on multi-circuit Wires will be comma separated. As shown in Figure 7.67, the Design Data of the selected home run indicates that it connects circuits 2 and 4, with loads of 900VA and 720VA, respectively.

CIRCUITING WHILE DRAWING WIRES

In this topic, we will assign Circuits to the connectors on Devices as we draw Wires. The result will be as shown in Figure .7.68.

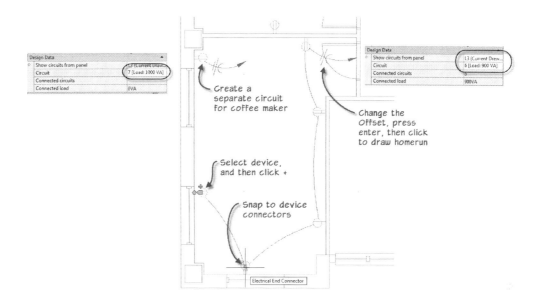

FIGURE .7.68

Assigning circuits as wires are drawn

1. In Object Snap Settings, disable all General snaps, and enable all AMEP snaps. The General snaps can interfere with accurately snapping to the electrical connectors on the Devices (see Figure .7.69).

FIGURE .7.69

Disable General snaps, and enable AutoCAD MEP snaps

2. In the southwest corner office, select the receptacle on the bottom/left, and then click the + (add Wire) grip.

3. On the Properties palette, set the following properties:

4. Wire Style: 1H 1N 1G

5. Segment: Arc

6. Height **6"**

7. Offset: Left

8. System: Standard

9. Show circuits from panel: L3

10. Circuit: 6

11. Going counter clockwise around the room, snap the Wire to the next 4 receptacles.

12. Set the offset to Right, and then press ENTER.

13. Click to place the home run.

14. Select the coffee maker receptacle, and then click the + grip.

   Notice that the Circuit incremented to 7, the next available free circuit, and the circuit load indicates 1000VA.

15. Press ENTER to set the wire to home run.

16. Click to place the home run.

WIRE DIMENSIONS

The Dimensions grouping on the Properties palette allows you to specify a size for each of the conductors in the Wire, with the assumption that all conductors of the same type (hot, neutral, ground) are the same size. The little calculator button is used to calculate the Wire sizes based on the estimated length, connected load, and certain assumptions about voltage drop (see Figure 7.70). See the next topic for a description of how the length is computed.

FIGURE 7.70
Dimension (Wire Size) properties

The sizes defined here also appear in the Wires column in Circuit Manager and are available to schedules using automatic properties that apply to Wires, as well as in Panel Schedules. The limitation with sizing wires is that if a Device is circuited to a Panel in another drawing, the length cannot be determined, and thus the Wire cannot be automatically sized.

Prior to sizing a wire, the Wire style must have some settings defined. If you click to calculate the Wire size without these settings in place, you will get a message like that shown in Figure 7.71.

FIGURE 7.71
Wire size calculation warning

MODIFY A WIRE STYLE TO ENABLE SIZING

1. Select the home run Wire.

2. On the Wire tab, click the Edit Style button.

On the Specifications tab, note that the Material for all the wire components (Hot, Neutral, Ground, and Isolated Ground) are set to <None>. Wire sizing is dependent on the material and temperature rating of the insulation; this is indicated by the large yellow warning icons. Note also that for this Wire Style, the Number of Isolated Ground conductors is set to 0. Thus, when the wire is selected in the Properties palette, there is no size listed for Isolated Ground.

3. On the Specifications tab, in the Hot section, set the following and then click OK.

- Material: **Copper**
- Temp. Rating: **60**

4. With the home run Wire still selected, on the Properties palette, click the Calculate sizes for the wire icon.

   Another message box appears, this time indicating that Neutral and Ground cannot be sized.

5. Click OK to dismiss the warning.

   The Hot was sized and the value (12 gauge in this case) appears on the Properties palette.

If you wish, you may further modify the Wire Style to specify the properties for the Neutral and Ground conductors so those may be sized as well.

6. Save 03 Power.

CIRCUIT LENGTH CALCULATION

AMEP calculates Circuit lengths when the panel and all connected devices are in the same drawing. The computation considers the connector location within the Device, and if the Panel has a connector, it takes that into account as well.

Panels and Devices are expected to have an insertion Z value of 0 in all cases; AMEP expects you to use the Elevation property to elevate panels and devices off the Z=0 plane. In Figure 7.72, even though the Panel indicates an elevation of 6'-0", the Z value of Panel's insertion point is 0. The Location window is opened by clicking the Additional information worksheet icon.

FIGURE 7.72

Device Elevation and Insertion Point properties

Since the Z coordinates are expected to be 0, AMEP ignores the Z coordinate of Devices and Panels when it calculates the circuit length; it does, however, consider the elevation value. The circuit length calculation ignores the Z component of the connector locations defined in the Device and Panel styles.

> Note: You should not use the MOVE command to move a device or panel away from the Z=0 plane; use the Elevation property instead. When objects or their connectors are at non-Z=0 coordinate, wires may appear strangely.

The basic calculation for the length of a circuit between a panel and a single device is the difference between the coordinate locations of the connectors. When multiple devices are connected to a circuit, the calculation sums the distance between the Panel and the closest Device, plus the distance to the next closest Device, plus the distance to the next closest Device, etc.

As an example, Device 1 (D1) and Device 2 (D2) are circuited to the same circuit in Panel (P). The locations (x, y, z coordinates) and their elevations (e) are tabulated as the Object Location in Figure 7.73. The Connector Positions in their respective styles are also tabulated. The Connection Points of each are based on the object location plus the connector position. Keeping in mind that the Z values in the object location and the connector position are ignored, only the elevation value is utilized to determine the connection point Z. The orthogonal distance between each object's x, y, and z components are computed. For example, the distance between P and D2 is figured as follows:

	Object Location				Connector Position			Connection Point			Ortho Distance		
	x	y	z	e	x	y	z	x	y	z	P	D1	D2
P	12	4	3	6	0	2	1	12	6	6	-	12	8
D1	15	10	9	6	0	1	10	15	11	2	-	-	18
D2	8	5	6	9	0	2	5	8	7	9	-	-	-

FIGURE 7.73
Device Insertion Point (Object Location) and connector coordinate example table for computing Wire length

The arrows indicate the summation. For example, in the P row, the x component is summed as 12 + 0 = 12. The x component of the Connection Point of D2 is computed as 8 + 0 = 8. The difference of the two is computed as, 12 - 8 = 4. This is simplified as follows:

```
x component: ABS((12+0) - (8+0)) = ABS(12 - 8) = 4
```

The y and z components are computed similarly. Note that the resulting y and z values would be negative if the absolute value was not considered (see Figure 7.74).

	Object Location				Connector Position			Connection Point			Ortho Distance		
	x	y	z	e	x	y	z	x	y	z	P	D1	D2
P	12	4	3	6	0	2	1	12	6	6	-	12	8
D1	15	10	9	2	0	1	10	15	11	2	-	-	18
D2	8	5	6	9	0	2	5	8	7	9	-	-	-

4   -1   -3

FIGURE 7.74

Example calculation for distance between x, y, and z coordinates of a Panel (P) and Device (D2)

```
y component: ABS((4+2) - (5+2)) = ABS(6 - 7) = 1
z component: ABS(6 - 9) = ABS (6 - 9) = 3
```

The total orthogonal distance is found by summing the three components:

```
orthogonal xyz distance = 4 + 1 + 3 = 8
```

Since D2 is closer to the panel than D1, this is the first segment of the circuit (distance of 8). The second segment is between D1 and D2 (distance of 18). The total circuit length is 8 + 18 = 26. In the case that the panel has no connector, the "Connector Position" values are all assumed to be 0, and thus the connection point is equivalent to the Object Location coordinates.

Of course, this computation is only an estimate. In some scenarios the installed length may actually be shorter if the wiring is installed more direct from point to point. In some scenarios the installed length may actually be longer as it does not consider any complexities in building geometry such as bends, columns, beams, chases, elevator shafts, stair towers, etc. The estimated length is simply there to give you an approximate overall length and help identify any circuits that are exceedingly long.

ELECTRICAL DISTRIBUTION BASICS

AMEP's support for panel to panel connections provides the ability to compute a total connected and demand electrical load for an entire project. Thus, all loads in the entire project, whether consisting of one main panel and one branch panel, or hundreds of panels consisting of multiple distribution panels, and branch circuit panels, may be connected using the AMEP

EPD functionality. This enables the electrical designer to account for every load in the project whether it is a single-story building or a high-rise tower.

Electrical Load Categories are used to compute electrical loads using some common demand/diversity rules. This functionality provides solutions to some load calculations that are difficult to manage, even in a spreadsheet. For example, keeping track of the largest motor load, or applying a different demand factor depending on the quantity of connected loads such as kitchen equipment (NEC 2008 220.56) or elevators (NEC 2008 620.14), can be cumbersome when spanning multiple panels using traditional spreadsheet methods.

Many electrical designers use spreadsheets to tabulate the connected and demand load. Using AMEP's functionality eliminates the error-prone process of transcribing loads. In a traditional spreadsheet/CAD workflow, if a load is added to a circuit on plan, this has to be accounted for on the spreadsheet. Likewise, if a load is removed from plan, it must be removed from the spreadsheet. If a light fixture type that requires two lamps changes to three, this must be changed on every circuit in the spreadsheet, whereas with AMEP this only needs to be changed in one place, the Device style, and ripples throughout the entire distribution system.

In this topic, we will explore the different types of demand factors that may be applied to Load Categories in AMEP.

DEMAND FACTOR TYPES

The load categories and demand factors are managed in Style Manager, under *Electrical Objects > Load Category Definitions*. The AMEP drawing templates come with a variety of load categories, each with preset demand factors. These can be modified to suit your specific needs.

There are four types of demand factors: Constant demand (i.e., lighting); Motors and largest motor (used to sub-categorize loads as motors); Varies depending on load total (receptacles); and Varies depending on quantity of objects (i.e., elevators). The functionality of the demand factors will be described using some common design scenarios so you can compare how this works to your traditional manual or spreadsheet workflows (see Figure 7.75).

FIGURE 7.75

Demand Factor types in a Load Category Definition in Style Manager

CONSTANT VALUE DEMAND FACTORS

Let's begin with constant value demand factors. As an example load scenario, let's consider lighting loads. Lighting loads may be considered continuous loads, and, as such, need to be rated at 125% of the actual load. The default Lighting Load Category Definition is defined in Style Manager as a constant value, with a 1.25 Demand Factor (Figure 7.76).

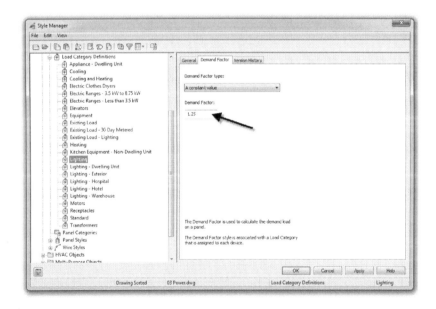

FIGURE 7.76

Lighting has a Demand Factor of 1.25 applied in its Load Category Definition

An example of this is depicted in Figure 7.77. Panel LP1 has a Connected load (C:) of 10kVA. Panel LP2 has 6kVA of lighting load. The total connected load that panel DP1 sees is 16kVA. Applying the 1.25 Demand Factor results in a total demand of 20kVA at panel DP1. You can verify the logic for the other panels.

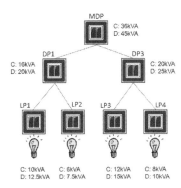

FIGURE 7.77

Connected and Demand lighting load in a sample distribution

VARIES DEPENDING ON TOTAL LOAD DEMAND FACTORS

The next type of Demand Factor is "Varies depending on load total." A common example of this is receptacle loads. In general, the effective total load in an electrical system is reduced to account for the fact that it is very unlikely that all receptacles in a building have something plugged in and turned on at the same time. This assumption permits the rating, which impacts the size and cost, of electrical equipment and feeders to be reduced.

For receptacle loads in the US, commonly the Demand Load is calculated according to these simple rules:

```
100% for the first 10,000VA
50% for load above 10,000VA
```

For example, if a panel has 50,000VA of receptacle load, this would result in the calculation:

```
10,000 VA+ 0.5*(50,000VA-10,000VA) = 30,000 VA.
```

A general simplified equation that you may find in a spreadsheet might state:

```
=IF(RECEPTACLELOAD>10000, RECEPTACLELOAD/2+5000,
    RECEPTACLELOAD).
```

In AMEP, this is implemented as a simple table in the default Receptacles Load Category Definition as shown in Figure 7.78.

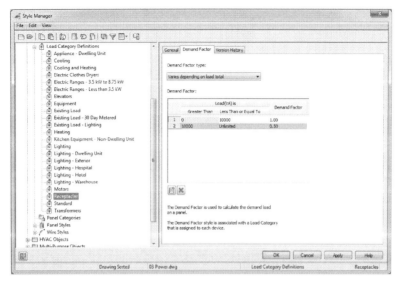

FIGURE 7.78

Default Receptacles demand factor definition

Let's look at an example employing a "Varies depending on load total" Demand Factor. In the distribution diagram in Figure 7.79, the C: indicates the total Connected load at the panel. For example, if each receptacle accounts for 200VA, panel LP1 would have 30 receptacles connected (200VA * 30 = 6kVA).

The D: represents the demand load, or the factored load, at each panel. For example, DP1 "sees" 6kVA from LP1 + 12kVA from LP2 for a total Connected load of 18kVA. Applying our Demand Load rule above, results in the following computation: 10000 + 0.5*(18000-10000) = 14kVA. You can follow the logic for the other panels.

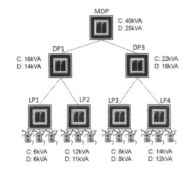

FIGURE 7.79

Example computation of receptacle demand factoring

VARIES DEPENDING ON QUANTITY OF OBJECTS

The next Demand Factor type is "Varies depending on quantity of objects." This is commonly used for elevators and non-dwelling kitchen equipment. This demand factor type is also implemented using a table as shown in Figure 7.80.

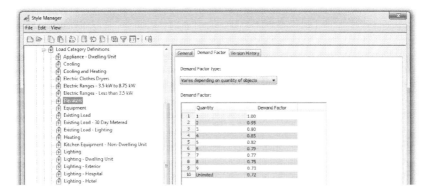

FIGURE 7.80

Demand factor table for elevators load type

As an example let's consider the distribution shown in Figure 7.81. Panel DP1 has three elevators connected to it, resulting in a demand factor of 0.90. Assuming that each elevator requires 10kVA, the connected load would be 30kVA, but the demand would be 0.90*(30kVA) = 27kVA. MDP *sees* seven elevators, and thus receives a 0.77 factor. Again, with elevators at 10kVA each, 0.77*(70kVA) = 53.9kVA.

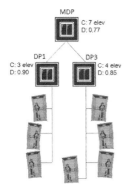

FIGURE 7.81

Example determination of demand factor for elevator loads.

MOTORS AND LARGEST MOTOR

According to the National Electrical Code (NEC), when sizing feeders, the largest motor has to be accounted for to incorporate additional demand related in the inrush current when the motor starts. The Motors and Largest Motor demand factor type facilitates identifying the largest motor value. As shown in Figure 7.82, in addition to the Load Category (cooling, fans, etc.) there is a property on a Device connector called Load Sub-category to identify if the connection is to a motor load.

Property Name	Value	Prevent Override
⊟ **Connector 1**		
Description		
System Type	Power and Lighting	
Connection Point X	0"	
Connection Point Y	11/16"	
Connection Point Z	0"	
Number of Poles	<Undefined>	No
Voltage	<Undefined>	No
Load Phase 1	0 VA	No
Load Phase 2	0 VA	No
Load Phase 3	0 VA	No
Load Category	Cooling	No
Load Sub-category	Motors	No
Maximum Overcurrent Rating (amps)	0	No
Power Factor	0.00	No

FIGURE 7.82

Load Category and Load Sub-category properties on a Device connector

The Load Sub-category is either specified as *None*, or any one of the Load Category Definitions that are defined as a "Motors and largest motor" type Demand Factor. The "Motors and largest motor" demand factor has two Demand Factors associated with it—one for the Largest Motor, and one for all other Motors. The default Motor load category applies a Demand Factor of 1.0, and the Largest Motor Demand Factor as 1.25 (see Figure 7.83).

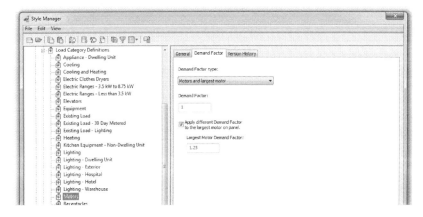

FIGURE 7.83

Default Motor demand factor definition

This functionality allows the electrical designer to sub-categorize a load as a motor load. This facilitates finding the largest motor at each Panel and applying an additional demand load to the load at the Panel's feeder. Say for example we have three loads classified as fans connected to panel LP1, and sub-categorized as motors. If the fans are 1000VA, 900VA, and 800VA respectively, the largest motor is 1000VA. Thus the total demand load for the fans on the Panel would be 2950VA (1000 + 900 + 800 + 0.25 * 1000).

OTHER ELECTRICAL SETTINGS

There are a few other electrical settings to be aware of in Electrical Preferences. These settings have an effect on how Circuits are named when they are created, how Devices may be circuited, and provides some feedback on overloaded circuits.

VOLTAGE DEFINITIONS

In the "Electrical Preferences" dialog, the Voltage Definitions tab has settings to control if a Device of a particular voltage and number of poles can connect to a circuit with the same number of poles. For example, you may have specified your 3-pole circuits to be 480v, however, the 3-phase motor Device you are connecting may be specified as 460v. Since the 3-pole 480v definition is defined with the range of 460 to 500 volts, the 460v motor will be allowed to connect.

You can add and remove voltage definitions to define the available voltages for 1-pole, 2-pole, and 3-pole circuits (see Figure 7.84).

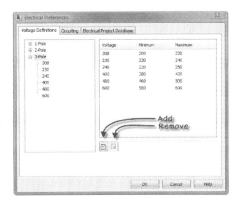

FIGURE 7.84

Voltage definitions in Electrical Preferences

CIRCUITING

The Circuiting tab settings effects how circuits are created, named, and sized (see Figure 7.85).

Require Unique Circuit Names per Drawing—Is not commonly used as this would preclude a circuit "1" from existing in two separate panels. This setting only has an effect when creating new Circuits, or renaming existing Circuits.

Prefix—Adds a prefix to the circuit name with the panel name or abbreviation as specified. However, typically "None" is used. Later in Chapter 15, when creating circuit Tags we will see how we can concatenate the Panel name with the circuit number. For example, if the panel name is L3, and the circuit number is 2, the Tag can be defined to display L3-1.

Numbering—These settings have an effect when creating Circuits. For example, if creating four 3-pole circuits, the names will be as follows using the different options:

> Use Sequential Numbers: 1, 2, 3, 4, etc.

> Group using Number of Poles: 1,3,5; 2,4,6; 7,9,11; 8,10,12; etc.

> Increase by Number of Poles: 1, 4, 7, 10, etc.

Typically, Group using Number of Poles is used. If the circuits are for a distribution Panel, they can always be renamed in Circuit Manager similar to the use sequential numbers option, instead of changing this setting prior to creating the Circuits. We saw an example of this in the "Add a Main Panel" topic above.

The start number simply specifies the starting point of circuit numbering. For example, if creating circuits numbered 43-84, changing this to 43 first simplifies the process.

FIGURE 7.85

Circuiting settings in Electrical Preferences

The Overload options provide warnings as indicated. The "Check Overload when circuiting" option provides a warning if the circuit is overloaded as you are circuiting (see Figure 7.86).

FIGURE 7.86

Overloaded circuit warning while circuiting.

The "Display Overload in Circuit Manager" setting highlights overloaded circuits in red (see Figure 7.87).

FIGURE 7.87

Overloaded circuit indication in Circuit Manager

CONTENT MIGRATION

There are several functions available in AMEP to help with the transition of an AutoCAD drawing to an AMEP drawing. These tools can help you convert your blocks and linework to objects that can be circuited. The tools are:

1. Convert to Device

2. Convert to Wire

3. Batch Convert Devices (Manage ribbon, MEP Content panel)

The first two will be covered below. The third will be briefly discussed but is out of the scope for this book.

CONVERT TO DEVICE

The Convert to Device functionality allows you to quickly convert all instances of a particular block to a Device, which you can then associate with electrical load information and a circuit. However, you have to keep in mind that the "align to geometry" functionality assumes that the block definition is oriented in a certain way. Thus, you may be able to quickly convert blocks to Devices, but reusing the Devices for other projects may not be practical unless the block definitions happen to be oriented in the right direction. Refer to Chapter 9 more information on creating Device styles.

CONVERT BLOCKS TO LIGHTING FIXTURES

In this exercise we will convert blocks representing lighting fixtures to Devices that may be circuited.

4. On the Project Navigator palette (on the Constructs tab) expand the *Electrical\Lighting* folder, and then double-click the *03 Lighting* Construct file.

5. Select one of the 2x4 lighting fixtures, right-click, and select **Convert To > Device**.

6. In the "Device Convert" dialog, type **2x4 Troffer - 3 Lamp** for the Name.

7. For the Type, choose **Lighting**.

8. For the Layer Key, click the browse button, select the E-DV-LIGHT layer key, and then click OK.

9. Clear the "Delete the original object" checkbox and then click Next.

   In the "Connectors" dialog only one connector (Connector 1) should be listed.

If necessary, you can use the Add Connector or Delete Connector icons to add or remove connectors as shown in Figure 7.87.

FIGURE 7.88
Connector settings in the Convert To Device wizard

10. Specify the following for Connector 1:

   - System Type: **Power and Lighting**
   - Connection Point X: **2'-0"**
   - Connection Point Y: **1'-0"**
   - Connection Point Z: **0"**
   - Number of Poles: for Value, choose: **1**; for Prevent Override, choose: **Yes**.
   - Voltage: for Value, choose: **<ByCircuit>**; for Prevent Override, choose: **Yes**.
   - Load Phase 1: for Value, type: **96**; for Prevent Override, choose: **Yes**.
   - Load Category: for Value, choose: **Lighting**; for Prevent Override, choose: **Yes**.

The connection point X, Y, and Z are relative to the block's origin at 0 rotation. In this case, the origin of the block is at a corner, and the center of the block is at 2'-0" in the x, and 1'-0" in the y. This is where Wires will connect.

The Number of Poles, Voltage, Load, and Load Category are all set to prevent override because we want to make sure that all instances of this Device Style contain the same electrical characteristics. The voltage is set to <ByCircuit> to allow this Device to connect to 277v or 120v circuits, such as may be the case with multi-volt ballast.

11. Click the Next button

12. Check the box "Convert all additional references to the selected block in the drawing" (see Figure 7.89).

FIGURE 7.89
Convert all additional references option

13. Click the Finish button.

All the 2x4 blocks have now been converted to Devices. However, since the original block was only 2D, there is no 3D component to the block. In Chapter 9 we will add a Model / 3D representation for the Device. In Chapter 15, we will define a lighting fixture Tag for these fixtures.

SPECIFY EPD AND CIRCUIT FIXTURES

In this exercise, we will associate the lighting drawing with the EPD we created earlier in this chapter. We will then quickly circuit the lighting fixtures.

14. On the Manage tab, on the Preferences panel, click the Electrical button.

15. On the Electrical Project Database tab, click the Open button.

16. Browse to select the *MAMEP Commercial.epd* file created earlier in this chapter.

17. Select the "Use relative path" and then click OK.

18. Select one of the lighting fixtures.

19. On the Device tab, on the General panel, click the Select Similar button.

20. On the Circuits panel, click the Circuit Properties button.

21. In the "Electrical Properties" dialog, specify the following:

- Show circuits from Panel: **H3 (03 Power)**
- Circuit: **21**

The Load on the circuit updates to reflect 3552 VA. The properties Load Phase 1, Voltage, Number of Poles, and Load Category are all disabled. This is because we specified Prevent Override in the Device style settings. This ensures that these settings will remain as expected, and won't be inadvertently overridden.

22. Click OK to dismiss the "Electrical Properties" dialog.

23. On the Home tab, on the Build panel, click the Wire button.

24. In the Properties palette, specify the following:

- Style: **2SL 1N 1G**
- Segment: **Arc**
- Height: **6"**

In the Advanced grouping, beneath Design Data:

- Show circuits from panel: **H3 (03 Power)**
- Circuit: **21**

25. In the drawing window, right-click, and then choose **Generate**.

26. Right-click again, and choose Cancel to exit the Wire command.

In this case, the homerun wasn't generated because the panel is in another drawing. You may manually draw a homerun using the Wire tool.

EXERCISE: CONVERT BLOCKS TO SWITCHES

In this exercise we will convert blocks representing light switches to Devices that may be circuited.

1. Select one of the double switch blocks in the drawing (see Figure 7.90).

FIGURE 7.90
Double Switch block

2. Right-click, and select **Convert to > Device**.

3. Specify the following, and then click Next:

- Name: **Double Switch**
- Type: **Switch**
- Layer Key: **E-DV-SWITCH**
- Delete the original object: Checked

4. Make sure there is only a single connector, specify the following, and then click Next:

- System Type: **General**
- Connection Point X: **0**
- Connection Point Y: **1'-0 3/8"**
- Connection Point Z: **0**

5. Check the box "Convert all additional references to the selected block in the drawing," and then click Finish.

6. Select one of the new switch Devices.

7. On the ribbon, click Select Similar.

8. On the ribbon click the Circuit Properties button.

9. From the Show circuits from panel list, choose **H3 (03 Power)**.

10. From the Circuit list, choose: **21** and then click OK.

> **Note:** Even though the switch has a "General" connector type, it is able to be connected to "Power and Lighting" Circuit types. Switches typically wouldn't have a "Power and Lighting" connector type because this implies that there is a load.

11. Save the file.

CONVERT TO WIRE

Lines, arcs, and polylines can be converted to wires. After you convert the linework to wires, you can select a wire style to assign tick mark and home run styles.

CONVERT ARCS TO WIRES

In this exercise, we will convert Arcs to Wires.

1. On the Properties palette, click the Quick select icon (see Figure 7.91)

FIGURE 7.91
Quick Select icon on the Properties palette

2. In the "Quick Select" dialog, specify the following, and then click OK:

   - Apply To: **Entire Drawing**
   - Object Type: **Arc**
   - Operator: **Select All**

3. In the drawing window, right-click, and select **Convert To > Wire**.

4. At the Erase layout geometry command line prompt, choose Yes.

This erases the original arcs and replaces them with Wires in the same shape. After you have converted the Arcs to Wires, the newly created Wires remain selected. This allows you to specify the Wire Style on the Properties palette.

5. On the Properties palette change the Style to 2SL 1N 1G.

6. Save 03 Lighting.

In real project drawings where you may be converting arcs, or other linework, to Wires, you may be more selective about the conversion process. For example, you could forego the Quick select and manually select objects to convert, or you could be more specific in the "Quick select" dialog, choosing only to select arcs on a particular layer.

BATCH CONVERT DEVICES

Batch Convert Devices allow you to create a script to convert multiple Blocks in a single drawing or multiple drawings into Devices. You can fine-tune the script before generating the Devices to specify many of the Device characteristics. However, you will likely need to fine-tune the Device style definitions after running the script.

You can launch the Batch tool from the Manage tab of the ribbon. Click on the MEP Content panel to expand it and reveal additional tools. Click the Batch Convert Devices tool. The detail of the Batch Convert tool is beyond the scope of this book; however, you are welcome to experiment with it on your own.

After creating Devices using the Batch Convert Devices, or the Convert To Device functionality, the Device styles may be added to your content library for use on future projects. Refer to Chapter 9 for more information on style-based content creation.

PANEL SCHEDULES

Panel Schedules may be created in any drawing associated with the EPD. Panel Schedules are actually based on AutoCAD table styles. The default US Imperial table styles are found in *C:\ProgramData\Autodesk\MEP 2011\enu\Styles\Imperial\Panel Schedule Table Styles (US Imperial).dwg*. There is a table style defined for single- and three-phase branch circuit panels. Additionally, the Switchboard and Distribution board styles only differ in their table heading. Finally, there is a style named Panel that resembles the panel schedule from earlier releases of AMEP and Autodesk Building Systems.

The easiest way to insert a panel schedule is to select one of the tool palette tools on the Electrical palette group on the Tag & Schedule tab. These tools are shortcuts to each of the styles available in the default table style file.

> **TIP:** If you create your own table styles, you may also want to define your own tool palette tools so you don't have to browse for the schedule style location file each time you need to place a schedule.

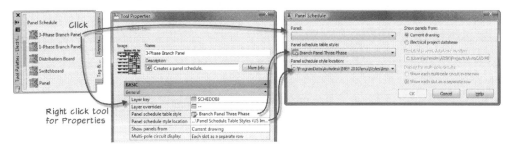

FIGURE 7.92
Panel schedule tool properties

Figure 7.92 shows the tool palette tools on the left. The middle of the figure shows the properties of the tool palette tool, and how it is configured with the table style and path. You could also define the tool to display Panels from the current drawing's electrical project database. The right side of the figure shows the "Panel Schedule" dialog where you specify from which Panel to create the schedule.

MODIFY THE TOOL PALETTE TOOL

In this exercise, we will first modify a tool palette tool to use the electrical project database option.

> If the Tool Palettes – Electrical are not already showing, on the Home tab, on the Build panel, click Tools.
>
> If the Electrical tool palette group is not active, right-click the Tool Palettes titlebar and choose **Electrical** to load it.

1. Click the Tag & Schedule tab.

2. Right-click on the 3-Phase Branch Panel tool and select Properties.

3. For the Show panels from option, select **Electrical project database**, and then click OK.

If you prefer, you can copy and paste this tool first and edit the copy instead. If you do so, be sure to rename the new copy logically as well.

INSERT A PANEL SCHEDULE

In this exercise, we will create a new drawing and associate it with the electrical project database. We will then use the tool palette tool created in the previous exercise to place a schedule for panel H3.

4. On the Project Navigator, click the Views tab.

5. Right-click the *Views* folder and choose **New Category**.

   A Category is another name for a folder in Project Navigator.

6. Name the Category **Electrical Schedules**.

7. Right-click on the new *Electrical Schedules* folder and choose **New View Drawing > General**.

8. On the General page, input **Panel Schedules** for the Name.

9. Check the "Open in drawing editor" checkbox, and then click Next.

10. Leave all Levels unchecked, and then click Next.

11. Leave all Construct folders unchecked, and then click Finish.

    The drawing will open.

12. On the Tool Palettes, click the 3-Phase Branch Panel tool modified in the previous exercise.

Since this drawing does not yet have an EPD specified, a message appears indicating that the EPD is missing or unavailable.

13. Click the "Open an existing EPD file" option (see Figure 7.93).

FIGURE 7.93
Click to Open an existing EPD file

14. Select the *MAMEP Commercial.epd* file created earlier in this chapter.

15. Specify the panel H3 (03 Power), and then click OK.

16. Click to place the schedule.

Zoom in on the entry for circuit 21. Note that the load indicates 3552 from the lighting load circuited earlier in this chapter. Note, however, that the Description indicates SPARE.

You may also want to place schedules for MDP and L3. Note that most of the schedule cell values are shaded in grey. These are AutoCAD Fields that will automatically update when the drawing is reopened. In the next exercise, we will update the description on circuit 21, and then refresh the schedule.

Update Panel Schedule

17. On the Project Navigator palette, on the Constructs tab, double-click to open the *03 Power* Construct.

We are defining the description on the lighting circuit. This circuit terminates in the panel in *03 Power*, and thus, this is the file we need to be in to edit the description.

18. Open Circuit Manager

19. Select Panel H3 in Circuit Manager, and double-click on the Description for circuit 21.

20. Change the Description to **LIGHTING** and then press ENTER.

    Leave Circuit Manager open.

21. Save *03 Power* to write the updated change to the EPD.

22. Switch back to the Panel Schedules drawing.

    A status notification indicates the EPD needs to be reloaded.

23. Click the icon to reload the EPD.

In Circuit Manager, circuit 21 in panel H3 should now indicate the new description, and the Field in the inserted panel schedule should update as well.

In this chapter, we demonstrated how to use the ribbon and settings on the Properties palette to place Devices. Although, for the most part, focus was on placing receptacles, you should know that Devices of any style may be placed and edited using the methods presented. This includes communication, lighting, fire alarm, general power, nurse call, security, switches, and any other type of Device Style you may require. Additionally, using the circuiting and scheduling functionality demonstrated how to report connected load information to avoid having to tabulate this information manually. Hopefully you are now em-"powered" and able to "energize" your electrical drawings in AMEP.

SUMMARY

- ✓ Devices may be placed using layout tools on Spaces or other drawing geometry

- ✓ Panels contain circuits to which Device connectors may be connected.

- ✓ Multi-View parts are used to physically represent electrical equipment

- ✓ Panels should be used (in lieu of Multi-View Parts) when the equipment is integral to the circuiting and load calculations of the distribution system.

- ✓ The Electrical Project Database (EPD) contains project-wide electrical load information for creating electrical panel schedules.

- ✓ Load Categories allow grouping of loads and application of Demand Factors.

- ✓ Wires are used to annotate circuiting, tick marks, and home runs.

- ✓ Existing AutoCAD geometry like Blocks and linework can be converted to Devices and Wires with simple right-click tools.

Chapter 8

Conduit Systems

INTRODUCTION

With AutoCAD MEP you can easily design and document Conduit Systems. We will cover the fundamentals on how Conduit works in AutoCAD MEP, describe the settings that control conduit and discuss the preferred workflow approaches when laying out conduit systems using Parallel Routing.

We will explore what you need to know about creating a new system definition for your Conduit system, how conduit systems are displayed, and how the routing tools work when you are laying out your conduit system. Routing tools incorporated into the ConduitAdd command allow you to create multiple parallel conduit routes using the ParallelRouting command. We will go through the best practices on how to lay out Conduit using both ConduitAdd and ParallelRouting.

OBJECTIVES

In this chapter you will create new system definitions and lay out a conduit system. We will create a conduit layout using the Parallel Routing abilities of AutoCAD MEP. By creating these systems, you will understand the fundamentals of routing, systems and routing preferences. Display configurations and how to control them will be explained and put into practice, and you will learn when to apply display overrides to create the desired look in construction documents. In this chapter you will:

- Learn the fundamentals of Conduit in AutoCAD MEP
- Learn about Settings and Controls
- Explore System Definitions
- Learn Conduit layout design
- Learn how to route multiple conduits using Parallel Routing

FUNDAMENTALS OF CONDUIT

Let's begin with an exploration of the fundamentals of 3D conduit by discussing what makes conduit work. 3D conduit has certain characteristics, features, and settings that allow AutoCAD MEP to be an efficient 3D conduit application. Like other 3D layout features such as duct and pipe, conduit uses system definitions to determine which settings the overall run will inherit. These include layers, display settings, system abbreviations, etc.

Much like piping, conduit requires content such as elbows, pipes, junction boxes (tees), etc. to assemble a run. These fittings are brought into the drawing from the conduit catalogs (see the Chapter 11 for more information on creating fittings). Unlike duct and cable tray, but similar to piping, the conduit feature leverages a style-based approach for storing the fittings that will be used while creating a run. This style is called a "Conduit Part Routing Preference," more commonly known as simply a "Routing Preference." A Routing Preference is a collection of fittings stored inside a style that assigns different fittings based on the routing preference selected. These fittings are automatically added to the run and multiple styles can be created based on your needs. See the "Creating a Routing Preference" topic below for more information.

Creating conduit runs leverages the auto routing functionality (built into AutoCAD MEP for all 3D objects) to automatically add fittings and required connections within the run. Conduit, like the other 3D layout objects, leverages the AecbCompass to restrict your cursor to the predefined angles stored within the AecbCompass dialog. See the "Auto Routing" topic in Chapter 6 for more information. Conduit runs can be displayed either as 2 line or 1 line using the display system.

As noted, System Definitions, Routing Preferences and Display are all explained in more detail in the tutorials later in the chapter. However, the fundamental feature that leverages all of these controls and settings is the "Auto Routing" feature. The Auto Routing feature is detailed in Chapter 6—Piping Systems. Routing preferences, the conduit fittings in the catalog and the AecbCompass directly impact the results of auto routing. Some items have been briefly explained, but you must understand their importance to auto routing before understanding how to create or configure the settings.

UNDERSTANDING ROUTING PREFERENCES

Conduit systems are created by accessing the Fittings stored inside the Conduit Catalogs supplied with AutoCAD MEP and specified in Conduit Part Routing Preferences (also referred to as simply "Routing Preference"). A Conduit Part Routing Preferences is a style that includes a collection of fittings. Like other styles, they are accessed through Style Manager (see Figure 6.10). The list of catalog fittings in the Routing Preference allows AutoCAD MEP to assemble a conduit system as it would be assembled during construction.

FIGURE 8.1

Style Manager Button on the Manage ribbon tab

You can open the Style Manager from the Manage ribbon tab on the Style & Display panel. On the left panel expand *Electrical Objects* and then expand *Conduit Part Routing Preferences*. All of the Routing Preference styles will appear both in the expanded list on the left side and on the right panel (see Figure 6.11).

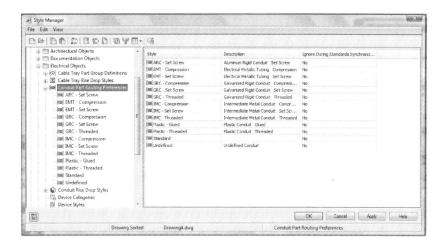

FIGURE 8.2

View available Conduit Part Routing Preferences in the Style Manager

A Conduit Part Routing Preference defines the part selection that will be used for a particular run and specifies the type of Conduit, Elbows, Tees, and Transitions. The parts available on the lists are stored in the Part Catalogs that are supplied with AutoCAD MEP.

In Style Manager, you can view the settings of any Routing Preference definition. The Type column specifies the types of fittings that are allowed to be stored in a Routing Preference and the Part Column specifies the selected fitting.

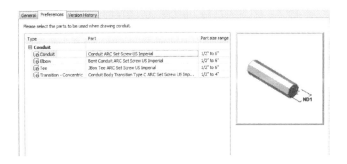

FIGURE 8.3
Specified parts for the selected Conduit Routing Preference

INSTALL THE DATASET FILES AND CREATE A DRAWING

In this exercise, we use the Project Browser to load a project from which to perform the steps that follow.

1. If you have not already done so, install the book's dataset files.

    Refer to "Book Dataset Files" in the Preface for information on installing the sample files included with this book.

2. Launch AutoCAD MEP 2012 and then on the Quick Access Toolbar (QAT), click the Project Browser icon.

3. In the "Project Browser" dialog, be sure that the Project Folder icon is depressed, open the folder list and choose your *C:* drive.

4. Double-click on the *MasterAME 2012* folder.

5. Double-click *MAMEP Commercial* to load the project. (You can also right-click on it and choose **Set Project Current**.).

> **IMPORTANT:** If a message appears asking you to repath the project, click Yes. Refer to the "Repathing Projects" heading in the Preface for more information.

> **Note:** You should only see a repathing message if you installed the dataset to a different location than the one recommended in the installation instructions in the preface.

6. Click the Close button in the Project Browser.

When you close the Project Browser, the Project Navigator palette should appear automatically. If you already had this palette open, it will now change to reflect the contents of the newly loaded project. If for some reason, the Project Navigator did not appear, click the icon on the QAT.

7. On Project Navigator, navigate to the *Electrical* Folder and then to the *Conduit* Folder.

8. Open the *01 - Electrical – Outdoor Power* Drawing.

LOAD THE ELECTRICAL WORKSPACE

Our first task will be to create a Routing Preference and a System Definition for the Power system. Make sure that the Electrical Workspace is active and that the Tool Palettes are displayed (see Figure 6.12).

Refer to the "Choosing your Workspace" topic in the Quick Start chapter if you are not sure how to load a Workspace and refer to the "Understanding Tool Palettes" topic in Chapter 1 for information on how to load and work with tool palettes.

FIGURE 8.4
Enabling the Electrical Workspace

CREATE A ROUTING PREFERENCE

Creating a Routing Preference is an important first step, but since we have already provided detailed steps for doing this in the piping Chapter, we will not reiterate the specifics here. Please refer to Chapter XX for details. For this exercise, we will use Routing Preference the **Plastic – Glued** routing preference supplied with AutoCAD MEP. The application has several different Conduit Routing Preferences predefined in the templates to meet most conditions on the project.

UNDERSTANDING SYSTEM DEFINITIONS

We will be using an existing System Definition for the Power conduit system. System Definitions assign the Layer Keys, Display System settings, Abbreviations, System Grouping, and Rise Drop styles. In addition, System Definitions determine whether items assigned to this system will follow the display system drawing default or will be assigned a system level override.

System definitions are used to separate objects based on their use such as a power conduit system versus a low voltage system. This allows you the flexibility to isolate the systems as well as assign uniquely different characteristics to the objects designated to this system such as layer, abbreviation, and display properties. For more information on how to create a System Definition refer to Chapter 6.

In this exercise we will duplicate an existing system definition for the underground conduit we will be adding to the building later on in the chapter.

9. On the Manage Tab, on the Style & Display Panel, click the Style Manager button.

10. Expand *Electrical Objects* and then expand *Electrical System Definitions*.

11. In the left panel, select the Cndt-Power - 480V system, right-click and choose Copy.

12. Select Electrical System Definitions, right-click and choose Paste (see Figure 8.5).

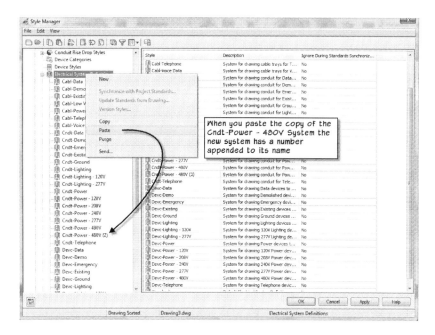

FIGURE 8.5

Enabling the Electrical Workspace

13. Select the Cndt-Power – 480V (2) system in the left panel.

   The right panel will update to show the system's settings.

14. Click the General Tab and change the name to Cndt-Power-480V Underground (see Figure 8.6).

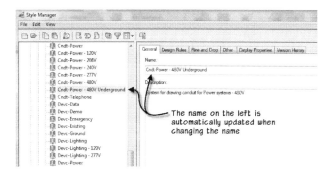

FIGURE 8.6

Rename the copied System

Next we will add a display override to change the linetype of the conduit to Hidden2 so it will appear dashed on the drawings.

DISPLAY PROPERTIES

When you configure a style, such as the System Definition in consideration here, you have the option to assign style-level display properties to it. Settings on the Display Properties tab are used to control the appearance of the style (in this case the System Definition). Before making any edits to the display properties of the System Definition, make sure you are comfortable with the display system hierarchy and definitions. Refer to the "Overview and Key Display System Features" heading in Chapter 2 and the "Display Properties and Definitions" topic in Chapter 13 for definitions of the key Display Control terms. You will also find detailed tutorials in Chapter 13 for working with the Display System.

CONFIGURE DISPLAY PROPERTIES

1. Click the Display Properties Tab.

Notice that there are several Display Representations listed. In this exercise we will focus on the Plan Display Representation.

2. Select the Plan Display Representation.

3. Place a checkmark in the style override box (see Figure 8.7).

    The override dialog will automatically open.

For the Cndt-Power – 480V Underground System Definition, we want to change the line type. We are adding a System Definition override because a style override applies to all objects assigned to this system.

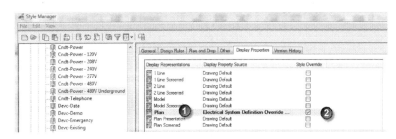

FIGURE 8.7
Adding a Display System Override

The "Display Properties" Dialog will appear. The titlebar will read: "(Electrical System Definition Override – Cndt-Power-480V Underground System) – Plan". This confirms that a System Definition override is now applied.

4. In the Linetype column next to the Contour Display Component, click on ByBlock (the current linetype designation) to change it.

5. Choose Hidden2 in the Linetype dialog then click Ok to dismiss the "Select Line type" dialog (see Figure 8.8).

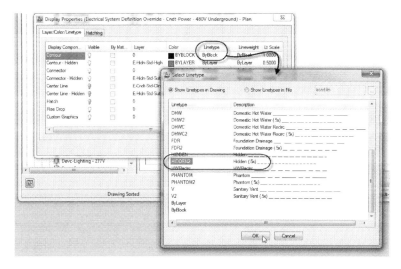

FIGURE 8.8

Change the Linetype for the Contour Display Component to Hidden2

6. Click OK to dismiss the "Display Properties" dialog.

7. Click OK again to dismiss the Style Manager and complete the System Definition setup.

The new System definition has been defined for the underground conduit system. This allows us to independently control the display properties of any conduit drawn on this system.

The setup is now complete and we are ready to begin laying out our equipment and conduit.

8. Save the drawing.

> **CAD MANAGER NOTE:** Routing Preferences, System Definitions and other kinds of styles can be stored in template files (DWT). This means that each new drawing created from the template will have those items already available by default. You can also create tool palettes to import systems, routing preferences and preferred settings as needed on a per drawing basis. For more information please refer to the online help.

EQUIPMENT AND CONDUIT LAYOUT

In general, layouts begin with determining the equipment location to allow others involved with the design to review access requirements, structural issues and additional information critical for making an informed decision.

The Equipment command on the ribbon allows you to add equipment (Multi-View Parts) from the AutoCAD MEP catalog. The catalog contains equipment that is modeled three-dimensionally at actual size and has specific information such as dimensions, required or optional connections, and 1 line symbols.

ADDING EQUIPMENT

When you click the Equipment button on the ribbon, it calls the MvPartAdd command. The MvPartAdd command opens the Multi-view Parts (MvParts) dialog which allows you to find specific equipment through the catalog's folder structure (see Figure 6.30).

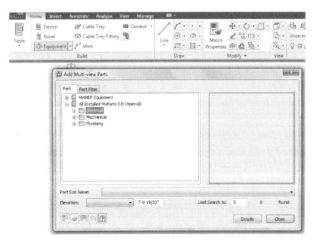

FIGURE 8.9

Adding Equipment starts the MvParts command

By expanding folders you can find parts listed in broad categories. Once you have selected a piece of equipment, you can click the Part Filter tab which allows you filter the list of parts based on specific criteria. Icons within the dialog will identify the content as either block-based or parametric. Refer to Chapter 9 for more information on types of content and for more information on how to filter part selections, refer to the "Pipe Layout" topic in Chapter 6.

ADDING AN OUTDOOR TRANSFORMER AND CONDUIT

Now that we have completed configuring the necessary settings, (we created a System Definition and explored the basics of adding equipment) we are ready to begin the layout of the outdoor transformer and the associated conduit runs to enter the building. For this exercise, we will work on the first floor of our commercial project.

ADDING THE TRANSFORMER

Let's add a Transformer outside the building. We will then connect conduits to the bottom of the Transformer and use the Parallel Routing command to create a conduit bank between the pieces of equipment.

> Continue in the *00 Electrical – Outdoor Power* construct.

Figure 8.10 shows the design we will be using for this exercise. We will be locating the transformer on the site to the left of the building. The Electrical Room is located on the back of the building in the center. We will add the Transformer in the site drawing and then connect it to the Control Center in the *01 – Electrical* drawing.

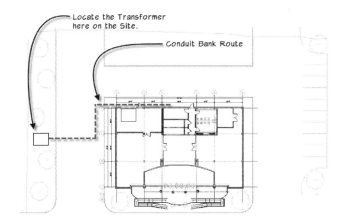

FIGURE 8.10
Site Plan Markup

1. On the Home Tab, Build Panel, click the Equipment button.

2. In the "Add Multi-view Parts" dialog, beneath the *Electrical* folder, expand the *Power Transformers* folder.

3. Select the Wet Type Transformer – 75-2000 kVa US Imperial.

    This is the transformer we are going to insert, but before we do, we want to change the size and elevation.

4. From the "Part Size Name" list, choose **2000 kVa Wet Type Transformer** (see Figure 8.11).

5. To allow a slab to be added below the transformer, change the Elevation to **6"**

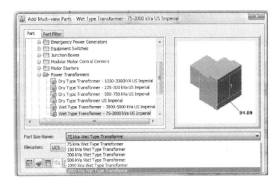

FIGURE 8.11

Select a Transformer

Do not click Close in the "Add Multi-view Parts" dialog.

6. In the drawing window, click to place the transformer between the 2nd and 3rd tree (counting from the bottom of the screen) on the left side (see Figure 8.10 above).

7. Move your cursor straight up on screen to rotate the part at 90 degrees (or type **90** at the command line).

8. Click Close to dismiss the "Add Multi-view Parts" dialog.

> Note: To add an Equipment pad, you can create a slab using the Slab tool in the Home tab of the Architectural workspace. For more information on how to create slabs, refer to the AutoCAD MEP Help.

There are two ways to set the system on the equipment connections on the Transformer. One option is you can modify the Equipment properties to assign the system to each connector as covered for piping in Chapter 6. An alternative is covered in the next topic. We are going to set the system through the Conduit command.

CONDUIT ROUTING

> **Note:** For this exercise we are going to assume the conduit will be located 4 feet below grade and will follow the path shown in Figure 8.10 above.

To assist us in the next sequence, let's divide the screen into two viewports.

9. On the View tab, on the Viewports Panel, click the Viewport Configuration List drop-down button and then choose: Two: Vertical (see Figure 8.12).

   This will divide your screen into two separate viewports oriented vertically.

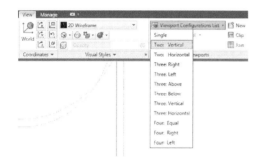

FIGURE 8.12

Divide the screen into two vertical viewports

10. In the right viewport, select the lower right-hand corner of the View Cube. This changes the view to a South West Isometric view (see Figure 8.13).

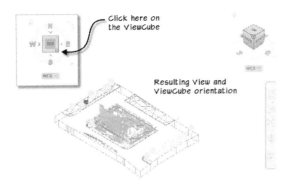

FIGURE 8.13

Changing the view with the View Cube and result

11. In the left (plan) viewport, zoom in closer so that the transformer and the location for the end of the run behind the stairs are in the view.

12. Zoom in on the transformer in the right viewport (see Figure 8.14).

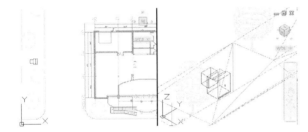

FIGURE 8.14

Zooming in both views

13. Select the Transformer in the right 3D view and click the upper right plus grip to start the Conduit command.

The "plus" grip starts the "Add" command associated with the connection type as defined in the part catalog. In this instance the connection type is defined as a Conduit connection and therefore the ConduitAdd command will start automatically.

14. On the Properties palette, change the System to **Cndt-Power – 480V Underground (PWR)** and then change the Routing Preference to **Plastic – Glued** (see Figure 8.15).

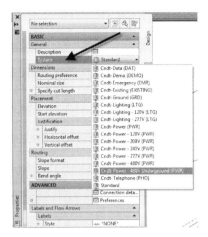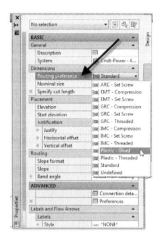

FIGURE 8.15

Choose the System and the Routing Preference

15. On the Properties palette, change the elevation (under Placement) to **-4'** and then press ENTER (see Figure 6.36).

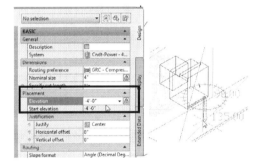

FIGURE 8.16

Change the elevation of the conduit

16. Press ENTER to exit the command.

17. On the Home tab, on the Modify panel, click the Copy tool. Select the newly drawn Conduit and then press ENTER.

    Zoom in close on the Conduit and Transformer connectors.

18. Using the Wireways End Connector (WCON) Osnap, make 7 copies of the Conduit being sure to snap from the end of the Conduit to the connections on the Transformer (see Figure 8.17).

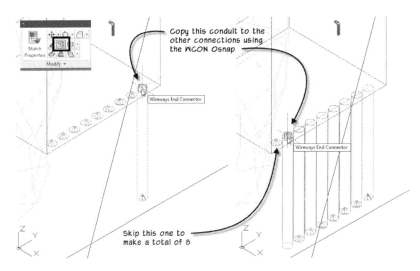

FIGURE 8.17

Copy the Conduit to the next 7 connectors

Now that we have a total of 8 Conduits, we are going to extend the 4 left most conduits down to an elevation of -5' 0" to allow us to create a 2 level bank of conduit.

> Continue in the 3D View.
>
> 19. Select the 4 leftmost conduits.
>
> 20. On the Properties palette type -5'-0" in the End elevation field (see Figure 8.18).

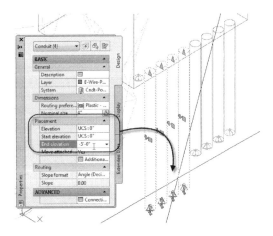

FIGURE 8.18

Changing the Elevation of the 4 left conduits

21. Press esc (or right-click and choose Deselect All) to deselect the Conduits and complete the operation.

The 8 conduits are now set at 2 different elevations to support a stacked conduit bank. Next we will route the conduit bank to the building.

PARALLEL ROUTING

AutoCAD MEP provides the ability to route multiple conduits, pipes, or a combination of both objects in parallel to create a bank. The command allows you to route a parallel conduit or conduit and pipe bank by selecting existing conduits or pipes in the drawing, then laying out the run in the same manner you would a single conduit or pipe run. The command used by this control is called Parallel Routing and works on both Conduits and Pipes. In this sequence we will use this functionality to route the conduit from the transformer to the building.

1. On the Home tab, on the Build panel, click the Conduit drop-down and choose Parallel Conduits (see Figure 8.19).

FIGURE 8.19
Choose the Parallel Conduit tool

Continue in the right 3D viewport.

2. At the "Select baseline object" prompt, select the upper right-hand Conduit.

This determines which object will be driving the layout.

3. At the "Select parallel conduits" prompt, select the next 3 Conduits and then press ENTER (see Figure 6.37).

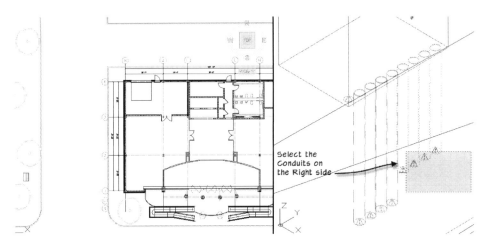

FIGURE 8.20

Select the Conduits for Parallel Routing

4. Click in the left view to route the conduit in the plan view. Position the cursor to the right of the transformer, and type 20' in the Dynamic Dimension and then press ENTER (see Figure 8.21).

> Note: You could also type **20'** at the command line if you do not have dynamic dimensions turned on.

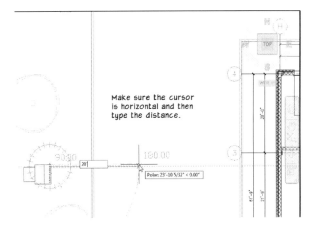

FIGURE 8.21

Use the onscreen dynamic dimension to route the first leg

5. On the Properties palette, under Routing, set Parallel bends to Concentric.

6. Move your cursor straight up onscreen and click to place the next run and add the elbows.

7. Press ENTER to exit the command.

We ended the command so the remaining 4 conduits can be routed under these conduits. The intent is to show how multiple levels of conduits can be routed simultaneously.

8. Click to activate the right 3D viewport.

Notice the elbow fittings that have been added automatically to our Conduits in the 3D view.

9. Press the SPACEBAR to repeat the ParallelRouting command.

10. In the right 3D view, select the upper right-hand conduit as your base conduit (this determines which object will be driving the layout).

11. At the "Select parallel conduits" prompt, select the next 3 Conduits and then press ENTER (see Figure 8.22).

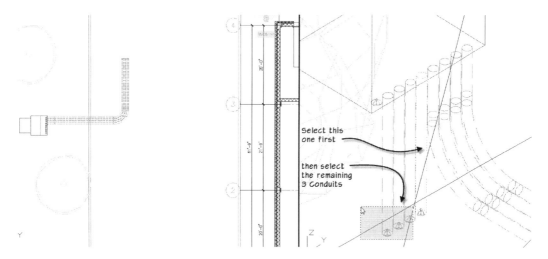

FIGURE 8.22
Selecting the Conduits for Parallel Routing

12. Click in the Plan viewport (left Viewport).

We want the conduits to be routed at elevation -5'-0" and we don't want them to attempt to connect to the other Conduits.

13. On the Properties palette, click the lock icon next to the Elevation (see Figure 8.23).

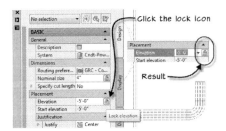

FIGURE 8.23

Locking the Elevation

Once the Elevation is locked, you can now snap to the objects above (or below) without inadvertently connecting to them.

14. Move the mouse horizontally to the right as before, type 20'-0" and then press ENTER to draw the Conduit parallel to the last route.

15. Using the WCON snap, click the end of the left Conduit from the previous run (see Figure 8.24).

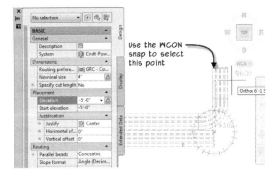

FIGURE 8.24

Snap the new run to the termination of the previous run

16. Press ENTER to end the command.

17. Start the ParallelRouting command again by pressing ENTER or SPACEBAR.

18. Select the Upper Left-hand conduit as your baseline in the plan view (see Figure 8.25).

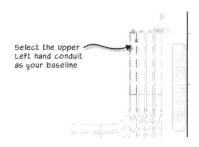

Select the Upper
Left hand conduit
as your baseline

FIGURE 8.25
Selecting the Baseline conduit for Parallel Routing

19. Use a crossing window to select the remaining 7 conduits and press ENTER.

Depending on where you terminated your vertical run, you may need to continue the run vertically now before turning toward the building.

If necessary, move the cursor straight up and click just above the point where the outside Wall of the building is.

20. On the Properties palette, beneath Routing, set Parallel bends to **Fixed radius**.

21. Extend the run towards the location shown in Figure 8.10 above, just beyond column line K and then press ENTER to end the command.

In this lesson, we have used both Concentric and Fixed radius Routing. Figure 8.26 shows an example of each. In your own projects, choose the routing that you prefer.

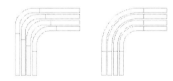

FIGURE 8.26
Conduit on left is routed using Fixed radius, conduit on right is Concentric

The final step is to turn the Conduit run up approximately where the main electrical gear will reside on the outside of the building.

22. Click in the 3D viewport and zoom in on the end of the Conduits.

23. On the Home tab, on the Build panel, click the Conduit drop-down button and choose Parallel Conduits.

24. Select one of the upper Conduits as the baseline object. Select the remaining three upper Conduits as the parallel Conduits and then press ENTER.

25. On the Properties palette, for Elevation, type 1' and then press ENTER.

> **Note:** If you prefer, you type E at the command line, and then specify the absolute elevation value instead.

26. Press ENTER to end the ParallelRouting command.

27. Press ENTER again to repeat the ParallelRouting command.

28. Select one of the lower Conduit as the baseline object. Select the remaining three lower Conduits as the parallel Conduits and then press ENTER.

29. Move the cursor straight along the X axis (using Polar or Ortho), type 6" and then press ENTER to extend the bottom run.

30. On the Properties palette, for Elevation, type 1' and then press ENTER (see Figure 8.27).

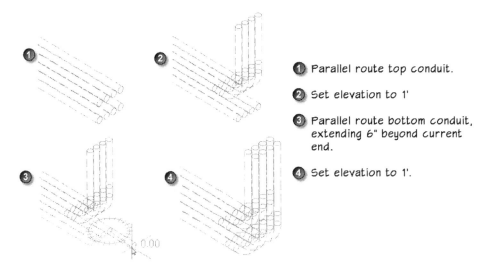

1. Parallel route top conduit.

2. Set elevation to 1'

3. Parallel route bottom conduit, extending 6" beyond current end.

4. Set elevation to 1'.

FIGURE 8.27

Routing the conduit up

We have now completed the routing of the conduit from the Transformer to the Switchgear equipment location using the Parallel Routing functionality. Feel free to experiment further using this functionality if you wish.

SUMMARY

Throughout this chapter, we have reviewed many key aspects of the 3D conduit tools in AutoCAD MEP. We have put these concepts into practice in our sample commercial office building. You should now have a good feel for how 3D conduit works as well as an understanding of how to route multiple conduit using the Parallel Routing feature.

✓ The Conduit Feature in AutoCAD MEP allows for accurate creation of models in a fast and accurate manner.

✓ Routing Preferences are used to establish the preferred fittings during layout.

✓ System Definitions define how the System is abbreviated in annotation, and how the conduit is displayed.

✓ System Definitions also control overrides to the default display, allowing for the Contour to be displayed with a different linetype.

✓ Leveraging AutoCAD commands like Copy helps us to create repetitive layouts quickly and easily while automatically maintaining relationships to other AutoCAD MEP objects.

✓ Parallel Routing allows us to create Conduit Banks quickly and accurately.

WHAT'S IN THIS SECTION?

This section explores the many ways to create content in AutoCAD MEP. Chapter 9 begins with a thorough look at styles and the Style Manager. Chapter 10 explores Equipment, block-based content and converting existing 3D blocks to Equipment. In Chapter 11, the process, rationale, and best practice techniques for building parametric parts is presented.

The basics of the Display System have been discussed and explored in previous chapters. Chapter 12 concludes this section with a deeper look at the Display System and its many nuances.

SECTION III IS ORGANIZED AS FOLLOWS:

- Chapter 9   Content Creation - Styles
- Chapter 10   Content Creation - Equipment
- Chapter 11   Content Creation - Parametric Fittings
- Chapter 12   Display Control

Section III

Content Creation—Styles

INTRODUCTION

In this chapter you will be introduced to the theory and application of Style Manager. Style Manager is the central repository for all AutoCAD MEP (and AutoCAD Architecture) style-based elements. The focus of this chapter is on the portions of Style Manager specific to the engineering disciplines served by AutoCAD MEP.

OBJECTIVES

In this chapter you will:

- Learn about the multi-view functionality of style-based content.
- Explore system definitions.
- Explore Rise Drop styles.
- Understand style-based content categories.
- Understand Part Group Definitions.
- Create Custom Fitting Styles.
- Understand Line and Wire styles.

TYPES OF STYLE-BASED CONTENT

There are many kinds of styles available in AMEP. Some of these are specifically architectural. Others are for the engineering disciplines, and still others are shared by all disciplines, architectural and engineering alike. In this chapter, we will focus on the Electrical, HVAC, Piping, Plumbing, and Schematic Objects (see Figure 7.1). Documentation objects such as Property Set Definitions, Tags (Multi-view Blocks), and Schedule Tables are also style-based; however, these items will be covered in Chapter 15.

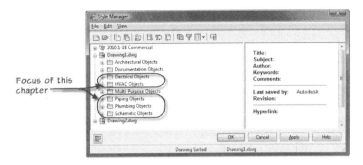

FIGURE 9.1

Electrical, HVAC, Piping, Plumbing and Schematic objects in Style Manager

Each of these five categories contains multiple styles types. However, there is much similarity across the disciplines. The following are the major groupings of styles that will be covered in this chapter:

Style-based Content—Panel Styles, Device Styles, Plumbing Fitting Styles, and Schematic Symbol Styles.

Categories—Device Categories, Panel Categories, Plumbing Fitting Categories, Schematic Symbol Categories

System Definitions—Electrical System Definitions, Duct System Definitions, Pipe System Definitions, Plumbing System Definitions, and Schematic System Definitions

Rise Drop Styles—Cable Tray Rise Drop Styles, Conduit Rise Drop Styles, Duct Rise Drop Styles, Pipe Rise Drop Styles, Plumbing Rise Drop Styles

Part Group Definitions—Cable Tray Part Group Definitions and Conduit Part Group Definitions. Duct Part and Pipe Part Routing Preferences are similar in concepts to Part Group definitions; however, these are discussed in Chapters 5 and 6, respectively.

Custom Fitting Styles—Duct Custom Fitting Styles, Pipe Custom Fitting Styles

Line Styles—Wire Styles, Plumbing Line Styles, Schematic Line Styles

There are a few remaining style types to consider: Load Category Definitions are covered in the Chapter 7. Pipe Single Line Graphics Styles are covered in Chapter 6. Finally, Fixture Unit Table Definitions, Sanitary Pipe Sizing Table Definitions, and Supply Pipe Sizing Table Definitions are covered in the AutoCAD MEP online help.

INSTALL THE DATASET FILES AND CREATE A DRAWING

The lessons that follow require the dataset files provided for download with this book.

1. If you have not already done so, install the book's dataset files.

    Refer to "Book Dataset Files" in the Preface for information on installing the sample files included with this book.

2. Launch AutoCAD MEP 2012.

3. On the Application Status Bar, set the Workspace to Electrical.

BLOCK-BASED STYLES

As we have already seen, AutoCAD MEP's display system allows a single object to have multiple representations depending on the orientation from which it is viewed. The most common application of this is giving an object a top (or plan) representation, and a 3D representation that is visible from other orientations. However, it is possible to give an object a unique view from any of the six orthogonal viewing angles (top, bottom, left, right, front, and back), plus a 3D view that is visible from non-orthogonal views. Typically, a simplified "flattened" view is used for orthogonal orientations because it reduces the overhead when AutoCAD displays the object.

Styles go way beyond simply offering unique display behavior. For example, styles may be used to ensure that all general purpose receptacles utilize 120v/1p connectors, have a load of 180va, have a Load Category of Receptacle. Further, if you define/utilize a separate style for each lighting fixture type in your project, it makes quantification for the purposes of energy tabulations or quantity take off a straightforward task, and can help ensure that a type F1 fixture isn't mislabeled or miscounted as a type F2.

CREATING A BLOCK-BASED STYLE

Perhaps the simplest style to understand and create is a multi-view block. Such an object can be considered a "block-based" style object. Up to seven blocks can be used in total for each Display Representation: one each for top, bottom, left, right, front, back, and 3D.

4. On the QAT, click the Open icon.

5. In the "Select File" dialog, browse to the *C:\MasterAME 2012\Chapter09* folder.

6. Open the file named: *Cube.dwg*.

This drawing has seven standard AutoCAD blocks oriented around the origin. The blocks are named: TOP, BOTTOM, FRONT, BACK, LEFT, RIGHT, and CUBE. All the blocks (except for CUBE) have a single text object that is the same as the block name. The text is oriented according to the view, is justified middle-center, and inserted at the origin. The cube is 1/8" square and centered on the origin. The drawing itself has an insert of each block, with the blocks offset to have them "float" around the cube (see Figure 9.2).

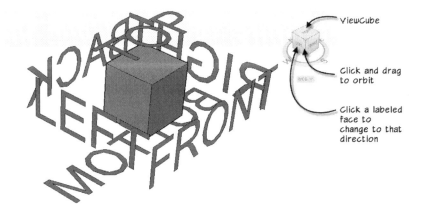

FIGURE 9.2
Standard AutoCAD blocks used as the basis of demonstrating multi-view components.

7. Click and drag the ViewCube to orbit the drawing and view the blocks from different angles.

8. Try clicking any of the labeled faces of the ViewCube to orient the view to that orthographic view (such as Top or Front).

As you change the viewpoint on screen using the ViewCube, notice that all the objects are visible regardless of your viewing direction. This is because AutoCAD blocks do not have awareness of the viewing direction. However, a benefit of block-based content is that only the necessary block is displayed depending on the view orientation. We will demonstrate this by creating a Device Style-based content component from these seven blocks.

> **Note:** Although we will be using a Device Style, the main concepts apply to all block-based styles, including Plumbing Fittings, Schematic Symbols, Panel Styles, multi-view Blocks and even block-based Multi-view Parts.

9. On the Manage tab, on the Style & Display panel, click the Style Manager button.

10. Expand the *Electrical Objects* category.

11. Select Device Styles, and then click the New Style icon (see Figure 9.3).

   This will create a new style named New Style.

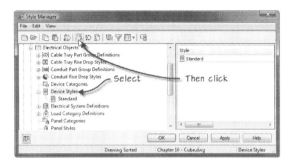

FIGURE 9.3

Select Device Styles, and then click New Style.

12. On the right side, click the General tab, and then change the Name to **Text Cube** (see Figure 9.4).

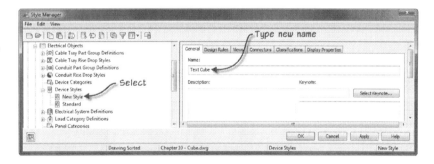

FIGURE 9.4

Specify the name for the newly created style

13. On the Views tab, add a new view (see Figure 9.5):

⇨ On the right, click the Add button.

⇨ In the View Name box, replace the default name of "New View" with **CUBE**.

⇨ From the View Block list, select **CUBE**.

⇨ From the Display Representation list, select **Model**.

⇨ From the View Directions list, uncheck everything, except for 3D.

⇨ Click the Apply button.

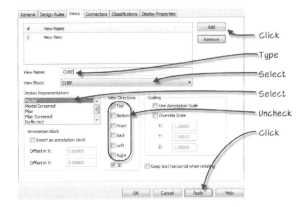

FIGURE 9.5

Add a view definition to the style using the 3D Block

14. Click the Add button to add another view:

⇨ In the View Name box, type **TOP**.

⇨ From the View Block list, select TOP.

⇨ From the Display Representation list, select **Plan**.

⇨ For View Directions, uncheck everything, except for Top.

⇨ Click the Apply button.

15. Repeat the process to create five more views (see Table 9-1):

TABLE 9-1—ADDITIONAL VIEWS TO CREATE

	View 3	View 4	View 5	View 6	View 7
View Name	**BOTTOM**	**LEFT**	**RIGHT**	**FRONT**	**BACK**
Block Name	BOTTOM	LEFT	RIGHT	FRONT	BACK
Display Representation	Model	Model	Model	Model	Model
View Direction	Bottom	Left	Right	Front	Back

16. Click OK to accept the changes and dismiss the Style Manager.

Note: If you get a "Views Invalid" message, make sure you have selected a View Block for each View in the list.

Now, we will test the Device we just created.

17. On the Home tab, on the Build panel, click the Device button.

18. On the Properties palette, (on the Design tab) click the Style image.

19. In the "Select a style" dialog, from the Drawing file list, choose <Current Drawing>.

20. Double-click on Text Cube.

21. On the Properties palette, specify the following:

   - Layout method: **One-by-one**.
   - Align to objects: **Yes**.

22. At the "Insert point" prompt, type **1,1,1** and then press ENTER.

23. Press ENTER to end the command.

   Right-click on the ViewCube, and make sure **Parallel** is selected.

24. Click on some of the labeled faces of the ViewCube such as Top, Front or Left (see the top of Figure 9.6).

Clicking on the labeled faces orients the view to that face. You can quickly orient to the six orthographic views in this way. When you do, you may need to zoom out a bit to see your new Device. For example, when you orient to Top view, only the "top" view block will appear. Compare this to the original inserted blocks that are still inserted at the origin of this drawing; they all appear superimposed on top of one another. When you click the various "hot spots" between faces of the ViewCube, you orient the view accordingly. In other words, if you select the long edge between two adjacent faces, the screen will orient along a 45° angle relative to the two adjacent surfaces. The corners of the ViewCube will orient the view to an isometric or perspective view. When choosing any orientation other than the six orthographic views, the 3D cube block will display instead of any of the text blocks.

25. Select the Device, and on the Properties palette, change the Rotation to **45** (see the lower left corner of Figure 9.6).

26. Use the ViewCube to orient to the different edge orientations (i.e., between RIGHT/BACK, RIGHT/FRONT, etc.).

Notice that the 3D cube is no longer displayed in these orientations; you now see one of the text blocks (see the bottom of Figure 9.6).

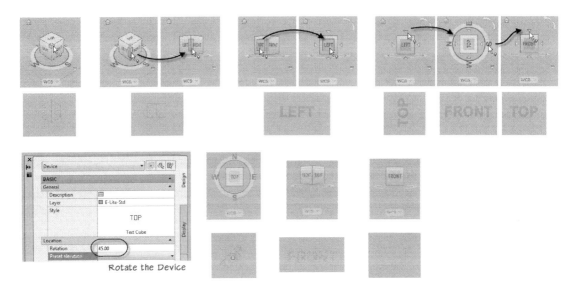

FIGURE 9.6

Cycle through various orientations and rotations to understand the behavior

27. Select the Device, and on the Properties palette, change the Rotation to **180**.

28. Use the ViewCube to orient to the different orientations again.

The view blocks now seem reversed, i.e., you see the FRONT block in the BACK orientation, LEFT in the RIGHT orientation, and so on. The concept to catch here is that the view block that is displayed for an object is based on that object's orientation with respect to the view, not just the view orientation itself.

The multiple views concept demonstrated here with Devices applies to Panels, Plumbing Fittings, Schematic Symbols, Multi-view Blocks, and even Multi-View Parts. Plumbing fittings, however, are intended only for 2D plans, and, as such, would not typically have a 3D view block. Schematic symbols are a bit different because they are designed to be used for Plan as well as 2D Isometric views. Thus, the Style editor for a Schematic Symbol contains additional settings for the various Isometric planes and rotations on the plane (see Figure 9.7).

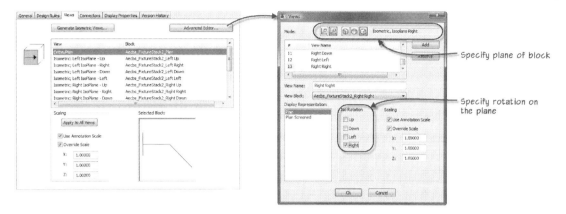

FIGURE 9.7

View editor for Schematic Symbols

SCALING AND ANNOTATION

Frequently, a block used in a style-based object needs to scale according to the drawing scale. For example, a block used for a tee fitting in a plumbing line layout needs to be half the size at 1/4" scale vs. 1/8" scale so that when it plots, the size is consistent. The "Use Annotation Scale" setting enables a block to scale according to the drawing annotation scale (see Figure 9.8). The Override Scale option (and associated x, y, and z values) provides the ability to scale the particular view block independent of the drawing scale. This is not commonly used; instead, the block geometry in the original block definition is typically resized if necessary.

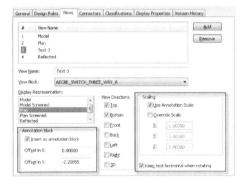

FIGURE 9.8

Annotation and scaling options for a view definition

Note: Blocks associated with style-based content should not be defined as annotative. Use the "Use Annotation Scale" setting in the AMEP style instead.

All the settings for a particular View Name are specific to that view. For example, the Model view typically does not have the "Use Annotation Scale" option enabled as the model block is typically created actual size. However, many plumbing line fittings, schematic symbols, and plan view blocks for devices are scaled per the annotation scale.

The Annotation Block option "Insert as annotation block" (see Figure 9.8) provides the ability for a particular view block to be offset and grip edited to move it independent of the location of the actual model geometry. This is typically used to offset the "3" in a 3-way switch plan representation, for example. This allows the 3 to be placed in a location independent of both the model geometry, and the "S" geometry in the plan representation of the switch. Note that the offset value is merely a default, and the actual location may be edited for each instance of the style (similar to grip-editing the location of an attribute in a block).

Finally, the "keep text horizontal when rotating" option keeps any text within the selected block oriented at 0 rotation, regardless of the rotation of the device. This works best if the text is middle or middle center justified, and defined at the origin.

SCALING OPTIONS

In this exercise we will inspect the annotation and scaling options of a block-based style.

29. On the View cube, orient to the Top view (see Figure 9.9).

FIGURE 9.9
Choose the top view orientation

30. On the Drawing Status Bar, make sure the drawing scale is set to 1/8" = 1'-0".

31. Select the Text Cube Device onscreen and then on the Properties palette, set the Rotation to **0**.

32. On the Device tab of the ribbon, on the General panel, click the Edit Style button (see Figure 9.10).

FIGURE 9.10
Edit the style of our Text Cube Device

33. Click the Views tab and create a new view definition:

- On the right, click the Add button.
- In the View Name box, replace the default name of "New View" with **PLAN SQUARE**.
- From the View Block list, select **UNIT_SQUARE**.
- From the Display Representation list, select **Plan**.
- From the View Directions list, uncheck everything except for Top.

34. Click OK.

There are now two separate blocks visible in the Plan representation (see Figure 9.11).

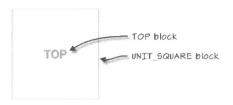

FIGURE 9.11
Two view definitions visible in the plan representation

SPECIFY AN ANNOTATION BLOCK

35. With the Device still selected, on the Device tab, click the Edit Style button.

36. Click the Views tab and create a new view definition:

⇨ On the right, click the Add button.

⇨ In the View Name box, replace the default name of "New View" with **THREE**.

⇨ From the View Block list, select THREE.

⇨ From the Display Representation list, select Plan.

⇨ From the View Directions list, uncheck everything, except for Top.

⇨ Select the "Inset as annotation block" checkbox.

⇨ In the Offset in X field, type **0.625**.

⇨ In the Offset in Y field, type **0.5**.

⇨ Select the "Keep text horizontal when rotating" checkbox.

37. Click OK.

The Device should remain selected and a grip should appear on the "3" block (see Figure 9.12). This is because the 3 block was configured as an Annotation Block. Using an Annotation Block with an offset is useful for cases such as 3-way switches and GFI receptacles. In such cases, the annotation block is used to annotate the Device, and is independently movable from the rest of the Device instance allowing you to position the annotation block wherever you want, but still keeping it as part of the Device.

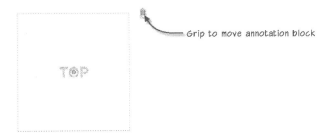

Grip to move annotation block

TOP

FIGURE 9.12
A grip appears on Blocks defined as Annotation

38. Click on the grip, and move it.

Notice that only the "3" block moves.

39. On the Properties palette, set the Rotation to **45**.

Note that the 3 rotates with the device, but maintains a horizontal orientation.

SPECIFY ANNOTATION SCALING

The Device should still be selected.

40. Set the device Rotation back to **0**.

41. On the Device tab, click the Edit Style button.

42. Click the Views tab.

43. Select the PLAN SQUARE view from the list.

44. Check the box "Use Annotation Scale" (see Figure 9.13).

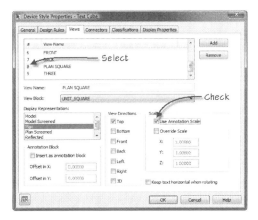

FIGURE 9.13
Set the PLAN SQUARE view definition to use Annotation Scale

45. Click OK.

The square resizes.

46. Zoom out to see the square.

The square has now scaled according to the Drawing Scale as well as the Annotation Plot Size. Annotation Plot Size is configured in the "Drawing Setup" dialog. For many Devices, such as receptacles and fire alarm devices, the Use Annotation Scale option is used for the Plan Display Representation. For Devices that are shown actual size, such as 2x4 light fixtures, you would not use the annotation scale option.

47. From the Application menu, choose **Drawing Utilities > Drawing Setup**.

48. Click the Scale tab (see Figure 9.14).

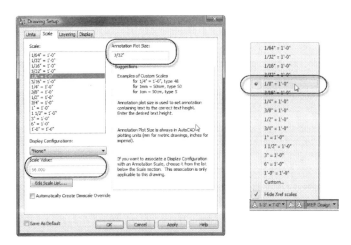

FIGURE 9.14

Annotation Plot Size in Drawing Setup (left) and Drawing Scale (right)

In the Figure 9.14, the annotation plot size is set to 3/32". A 1/8" scale drawing is scaled by 1:96 (indicated by the Scale Value). When an annotation block is displayed in a drawing, it is scaled to maintain a consistent size when plotted regardless of the drawing scale. With a drawing scale of 1/8"=1'-0" (1:96), and the Annotation Plot Size 3/32", the block displays as 9" square (3/32 * 96 = 9).

49. Close the "Drawing Setup" dialog.

50. From the Scale pop-up menu on the Drawing Status Bar, choose 1/4"=1'-0".

The block will resize, and is now 3/32" x 48 = 4.5" square. Use the DIST command to measure the square to confirm.

51. Set the scale back to: 1/8"=1'-0".

52. Select the Device and edit the style again:

⇨ Select the TOP view.

⇨ Check the box for Use Annotation Scale, and Keep text horizontal when rotating (see Figure 9.15).

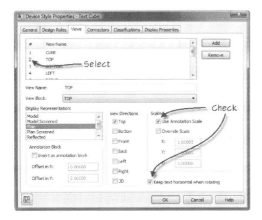

FIGURE 9.15

Set the TOP view to use Annotation Scale and Keep text horizontal when rotating

⇨ Select the THREE view.

⇨ Check Use Annotation Scale.

53. Click OK.

54. Rotate the Device and change the scale of the drawing.

Notice how the TOP and "3" blocks maintain their relative positions and horizontal orientation, and how the square and "3" scale do keep their relative distance.

55. Save the drawing if you wish, and then close it.

MODIFY A DEVICE STYLE TO MAKE IT ANNOTATIVE

In Chapter 7, we created switches by converting existing blocks to Devices. However, these switches are not annotative; that is, if we change the drawing scale, the switches don't update accordingly. In this exercise, we will make the necessary changes to make the switches annotative.

> Make sure the Project is loaded and the Project Navigator palette is displayed onscreen (see the "Install the CD Files and Open a Project" topic above).

1. On the QAT, click the Open icon.

2. In the "Select File" dialog, browse to the *C:\MasterAME 2012\Chapter09* folder.

3. Open the file named *03 Lighting.dwg*.

4. On the Insert tab, on the Block panel, click the Edit Block button.

5. Scroll to the bottom of the list, select Double Switch, and then click OK.

6. On the Home tab, expand the Modify panel and click the Scale button.

   If you prefer, you can type SCALE at the Command Line instead.

⇨ At the "Select objects" prompt, type **all,** and then press ENTER twice.

⇨ At the "Specify base point" prompt, type **0,0,** and then press ENTER.

⇨ At the "Specify scale factor" prompt, type **1/9,** and then press ENTER.

When defining blocks that use the Device annotation scale, the block definition needs to be defined to be the size you intend for it when plotting. In this case, the block was defined for the size it would need to be in a 1/8" scale drawing. However, to make it more flexible and usable at any scale, we scale the block to the desired plot size, and the Annotation Scale functionality will scale it up as appropriate.

7. On the Block Editor tab, on the Close panel, click Close Block Editor button.

8. When prompted, click Save the changes to Double Switch.

9. Select one of the Double Switch Devices.

10. On the Device tab, on the General panel, click the Edit Style button.

11. Click the Views tab and make the following changes:

⇨ Select the Plan view.

⇨ Beneath Scaling, check the "Use Annotation Scale" checkbox (see Figure 9.16).

FIGURE 9.16

Set the Plan view definition to Use Annotation Scale

12. Click the Connectors tab.

13. Change the Connection Point Y to **1.375,** and then click OK.

Just as with the block scaling, the Connector location will scale as well. In this case, the Connector was at 12.375, scaling by 1/9 results in 1.375.

14. On the Drawing Status Bar, change the drawing scale from 1/8"=1'-0" to **3/32"=1'-0"**.

Note that changing the scale resizes the switch Device accordingly, and the Wires stay connected. This is a huge timesaver if, for example, you need to change the drawing scale at the last minute. Feel free to experiment with other scales if you like.

15. Save the drawing and then close it.

CONNECTORS

Connectors are what allow components to interconnect. Devices, Plumbing Fittings, Schematic Symbols, and Panels are all style-based components that have connectors. The connectors define the point of connection, and except for electrical connectors, also define the direction of the connection (*not* the direction of flow). The connection direction makes sure that a connected object (Plumbing Line or Schematic Line) is drawn in the correct direction from a Fitting, Symbol, or Multi-View Part.

CREATE A PLUMBING FITTING WITH CONNECTORS

In this exercise, we will create two Plumbing Fitting Styles: a Tee, and a Cross.

1. On the QAT, click the Open icon.

2. In the "Select File" dialog, browse to the *C:\MasterAME 2012\Chapter09* folder.

3. Open the file named *Plumbing Fitting.dwg*.

4. On the Manage tab, click the Style Manager button.

5. Expand the *Plumbing Objects* category.

6. Right-click *Plumbing Fitting Styles* and choose **New**.

7. Enter the name **Donut Tee,** and then press ENTER.

8. On the right side, click the Details tab.

9. From the Type list, choose **Tee**.

10. Click the Views tab and configure the following settings (see Figure 9.17):

   ⇨ Click the Add button.

   ⇨ In the View Name box, replace the default name of "New View" with **General**.

   ⇨ From the View Block list, select DONUT.

   ⇨ From the Display Representation list, select General.

   ⇨ From the View Directions list, leave all boxes checked.

   ⇨ Check the "Use Annotation Scale" box.

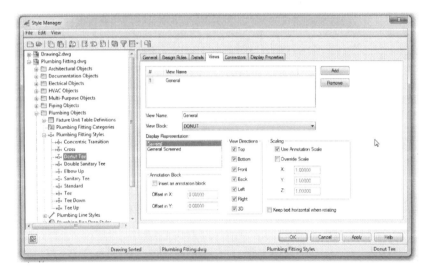

FIGURE 9.17

Define the General view

11. Click the Connectors tab:

   ⇨ Click the Add connector icon three times to create three connectors (see Figure 9.18).

   ⇨ Specify the values for the three connectors as shown in Table 9-2.

TABLE 9-2—*CONNECTOR SETTINGS*

	Connector 1	Connector 2	Connector 3
Connection Point X	0.5	-0.5	0
Connection Point Y	0	0	0.5
Connection Point Z	0	0	0
Connection Direction X	1	-1	0
Connection Direction Y	0	0	1
Connection Direction Z	0	0	0

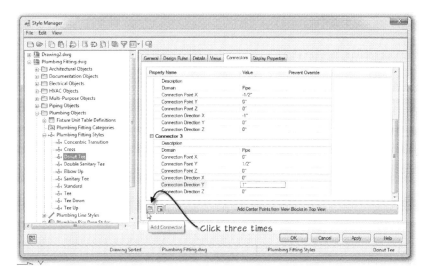

FIGURE 9.18
Add three connectors and configure the settings

12. Select Donut Tee on the left side of Style Manager.

13. From the Style Manager menu, choose Edit > Copy.

14. From the Style Manager menu, choose Edit > Paste.

15. Select Donut Tee (2) on the left side of Style Manager.

16. Click the General tab and change the name to **Donut Cross**.

17. On the right side, click the Details tab.

18. From the Type list, choose **Cross**.

Since this style is a copy of Donut Tee, the Views tab should already be configured properly. Feel free to verify it if you wish.

19. Click the Connectors tab:

The first three connectors came from the Donut Tee style we copied. We need one more.

⇨   Click the Add Connector icon.

Connection 4 will appear with its settings copied from one of the other connectors.

⇨   Specify the values for Connector 4 as shown in Table 9-3.

TABLE 9-3—*CONNECTOR SETTINGS*

	Connector 4
Connection Point X	0
Connection Point Y	-0.5
Connection Point Z	0
Connection Direction X	0
Connection Direction Y	-1
Connection Direction Z	0

20. Click OK to complete the settings and dismiss the Style Manager.

TEST THE NEW FITTINGS

21. On the Application Status Bar, make the Plumbing Workspace current.

22. On the Home tab, on the Build panel, click the Plumbing Line button.

23. On the Properties palette, beneath the Labels and Flow Arrows grouping, change the Style for both Labels and Flow Arrows to *NONE* (see Figure 9.19).

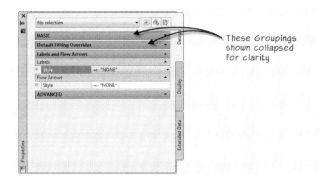

FIGURE 9.19

Set the Labels and Flow Arrows styles to *NONE* (some groupings collapsed for clarity)

24. Create a single Plumbing Line that is approximately 10'-0" long, and then cancel the Plumbing Line command.

25. Select the Plumbing Line you just created.

26. On the Plumbing Line tab of the ribbon, on the General panel, click the Edit System Style button.

27. Click the Defaults tab.

28. From the Tee drop-down list, select **Donut Tee**.

29. From the Cross drop-down list, select **Donut Cross**.

30. Click OK.

31. Click the plus (+) grip in the middle of the plumbing line (see Figure 9.20).

FIGURE 9.20

Click + grip to add a fitting and new pipe section

32. Click to route a Plumbing Line perpendicular to the first. Note that the Tee is automatically inserted.

33. Cancel the Plumbing Line command.

34. Select the Tee that was inserted.

35. Click the plus (+) grip.

36. Move your mouse in the opposite direction and then click to place a Plumbing Line segment (see Figure 9.21).

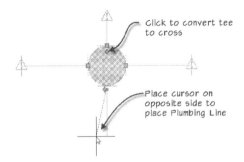

Click to convert tee to cross

Place cursor on opposite side to place Plumbing Line

FIGURE 9.21
Route a Plumbing Line from the converted cross

37. Cancel the Plumbing Line command.

If you wish, you can reopen the DONUT block to explore how it is defined. The outside diameter of the donut is one unit, and centered on the origin. This results in the four quadrants of the donut being at the coordinates: (0.5, 0), (-0.5, 0), (0, 0.5), (0, -0.5) which coincides with the connection points we defined above. For each connection point, there is a connection direction, which is a vector in the direction away from the connection. For this simple case, the vectors corresponding to the connector locations listed above are positive x, negative x, positive y, negative y. In the definition, we specified +1 or -1; however, the scale doesn't really matter, +0.1 or -0.1 would work just as well. We will see how these vectors are derived in a more complicated case later in this chapter.

In this topic, you saw how Connectors on Plumbing Fittings define the connection points and directions. Below, you will see how Devices utilize connectors to define electrical loads.

38. Save the drawing if you wish, and then close it.

DEVICES

This topic describes some characteristics unique to Devices that you should consider when creating them. Devices (and Panels) can automatically orient to other geometry, reducing the need to manually rotate Devices as you place them. This is conceptually similar to the

Alignment Grip within a Dynamic Block Definition. For this to work properly, the blocks making up the Device must be oriented in a certain direction, otherwise, the rotation may appear off by 90 or 180 degrees.

When a device is inserted, its origin coincides with the origin of the blocks that make up the Device Style. For a typical wall-mounted device, the block should be oriented to point "north". As shown in Figure 9.22, the Device on the short Wall segment is placed with 0° rotation. With the block defined in this manner, the Device will insert perpendicular to the wall as expected, regardless of the rotation of the wall or other geometry.

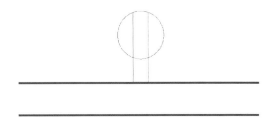

FIGURE 9.22
Device 0° orientation

MODIFY A BLOCK DEFINITION TO PROPER ORIENTATION

The "align to geometry" functionality has not always been a part of AMEP. As such, a handful of the electrical Devices that ship with AutoCAD MEP have not been updated to properly take advantage of this feature. Additionally, it is possible that some of your company standard blocks may have been converted to Devices but are likewise not oriented properly either. Use the following procedure to correct the block definitions so that when you use the Devices in the future, they function as expected.

1. On the QAT, click the Open icon.

2. In the "Select File" dialog, browse to the *C:\ProgramData\Autodesk\MEP 2012\enu\MEPContent\USI\Electrical\Devices* folder.

3. Open the file named *Fire Alarm (US Imperial).dwg*.

This folder is the default location indicated for Electrical Devices in the "Options" dialog. If your CAD Manager has relocated this folder, check with them to see where the location is in your installation and get permission to edit this file. Alternatively, if this is not possible, a copy of the *Fire Alarm (US Imperial).dwg* has been provided in the *C:\MasterAME 2012\Chapter09* folder for your convenience.

4. On the Drawing Status Bar, change to the Electrical Workspace.

5. On the Insert tab, on the Block panel, click the Edit Block button.

6. Expand the size of the window so you can see the full block names in the list.

7. Select the AECB_SCM_FA_SpeakerHornElec_P block, and then click OK (see Figure 9.23).

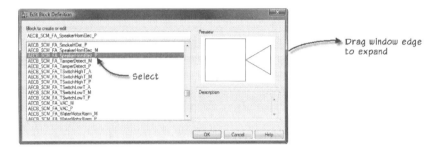

FIGURE 9.23

Select the block from the Edit Block Definition window

8. On the Home tab, on the Modify panel, click the Rotate button (see Figure 9.24).

FIGURE 9.24

Rotate tool on the Home ribbon, Modify panel

- At the "Select objects" prompt, type **all** and then press ENTER twice.
- At the "Specify base point" prompt, type **0,0** and then press ENTER.
- At the "Specify rotation angle" prompt, type **90** and then press ENTER.
- On the ribbon, click the Close Block Editor button.

9. When prompted, click Save the changes to AECB_SCM_FA_SpeakerHornElec_P.

10. Repeat the above procedure to rotate the 3D model block named: AECB_SCM_FA_SpeakerHornElec_M.

11. Close the drawing, saving your changes.

> **Note:** If you edited the file provided in the *C:\MasterAME 2012\Chapter09* folder instead of the one in the out-of-the-box *MEPContent* folder, you will either have to test the Device in the current drawing, or temporally change the path for Electrical Devices in the "Options" dialog to the *C:\MasterAME 2012\Chapter09* folder. If you do change the path, be sure to change it back to its original setting later. Ideally, the change detailed here should be saved permanently over the original file in the default location.

12. On the Home tab, on the Build Panel, click the Device tool.

13. On the Properties palette, click the Style image.

14. From the Drawing File list, choose **Fire Alarm (US Imperial)**.

Notice that the Speaker-horn (electric horn) Device Style is now oriented upright (see Figure 9.25).

15. Add one to the file if you wish. It should behave as expected.

FIGURE 9.25

Resulting preview image in the Select a style window

If you do not want to make edits to the original out-of-the-box content files, you can save the modified file with a new name to the Electrical Devices path (configured in the "Options" dialog). All drawings located in that folder will appear on the Drawing file list shown in the figure.

CLASSIFICATION

Many styles have a Classifications tab. Classifications are used within Display Sets to control which objects are visible in various circumstances (see the "Understanding Sets" topic in Chapter 2 for more information) and in defining Property Sets (refer to Chapter 15). This primarily applies to Multi-View Parts and Devices, as other style-based content styles (Panels, Wires, Plumbing Fittings, and Schematic Symbols) don't have a Classifications tab.

With Devices, there are two potentially conflicting settings for classification in the Device Style. On the Design Rules tab is a "Type" setting and on the Classifications tab is a "Device Type" classification. The Design Rules Type is a hard-coded list of device classifications consisting of: Undefined, Receptacles, Lighting, Switch, Junction Box, Communication, Fire Safety, Other Power, and Security. When inserting a Device into a drawing for the first time, the Type setting on Design Rules will determine what the Device Type Classification (on the Classifications tab) will be. For example, if the Design Rules Type is set to Receptacles, the Device Type Classification will also become Receptacles, even if in the content drawing the Classification is set to something else. In other words, the Design Rules setting will take precedence. The exception is if the Design Rules Type is set to Undefined; in this case, the Classification setting will be used. This allows you to add additional device types for the purposes of schedule and visibility filtering when necessary.

> Note: The Types on the Design Rules tab correspond to commands such as ReceptacleModify, FireSafetyModify, etc. These commands will filter for the particular Type of Device before showing the selected objects in the Properties palette. This functionality is not commonly used.

IFC MEP TYPE CLASSIFICATION

Device Styles also have an IfcMEPTypeClassification. Since a Device is a rather generic term, the Classification is used to categorize elements more specifically as Lighting Fixtures, Outlets, and the like, when using the IFCEXPORT command to create an IFC file from your drawing. For example, the Duplex Receptacle has its IfcMEPTypeClassification set to IfcOutletType.POWEROUTLET (see Figure 9.26).

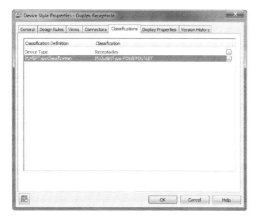

FIGURE 9.26
Device Classifications

LAYER KEY

The Layer Key is used to determine on what layer a Device will be inserted. The Layer Key for a Device Style is assigned on the Design Rules tab (see Figure 9.27).

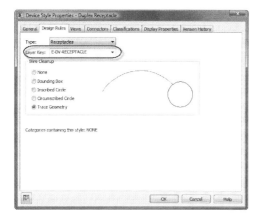

FIGURE 9.27
Layer Key setting for a Device Style

In the default template, the Layer Keys for devices generally begin with E-DV (for Electrical Device). The keys reflect the type/classification of the Device, such as E-DV-RECEPTACLE.

If the Layer Key on a Device style is left empty, when the Device is inserted it will insert on the layer associated with the current System. This is used in scenarios where you may use the same

Device styles on multiple systems, such as specialty fire alarm systems. If the selected System has no Layer Key assigned to it, the Device will insert on the layer associated with the DEVICE layer key (see Figure 9.28).

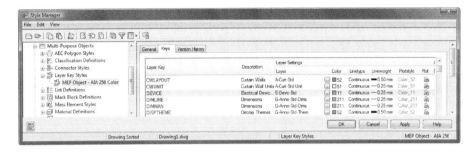

FIGURE 9.28
The DEVICE Layer Key in the MEP Object - AIA 256 Color Layer Key Style

> **Note:** As you can see from the figure, Layer Key Styles also are styles and are accessed via Style Manager like any other style.

Typically, however, Devices are layered according to the key assigned to the Device Style, as this minimizes any necessity to select a System when placing the Device. By modifying the Layer Key Style, it is possible to define additional keys or edit the existing ones if desired to comply with your firm's layering requirements.

Electrical Panels, Schematic Symbols, and Plumbing Fittings do not have layer keys associated with their styles. The layering of these elements is typically controlled by the System selected when creating such objects.

WIRE CLEANUP

The Wire Cleanup setting on the Design Rules tab (see Figure 9.27, above) controls how Wires trim to a Device. Typically, the Trace Geometry or the None options are used. The Bounding Box, Inscribed Circle, and Circumscribed Circle options can result in some strange effects for non-rectangular Devices (see Figure 9.29).

NONE   BOUNDING   INSCRIBED   CIRCUMSCRIBED   TRACE
            BOX   CIRCLE   CIRCLE   GEOMETRY

FIGURE 9.29
Wire Cleanup samples

ELECTRICAL CONNECTORS

Connectors on electrical Devices provide the capability for Devices to be circuited to Panels. The connector location properties define where a Wire will land on the device. There are three System types that can be assigned to an electrical connector which correspond to the three types of Circuits (refer to Chapter 7 for more information on Circuits). The Power and Lighting System type connector provides properties for voltage/poles, load, and load category that allow total connected and demand load computations to occur within AutoCAD MEP. Additionally, there are Maximum Overcurrent Rating and Power Factor properties that exist solely for third-party applications (AutoCAD MEP does nothing with these values other than store the data). The Other and General System type connectors do not provide load information; they are used for grouping or zoning devices, such as for fire alarms. The General System type connectors are used for some lighting and power devices, such as switches, that themselves contribute no load.

ASSOCIATE A 3D MODEL BLOCK WITH A DEVICE STYLE

In Chapter 7, we created a lighting layout. Since the lighting fixtures in that drawing were created by converting 2D AutoCAD blocks to Devices, there is no 3D representation to these lighting fixtures. In this exercise, we will define a Model/3D view for the 2x4 Troffer fixture style. Additionally, we will modify the load information on the connector to define different conditions.

> The Electrical Workspace should still be current from the previous exercise. If it is not, please make it current now.

1. On the QAT, click the Open icon.

2. In the "Select File" dialog, browse to the *C:\MasterAME 2012\Chapter09* folder.

3. Open the file named: *03 Lighting.dwg*.

4. On the View tab, on the Appearance panel, from the view list click SW Isometric.

5. On the Manage tab, on the Style & Display panel, click the Style Manager drop-down button and choose Electrical Device Styles (see Figure 9.30)

FIGURE 9.30

Select Electrical Device Styles from the Style Manager drop-down button

6. Select 2x4 Troffer - 3 Lamp style on the left side.

7. On the right side, click the Views tab and configure the following settings (see Figure 9.31):

⇨ Click the Add button.

⇨ In the View Name box, replace the default name of "New View" with **Model**.

⇨ From the View Block list, select **2x4_troffer_m**.

⇨ From the Display Representation list, select **Model**.

⇨ From the View Directions list, uncheck Top and Bottom. Make sure Front, Back, Left, Right, 3D are all checked.

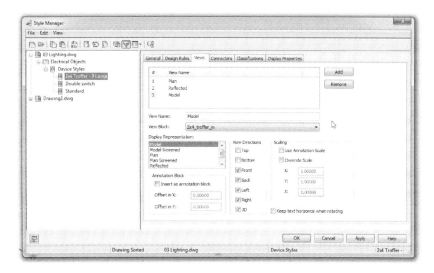

FIGURE 9.31
Add a Model block to the existing light fixture

CREATE 2 AND 4 LAMP TROFFERS FROM THE 3 LAMP TROFFER

8. With the 2x4 Troffer - 3 Lamp still selected, click Copy and Paste twice to create two duplicates.

   The duplicates will be named 2x4 Troffer - 3 Lamp (2) and 2x4 Troffer - 3 Lamp (3) (see Figure 9.32).

FIGURE 9.32
Create copies of the selected Device Style

9. Select 2x4 Troffer - 3 Lamp (2) and on the General tab, rename it to **2x4 Troffer - 2 Lamp**. Update the Description accordingly.

10. Repeat for 2x4 Troffer - 3 Lamp (3) naming it **2x4 Troffer - 4 Lamp**.

11. Select 2x4 Troffer - 2 Lamp and then click the Connectors tab.

12. Set the Load Phase 1 to **64**.

13. Select 2x4 Troffer - 4 Lamp.

14. Set the Load Phase 1 to **128**.

15. Click OK to accept all changes and close Style Manager.

Since we switched the drawing to a 3D view before opening the Style Manager, you should immediately see all of the lighting fixtures now displayed in the 3D view. We will take advantage of the two and four lamp styles in Chapter 14 where we will change some of the 3 Lamp fixtures to 2 Lamp or 4 Lamp fixtures, to have lighting fixture tags and circuit loads update accordingly.

16. Save and close the file.

PANEL STYLES

Panel objects provide the capability to create circuits. The style-based definition of Panels is similar to Devices with the following exceptions:

- Panels don't have Type or Classification properties.
- Panels don't have a Layer Key property; a panel is layered according to the System that is current when the panel is inserted.
- Panel connectors don't allow you to assign a load. This stands to reason as panels feed loads, they aren't loads themselves; loads come from Circuits to which Devices are connected.
- Connectors can be defined to be for Conduit and Cable Tray. However, the functionality is not similar to that for Conduit/Cable Tray connectors on MvParts. As such, they are not commonly used.

CREATE A PANEL STYLE

IN THIS EXERCISE, WE WILL UTILIZE BLOCKS THAT ARE ALREADY IN THE DRAWING TO DEFINE A SWITCHBOARD PANEL STYLE.

1. On the QAT, click the Open icon.

2. In the "Select File" dialog, browse to the *C:\MasterAME 2012\Chapter09* folder.

3. Open the file named *Switchboard.dwg*.

4. From the Manage tab, click the Style Manager button.

5. Expand the *Electrical Objects* folder.

6. Right-click *Panel Styles*, and choose New.

7. On the General tab, change the name to **800A Switchboard**.

8. Click the Views tab and configure the following settings:

⇨ Click the Add button.

⇨ In the View Name box, replace the default name of "New View" with **Plan**.

⇨ From the View Block list, select Switchboard_800A_P.

⇨ From the Display Representation list, select **Plan**.

⇨ From the View Directions list, uncheck all View Directions, except for Top.

9. Click Add again, and then specify the following:

⇨ Change the View Name to **Model**.

⇨ From the View Block list, select Switchboard_800A_M.

⇨ From the Display Representation list, select **Model**.

⇨ From the View Directions list, uncheck Top.

10. Click OK.

If you open the "Options" dialog (Application menu) and click the MEP Catalogs tab, the default location for Electrical Panels is: *C:\ProgramData\Autodesk\MEP 2011\enu\MEPContent\USI\Electrical\Panels*. This next sequence requires us to save the file to this location or whatever location is indicated for Electrical Panels in your "Options" dialog. If you are unable to save files to this location, you can leave the file open and use the <Current Drawing> option to test your work.

11. From the Application menu, choose Save As > AutoCAD Drawing.

12. In the "Save Drawing As" dialog, browse to the *C:\ProgramData\Autodesk\MEP 2012\enu\MEPContent\USI\Electrical\Panels* folder (or whatever folder is listed in your "Options" dialog).

13. Name the file **My Switchboards.dwg** and then click the Save button.

14. Close the *My Switchboards.dwg* drawing file.

INSERT THE SWITCHBOARD

15. On the QAT, click the Open icon.

16. In the "Select File" dialog, browse to the *C:\MasterAME 2012\Chapter09* folder.

17. Open the file named *01 Power.dwg*.

18. Select the Panel on the north outside wall by the stairway (see Figure 9.33).

19. On the Properties palette, click the Style preview image.

20. From the Drawing file list, choose **My Switchboards**.

21. Double-click 800A Switchboard.

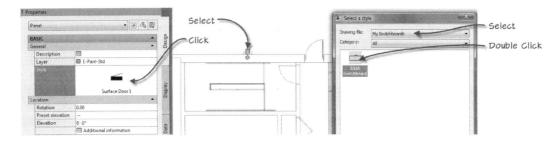

FIGURE 9.33

Change the style of the specified panel

22. On the Properties palette, set the Elevation to 4".

Changing the Panel Style using the method above is preferable to erasing and reinserting a new Panel. First, it is fewer clicks; second, if the Panel were erased, any association with Circuits and panel schedules would be disrupted. When a Panel is erased, the Circuits are orphaned under the <Unassigned> node in Circuit Manager. These Circuits can be dragged and dropped to another panel; however, such efforts are not necessary if the above procedure is adopted. Any panel schedules that refer to Panels that have been erased will cease to update (all Fields will indicate ####). Even if a new Panel with the same name is created, the schedule will need to be erased and replaced. Swapping the Panel in this way instead keeps the link to the Schedule intact.

23. Save and close the *01 Power* drawing.

PLUMBING FITTINGS

Plumbing Fittings are similar to Devices, with the following exceptions:

- Plumbing Fittings don't have a Layer Key property. Plumbing Fittings are layered according to the System that is current when the fitting is inserted.

- Plumbing Fittings don't have a Classification property. There is a Type property, but this has specific functionality for Plumbing Fittings, and will be detailed below.

Earlier in this chapter, in "Create a plumbing fitting with connectors" topic, we created a Plumbing Fitting. In this topic, we will discuss more about the Type and Sub Type properties of Plumbing Fitting styles. We will also discuss a little more about how plumbing fitting connector directions work.

If you return to the Style Manager and click the Details tab for a Plumbing Fitting style, you will find the Type drop-down list. This list of Plumbing Fitting Types comes from a built-in list that cannot be edited (see Figure 9.34).

FIGURE 9.34

Type options for a Plumbing Fitting style

The Type of fitting determines what plumbing fitting styles are available for the fitting options in a Plumbing System Definition (see Figure 9.35).

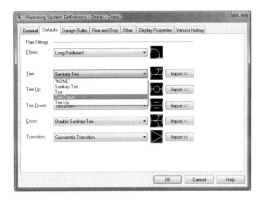

FIGURE 9.35

Plumbing System Definition fitting Defaults; the Tee box lists the fittings specified as type Tee

In the System Definition, only a few fitting types are used: tees, crosses, and transitions. These fitting types are automatically inserted as your route pipe. Elbows of any angle are automatically placed as well, and their styles are generated dynamically as needed.

DYNAMICALLY GENERATE ELBOWS

In this exercise, we will dynamically create some elbow fittings to inspect how they are created. First, we will set the units settings to high precision to be able to see exactly what is going on.

1. Create a new drawing based on the *Aecb Model (US Imperial Ctb).dwt* template.

> Note: If for some reason you do not have this template file, a copy has been provided in the *C:\MasterAME 2012\Template* folder.

2. On the Application Status Bar, change the Workspace to **Plumbing**.

3. From the Application menu, choose **Utilities > Drawing Setup**.

4. Click the Units tab.

5. Set the Length Type to **Decimal**, select the highest level of precision and then press OK (see Figure 9.36).

FIGURE 9.36

Modify the Units settings

6. On the View tab, on the MEP View panel, click the Compass button.

7. Uncheck Enable Snap and then click OK.

8. On the Manage tab, click the Style Manager button.

9. Expand the *Plumbing Objects* category and then *Plumbing Fitting Styles* (see Figure 9.37)

FIGURE 9.37

Plumbing Fitting Styles in the default template

10. Take note of the list of fittings available and then click OK.

11. On the Home tab, on the Build panel, click the Plumbing Line button.

12. Click in the drawing area to pick a start point.

13. At the "Specify next point" prompt, type **@120<0** and then press ENTER.

14. At the "Specify next point" prompt, type **@120<41.222** and then press ENTER.

15. At the "Specify next point" prompt, type **@120<12.345** and then press ENTER.

16. Re-open Style Manager, and expand *Plumbing Fitting Styles* again.

17. Note the two new additions at the top of the list (see Figure 9.38).

FIGURE 9.38

Dynamically created Plumbing Fitting Styles appear automatically in Style Manager

> **Note:** You may have different elbow types depending on the system you had current when routing the plumbing line.

18. Select the 41.222 elbow style.

> **Note:** If you have the compass snap turned on, you will be limited to more "standard" elbow sizes.

19. Click the Connectors tab.

Connector 1 has an X,Y,Z direction of -1,0,0, at a location of -0.75,0,0 That is, a pipe that connects to it goes straight to the left (negative X direction). Connector 2 has a more odd set of values for the direction as shown in Figure 9.39.

FIGURE 9.39

Connector 2 settings for the 41.222 elbow style

20. Click OK to close Style Manager.

Let's take a look at where these values come from:

21. At the Command Line, type **CAL** and then press ENTER.

22. At the ">> Expression" prompt, type **COS(41.222)** and then press ENTER.

The result is: 0.752161935. Take another look at Figure 9.39; this is the Connection Direction X shown for Connector 2.

23. Repeat CAL and at the ">> Expression" prompt, type **0.75*COS(41.222)** and then press ENTER.

This time the result is: 0.564121451. This is the Connection Point X shown for Connector 2.

24. Repeat CAL and at the ">> Expression" prompt, type **SIN(41.222)** and then press ENTER.

The result is: 0.658978318. This is the Connection Direction Y shown for Connector 2.

25. Repeat CAL and at the ">> Expression" prompt, type **0.75*SIN(41.222)** and then press ENTER.

The result is: 0.494233739. This is the Connection Point Y shown for Connector 2.

The 0.75 is the length of the "leg" of the fitting, regardless of the angle. Of course, this is scaled by the Drawing Scale and the Plot Annotation Size. Thus, for the standard 3/32" Annotation Plot Size at a 1/8"=1'0" (1:96) scale, this works out to: 3/32 * 96 * 0.75=6.75 (see Figure 9.40).

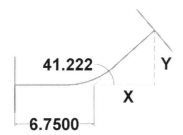

FIGURE 9.40
Plumbing fitting style showing key dimensions

As indicated, such elbow fittings are created dynamically, and thus the theory is trivial. However, understanding how these elbow fittings work is more important if you find yourself defining other fitting types that aren't created automatically.

26. On the View tab, on the MEP View panel, click the Compass button.

27. Check the Enable Snap option, and then click OK.

28. Close the drawing—you don't need to save it.

SCHEMATIC SYMBOLS

Schematic symbols are found on the Tool Palettes when the Workspace is set to Schematic. These symbols are slightly different from the other style-based content we have seen thus far. Instead of having a model representation and a plan representation, these symbols have a single Ortho (plan) representation, and isometric representations for each of the various isometric rotation directions (see Figure 9.41).

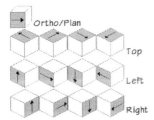

FIGURE 9.41
The possible block plan and isometric orientations and rotations

As you can see, each schematic symbol has up to 13 separate blocks. Including all gives you the most flexibility in orienting the symbol to create a flat or isometric diagram. It could potentially be a lot of work to generate a block for each orientation; however, AutoCAD MEP provides an Isometric View Generator that will create the 12 isometric blocks from a single flat block.

CREATE AN ISOMETRIC SYMBOL

In this exercise, we will create an isometric symbol based on the block shown in Figure 9.42.

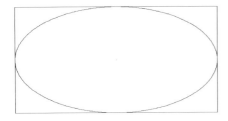

FIGURE 9.42
The block that will be the basis of a new Schematic Symbol

29. Create a new drawing based on the *Aecb Model (US Imperial Ctb).dwt* template.

30. On the Application Status Bar, make the **Schematic** Workspace current.

31. On the Insert tab, on the Block panel, click the Edit Block button.

32. In the "Block to create or edit" field, type **My Schematic Symbol Block,** and then click OK.

> **Note:** When defining blocks that will be used for AutoCAD MEP objects, avoid using Dynamic Block and Parametric functionality. Such features are ignored when used in AutoCAD MEP objects.

33. On the Home tab, on the Draw panel, click the Rectangle icon (see Figure 9.43).

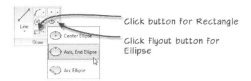

Click button for Rectangle

Click Flyout button for Ellipse

FIGURE 9.43

Rectangle and Ellipse tools on the Draw panel

34. At the "Specify first corner point" prompt, type **-2,-1**, and then press ENTER.

35. At the "Specify other corner point" prompt, type **2,1,** and then press ENTER.

36. Zoom Extents and then zoom out a little more.

37. On the Home tab, on the Draw panel, click the Ellipse drop-down button, and choose Axis, End Ellipse (see Figure 9.43).

38. Use the midpoint snaps to pick the points in the order indicated in Figure 9.44.

FIGURE 9.44

Use the midpoint snaps, and pick in the indicated sequence

39. Click Close Block Editor on the right end of the Ribbon.

40. When prompted, select the "Save the changes to My Schematic Symbol Block" option.

41. On the Manage tab, on the Style & Display tab, click the Style Manager button.

42. Expand the *Schematic Objects* category.

43. Right-click the *Schematic Symbol Style* item and choose **New**.

44. On the General tab, change the Name to **My Schematic Symbol**.

45. Click the Design Rules tab and then select the Trace Geometry option.

46. Click the Views tab and then click the Generate Isometric Views button (see item 1 in Figure 9.45).

Read the information presented in the "Generate Isometric Symbol Blocks" dialog. This will give you a better understanding of the process.

47. Select the "In-Line symbol" option.

48. From the Plan Block list, select **My Schematic Symbol Block**.

49. Make sure the "Use new blocks in the style definition" checkbox is selected, and then click OK (see item 2 in Figure 9.45).

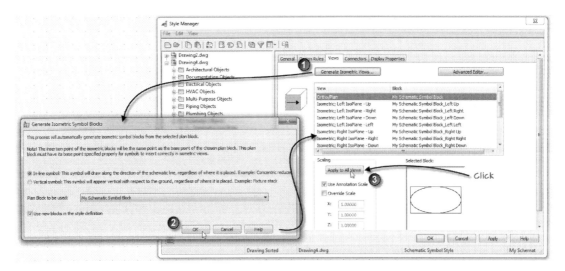

FIGURE 9.45

Generate the Isometric Blocks automatically

View blocks are created for each isometric side and orientation combination, all block names are prefixed with the original block name.

50. Select the "Use Annotation Scale" option, and then click the Apply to All Views button (see item 3 in Figure 9.45).

51. Click the Connectors tab.

Note that a default "Connector 1" has been created, with the Point at 0,0,0, and a direction of 1,0,0.

In schematic symbols, the Direction isn't taken into account when a schematic line is drawn from a connector.

52. Click OK to complete the symbol and close Style Manager.

53. On the Home tab, on the Build panel, click the Schematic Symbol button.

54. On the Properties palette, click the Style preview image to open the "Select a style" dialog.

55. From the Drawing file list, choose <**Current Drawing**>.

56. Double-click on My Schematic Symbol.

57. On the Properties palette, set the Drawing Mode Orientation to **Isometric**.

58. Click to place the block, then move the cursor to rotate it.

Notice how "Rotation in isoplane" (under Location) changes automatically as you move the mouse.

59. Click to set the rotation and then press ENTER to complete the command.

60. Select the inserted symbol, and click the grips to see the effect of each (see Figure 9.46).

FIGURE 9.46
Grips on a selected Schematic Symbol shown isometrically

61. Close the drawing—you don't need to save it.

CATEGORIES

Categories are used with Devices, Panels, Plumbing Fittings, and Schematic Symbols. Categories are used to organize and to make finding a particular style easier. For example, when selecting a Device Style from the *Lighting - Fluorescent (US Imperial)* content file, a list of categories is available (see Figure 9.47).

FIGURE 9.47

Categories of Devices in the Lighting - Fluorescent (US Imperial) content file

This list of categories exists in the content file itself. For example, if you open the *Lighting - Fluorescent (US Imperial).dwg* file, open Style Manager and expand *Electrical Objects*, you will see the *Device Categories* item (see Figure 9.48).

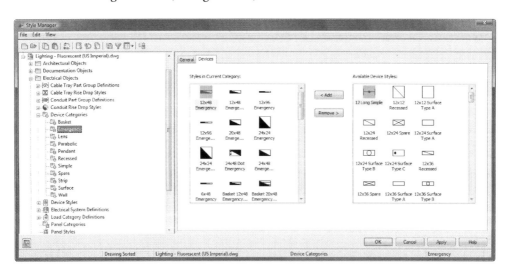

FIGURE 9.48

Device Category definitions in the Lighting - Fluorescent (US Imperial) content file

On the left side of Style Manager, you see the list of Device Categories (Basket, Emergency, Lens, etc.). You may create and name a Category anything you want. On the right-hand side, there are two boxes, one labeled "Styles in Current Category" and the other "Available Device Styles." As shown above, there are a variety of Devices in the Emergency category. A Style may belong to more than one Category. For example, a 24x48 Recessed Emergency Lensed Troffer may belong to "Recessed," "Emergency," and/or "Lens" Categories.

You can delete a Category by right-clicking on the Category name, and selecting Purge. Deleting a Category has no effect on the Styles in that Category. A new category may be added by right-clicking Device Categories and selecting New.

To add a Style to a Category, simply select it from the "Available Device Styles" list and click the < Add button. To remove a Style from a Category, select it in the "Styles in Current Category" list and click the Remove > button.

When you insert a Style-based content object in a drawing (such as a Device), the associated Category does not come along. Categories are only intended to organize content files.

SYSTEM DEFINITIONS

System definitions are primarily used to define what layer components will be inserted. For example, the Plumbing System Definition shown in Figure 9.49 below is keyed to the P-SY-PIPE-DEMO layer key. In the default templates, this key is associated with the layer P-Pipe-Std-Demo.

Note, however, that this particular system definition has a Layer Key Override of -San on Major 1, thus the resulting layer name will be P-Pipe-San-Demo. Refer to the AutoCAD MEP documentation for more information on Layer Keys and Layer Key Overrides.

In addition to the Layer Key, all System Definitions have an Abbreviation. This abbreviation is used when labeling elements; for example, the Chilled Water System Definition has a CHW abbreviation.

As indicated in the System Definition window, the System Group provides a way for different systems to interconnect.

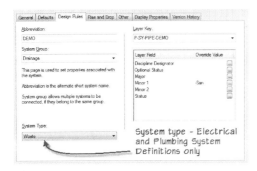

FIGURE 9.49
System Type setting for Electrical and Plumbing System Definitions

Plumbing and Electrical systems have a System Type property (see Figure 9.49) that defines some additional functionality for these systems.

For Plumbing, available system types are Cold Water, Drain, Fire Protection, Gas, Hot Water, Undefined, Vent, and Waste. Plumbing system definitions facilitate some sizing functionality:

Cold Water—Plumbing lines in this system accumulate fixture units based on the Cold Water column in the Fixture Unit Table definition.

Drain—Plumbing lines in this system will not accumulate fixture units when connected to a Waste system.

Fire Protection—Plumbing lines in this system will not accumulate fixture units when connected to a Water system.

Gas—Plumbing lines in this system will not accumulate fixture units.

Hot Water—Plumbing lines in this system accumulate fixture units based on the Hot Water column in the Fixture Unit Table definition.

Undefined—Plumbing lines in this system will not accumulate fixture units.

Vent—Plumbing lines in this system will not accumulate fixture units when connected to a Waste system.

Waste—Plumbing lines in this system accumulate fixture units based on the Waste column in the Fixture Unit Table definition.

For Electrical systems, available system types are Cable Tray, Conduit, General, Other, Power, and Lighting. When routing Conduit, only an electrical system with the Conduit System Type may be selected. Similarly Cable Tray, Devices, Panels, and Wires may only be placed on systems defined as General, Other, or Power and Lighting system types. There is no special sizing functionality related to the electrical system types.

Duct systems have a Design Parameters tab used for defining the duct and air characteristics, and it is used for sizing duct. For more information refer to the "Duct System Definitions" topic in Chapter 5.

Pipe systems have Single Line Graphics settings that are used to display pipe under a specified size as single line. Refer to Chapter 6 for more information.

Plumbing Systems have a Defaults setting that defines what type of fittings to place (see Figure 9.35).

Excluding the Schematic System Definitions, all System Definitions have an Other tab with a single setting to optionally exclude the Shrinkwrap display in generated 2D sections. Refer to Chapter 13 for more information on sections.

All System Definitions have Display Properties settings. Refer to Chapter 2 for more information on Display Overrides.

RISE DROP STYLES

All systems (except for schematic) have a Rise and Drop setting that defines the symbology used for vertical segments. Separate symbols may be used for 1-line vs. 2-line, and for rise vs. drop. The "rise" symbol is used when the object is "open" at the top, such as when a duct continues to the level above. The "drop" symbol is used when the vertical section is obscured, such as when below a turned-down elbow. In addition to the rise/drop graphics themselves, a Rise Drop Style also allows you to define a block that is displayed when center line graphics are turned on. For more information on Rise Drop Styles, refer to Appendix A.

For duct, separate blocks are used for Exhaust, Supply, and Return Duct Rise Drop styles. Of course, these may be expanded and modified if desired. Separate blocks exist for conduit, cable tray, pipe, and plumbing rise drop styles, though typically, for these domains, all systems use same rise drop style and thus the same blocks for all systems.

A Rise Drop Style consists of several views, generally named to reflect the shape, system, and display representation. Each view requires a Name, Display Representation (2-Line or 1-Line), Block, and Center Line Block. Additionally, each view specifies whether it is for Rise or Drop, and, if for a drop, whether for tee/takeoff (vs. elbow). For ducts only, the shape is also specified.

> **Note:** Conduit, Cable Tray, and Pipe Rise Drops all use the 2-Line Display Representation; however, Duct Rise Drops use the Plan representation in lieu of the 2-Line Display Representation. Plumbing Rise Drop Styles use the General display representation since they are always 2D/single line.

The rise/drop blocks, when displayed in a layout, are scaled according to the size of the object (i.e., a drop in a rectangular 18x12 will scale accordingly). Typically, the Annotation and Override scale options are not used, though you may want to use these options for small diameter piping. For consistency, all the rise drop styles and views use the same block for the Center Line Block, with the same Center Line Block Scale factor. However, you may deviate from this if desired.

THE RISE AND DROP BLOCK DEFINITION

A "Rise Block" should be built on Layer 0 with the color, Linetype, and Lineweight all set to ByBlock. This ensures that the graphics will display according to the display control hierarchy, instead of being hard-coded within the definition of the block.

A "Drop Block" should be built on a specific layer like "G-Risr-Std-High" and its color, Linetype, and Lineweight set to ByLayer. This makes control of the "hidden components" of the block easily editable simply by modifying the associated layer's definition. The perimeter components of the 1-line blocks should be on Layer 0 with the color, Linetype, and Lineweight set to ByBlock to ensure that they inherit the characteristics of the system/layer they are associated with, according to the display control hierarchy.

DISPLAY REPRESENTATIONS

The Display Representation selected is dependent on whether the symbol is for 2-Line Representation or 1-Line Representation. Selecting other Display Representations may not provide desired results, except as noted above where duct 2-Line Rise Drop blocks are specified to use the Plan Representation.

Use the 1-Line Display Representation when selecting the block to display for Graphical 1-Line (pipe only), Component 1-Line (pipe only), and the Display Representation 1-Line (Pipe, Conduit, Cable Tray, and Duct).

Use the 2-Line (or Plan) Display Representation when selecting the block to display in all Plan Display Representations.

Center Line block display is controlled by the Center Line Display Component, which is the same regardless of 1-Line or 2-Line display.

RISE/DROP

The Rise/Drop options show how the software determines which block to display for a variety of conditions. Select the Rise option to display the view when the vertical segment is "open," such as when passing to the level above. Select the Drop option when the segment is obscured by a fitting or MvPart component above.

The two options under Rise provide additional flexibility in defining how the Rise Drop graphics display when a vertical segment is under an MvPart or an Endcap. Some conventions show a solid line Rise under MvParts (such as a vertical duct dropping out of the bottom of a roof top air handling unit), whereas others prefer to show hidden line Drops. The same option holds for Endcaps.

Drop Tees/Takeoffs only: This check box specifies that the current view block is to be used for a Tee or a Takeoff in a drop position. With this option unchecked, the Rise Drop view displays a block suited for a down-turned elbow condition.

SCALING

The Scale of the Rise Drop block and Centerline Block are based on the physical dimensions of the vertical segment (in the case of pipes, this is the nominal diameter).

Using annotation scale allows the Rise Drop block size to be controlled by the drawing scale. This can be handy for displaying small diameter pipes with an exaggerated size.

Setting a specific override on a Rise Drop block will scale the block to that size for all sizes of the associated vertical segments. For example, if you override the scale to 6", pipes of all sizes (whether 1/8", 6", or 18") will show the same 6" graphic.

> Note: Checking "Override Scale" overrides the "Use Annotation Scale" option.

A Rise Drop Style is assigned to a System Definition. Each system can have its own Rise Drop Style. The Style should contain a Symbol for each of the basic five conditions.

- Drop 2 Line
- Rise 2 Line
- Drop 1 Line
- Rise 1 Line
- Drop 1 Line Tee

For Duct Rise Drops, this extends to a definition for each of the three (rectangular, round, oval) shapes, resulting in a total of 15 separately defined blocks (5 conditions x 3 shapes = 15 definitions). Refer to Chapter 5 for a tutorial on modifying Duct Rise Drop Styles.

PART GROUP DEFINITIONS

Part Group definitions are used for Cable Tray and Conduit. Plumbing layouts rely on the Defaults settings within the system definitions. Pipe and Duct layouts rely on Pipe and Duct Part Routing Preferences, which are discussed in their respective chapters.

The Part Groups define the specific fittings that will be used as you route Cable Tray or Conduit. For example, you can define a part group to use mitered elbows, and define another part group for radius elbows.

Within Style Manager, you only copy, paste, and define the name and description of a part group definition. The actual settings for the part group are configured on the Parts tab of the associated Preferences window.

The sole tab, General, is where you specify the name and description of the Part Group.

The settings for a Part Group definition are found on the Manage tab on the Preferences panel. For the Cable Tray and Conduit Preferences windows, the Electrical Workspace must be current. Figure 9.50 shows the Parts tabs for the Conduit Layout Preferences. Cable Tray Layout preferences (not shown) are similar.

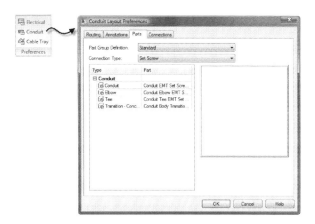

FIGURE 9.50

Conduit Part Group Definition (left) and Duct Part Group Definition (right) settings

You can modify the Part Group Definition settings via the drop-down menu in the "Add" dialogs. Figure 9.51 shows the "Add Duct" dialog. However, in some cases it may be beneficial to set up Part Group Definitions to accommodate multiple routing scenarios.

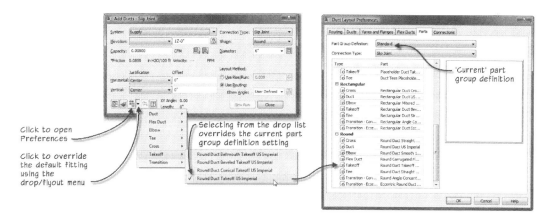

FIGURE 9.51

Modifying the Part Group Definition from the drop-down on the "Add" dialog

CUSTOM FITTING STYLES

Custom Duct and Pipe fittings, when generated, show up in Style Manager. These objects may be generated from 2D linework using the tools on the ribbon. Refer to the product documentation for more information on creating such fittings. However, rather than using Custom Fitting Styles, it is recommended that you create the part definition using the Content Builder. Refer to Chapter 10 for more information.

WIRE STYLES

Wire styles allow you to define the tick marks and home run arrows that are used to annotate a wire segment. The Specifications tab of the wire style editor allows you to define the number of hot, neutral, ground, and isolated ground conductors (see Figure 9.52). These settings control the number of each type of tick mark displayed on the wire segment.

FIGURE 9.52

Wire style Specifications define the quantity and type of each conductor in the Wire

For each type of conductor, you may also define the material, insulation, and temperature rating of the conductor. These settings affect the wire sizing functionality (refer to Chapter 7 for more information).

The Annotation tab lets you select the blocks to use for the various conductor ticks, and specify how they are oriented along the wire (see Figure 9.53). The home run arrow options allow you to specify the arrow block, and whether you show multiple arrows for multi-circuit wires or just a single arrow. The crossing options allow you to specify that wires automatically "gap" when they cross one another.

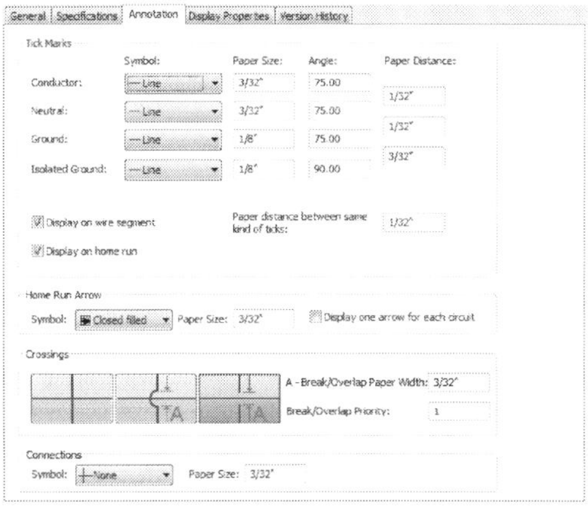

FIGURE 9.53

The Wire Annotation tab allows you to specify the block symbol to use for each type of conductor

On the Display Properties tab, you can modify the various display components of the wire. There are three main display components used for the display of wires: Linework, Tick Marks, and Home Run (see Figure 9.54).

Linework   Tickmarks   Homerun

FIGURE 9.54

A home run wire, showing each display component

Refer to Chapter 1 for more information on display control hierarchy. Refer to the Electrical Layout chapter for more information on creating Wire Styles.

LINE STYLES

Plumbing Lines and Schematic Lines are based on Styles, similar to Wires. Schematic Line Styles have a Designations tab that pre-populates the Design Data ID field on the Properties palette (see Figure 9.55).

FIGURE 9.55

Designations settings in the Schematic Line Style

The Annotation tab includes settings that specify how lines break one another, and how connections are depicted (see Figure 9.56). The Break/Overlap Priority setting controls which lines break which. Schematic Lines of the same priority break based on the drawing order.

FIGURE 9.56

Schematic Line Style Annotation tab settings

Plumbing Line Styles have a Standard Sizes list that provides a selection list of Nominal sizes on the Properties palette for Plumbing Lines (see Figure 9.57).

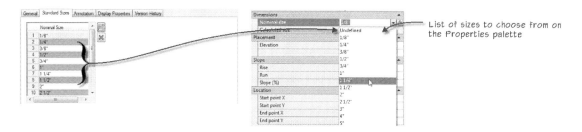

FIGURE 9.57
Standard Sizes settings for Plumbing Line Style

The Annotation tab provides options for how plumbing lines will break when crossing one another (see Figure 9.58). Since Plumbing Lines have an elevation property, there is no Break/Overlap Priority setting..

FIGURE 9.58
Plumbing Line Style Annotation settings

We covered a lot of ground in this chapter. However, there is a lot of overlap between the different domains (Duct, Pipe, Conduit, Cable Tray, Schematic, and Plumbing) which should help solidify some of the concepts you have learned in earlier chapters. Additionally, understanding the foundations of Block-based Styles sets the stage for building Block Based MvPart content, which will be covered in the next chapter.

SUMMARY

✓ You learned about style-based content, and how to modify it within Style Manager.

✓ Standard AutoCAD blocks are used to define Block-based Styles.

✓ Connectors are used in Devices to define load characteristics and where Wires land; in Plumbing fittings, they're used to define where and how Plumbing Lines connect.

✓ Devices typically have both Plan and Model representations, and may also have an annotation block. The Plan blocks may be defined to scale per the drawing's annotation scale.

✓ Panel Styles are used to graphically depict electrical distribution elements that may be used for hosting circuits.

✓ Plumbing fittings are 2D elements used for schematic plumbing layouts.

✓ Schematic Symbols are used to define 2D blocks that are used for flat and isometric schematic layouts.

✓ Categories are used to organize Devices and Schematic Symbols within their content files.

✓ System definitions are used to specify the layers of routed components. Each domain (Duct, Pipe, Conduit, Cable Tray, Schematic, and Plumbing) has some unique settings.

✓ Rise Drop Styles are used to automatically annotate vertical segments, and are associated with Systems.

✓ Part Group Definitions establish default fittings to insert as you route duct, conduit, and cable tray.

✓ Wire styles are used to annotate electrical designs.

✓ Line styles are used to facilitate annotation.

Chapter 10

Content Creation—Equipment

INTRODUCTION

Equipment creation in AutoCAD MEP is done by different methods depending on how the equipment is intended to be used. Equipment can be created as a single unique object in a drawing using the Multi-view Part convert utility, it can be commonly used equipment that is supplied in limited sizes that should be available in the Multi-view Part catalog (block-based) using content builder, or the equipment can have customizable sizes that can be based on parametric dimensions. This chapter will explain the differences between each type, discuss how to determine which kind should be created, and how to define equipment for various situations including one-of-a-kind equipment and limited-size equipment stored in the equipment catalog.

OBJECTIVES

In this chapter you will learn how to create Multi-view Parts from an AutoCAD block using both MvPart Convert and Content Builder. Block-based content is similar to style-based content covered in chapter 9, which allows for content to be created quickly using AutoCAD solids. AutoCAD 3D solids are recommended for equipment since most equipment is complex in nature, only available in limited sizes and can have different requirements for each available size.

The main difference between style-based content (Devices, Panels, Schematic Symbols, and Plumbing Fittings) and Multi-View parts are the types of connectors they support. Additionally, the other key difference is that style-based content resides as styles within drawing files, and are managed in Style Manager, whereas Multi-view Parts are stored in Catalogs and are managed through Content Builder and Catalog Editor. We will cover:

- The different commands for creating equipment
- Create unique equipment using Multi-view Part convert utility
- Create block-based catalog equipment using Content Builder

- Customize block-based equipment

INSTALLING TUTORIAL CATALOGS

We need first to install the fitting and equipment part catalogs from the CD to store the new equipment and access the completed version.

INSTALL THE DATASET FILES AND CREATE A DRAWING

The lessons that follow require the dataset files provided for download with this book.

1. If you have not already done so, install the book's dataset files.

   Refer to "Book Dataset Files" in the Preface for information on installing the sample files included with this book.

2. Launch AutoCAD MEP 2012.

3. From the Application menu choose **Options** (this is shown in Figure 5.1 in Chapter 5).

4. Click the MEP Catalogs Tab.

5. In the Catalogs area, select the *Multi-view Part* Folder and then click the Add button.

6. Browse to the *C:\MasterAME 2012\MAMEP Equipment* folder, select the *MAMEP Equipment.APC* file and then click Open.

7. Back in the "Options" dialog, click the Move Up button to place the MAMEP catalog at the top of Multi-view Part list (see Figure 10.1).

FIGURE 10.1

Adding the MAMEP Equipment Tutorial Catalog

8. Click OK to exit.

9. On the Manage tab, expand the MEP Content panel and then click the Regenerate Catalog button.

FIGURE 10.2

Regenerate Catalog

10. At the Command Line type M (for the Multi-view Parts Catalog) and then press ENTER.

11. In the Catalog Regen dialog that appears, click OK.

12. Press ENTER to complete the command.

The Tutorial Catalog is now accessible.

EQUIPMENT CREATION

There are three methods for creating equipment in AutoCAD MEP; Single instance equipment (Multi-view PartConvert), block-based equipment using Content Builder and parametric equipment using Content Builder.

Multi-view Part Convert (MVPARTCONVERT) is a command that creates a single instance of the equipment within the current drawing from an AutoCAD block, a 3D solid or a Multi-view Block. This type of part is not saved in the catalog. This is the easiest way to create a unique piece of equipment. However, it should only be used for equipment that will not be reused in other projects.

Content Builder is a tool that can convert an AutoCAD block containing either 3D solids or Mass Elements into a Multi-view Part and save it to the catalog. This type of part is limited to AutoCAD blocks based on 3D solids, but is commonly used in more than one project. Additional sizes can be added to a part created this way by editing the content builder drawing file. Each size is created using a unique set of blocks specific to that part size.

Parametric parts are also created inside Content Builder using constraints and dimensions to define the sizes of the part. The constraint and dimension system used in Content Builder is unique and independent from the parametric constraint and dimension tools found in more recent versions of AutoCAD. Content Builder is best described as an application that runs on top of AutoCAD MEP. Content Builder supplies its own commands and controls, it creates drawings and associated storage files and indexes them into a catalog structure. The content built can also be based on parametric constraints, similar to Autodesk Inventor or other parametric modeling programs. The parametric constraints allow simple parts to supply hundreds or thousands of possible sizes. This type of part is used for items that have multiple combinations for the equipment or to define fittings. Creating parametric parts is covered in Chapter 11.

Multi-view PartConvert should be limited to parts that will not be reused in other projects. Equipment created from this command cannot be added to the Multi-view Part catalog after creation. The preferred method is to use the Content Builder to create a block-based catalog part for the equipment. AutoCAD MEP supports multiple catalogs for each domain, which allows you to create a unique project catalog or a company specific catalog in addition to the standard AutoCAD MEP catalogs.

When creating content using MVPARTCONVERT or when using Content Builder to build block-based parts, you should have good working knowledge of solid modeling in AutoCAD. Significant changes were introduced in 2007, 2010, and 2011. So if you historically have primarily used 2D functionality in AutoCAD, it may be worthwhile to find some tutorials or books focusing on 3D modeling in AutoCAD. The large topic of 3D modeling in AutoCAD is outside the scope of this book.

> **TIP:** Search YouTube for AutoCAD 2011 Convert 2D Objects to 3D Objects

MULTI-VIEW PART (MVPART) CONVERT

Let's begin with an exploration of the fundamentals of Multi-view Part Convert. This command should be used when the part is only needed for a single project/drawing and will not be reused frequently for other projects. The command takes an AutoCAD 3D solid, an AutoCAD Architecture Multi-view Block or an AutoCAD block (that contains 3D Solids or Mass Elements) and converts it to a Multi-view Part. During this conversion the different views of the block (Top, Bottom, Front, Back, Left, Right, 3D model) are created. The resulting Multi-view Part definition adheres to the display system settings.

We will begin with creating a series of multi-view parts using the Multi-view Part convert command by converting a single AutoCAD Block containing several 3D solids to define the equipment.

LOAD A PROJECT

1. On the Quick Access Toolbar (QAT), click the **Project Browser** icon.

2. Click to open the folder list and choose your *C:* drive.

3. Double-click on the *MasterAME 2012* folder.

4. Double-click *MAMEP Commercial* to load the project. (You can also right-click on it and choose **Set Project Current**.) Then click Close in the Project Browser.

IMPORTANT:

If a message appears asking you to repath the project, click Repath the project now. Refer to the "Repathing Projects" heading in the Preface for more information.

USING MULTI-VIEW PART CONVERT

5. On the Project Navigator palette, click the Constructs tab.

6. Click the *Elements* folder, expand *Equipment Creation* folder, and then double-click the *Mvpart Convert* file (see Figure 10.3).

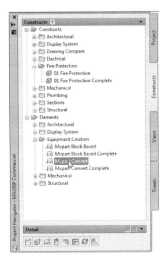

FIGURE 10.3
Open the MvPart Convert drawing from Project Navigator

In this drawing is an AutoCAD Block that represents a Recessed Impeller Pump. The block contains three AutoCAD 3D solids. In the next sequence we will convert it to a multi-view part using the Multi-view Part convert command.

7. Select the Recessed Impeller Pump block instance, right-click, and choose **Convert to >** **Multi-view Part** (see Figure 10.4).

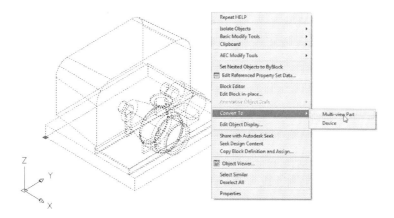

FIGURE 10.4

Converting an AutoCAD Block to a Multi-view Part

8. In the "Multi-view Part Convert" dialog, for the Name, type **8 x 4 in Recessed Impeller Pump Belt Driven**.

9. From the Type drop-down list, select **Pump** (see Figure 10.5).

FIGURE 10.5

Defining the Multi-view Part behavior

Note: Once the Pump Type is selected, the Subtype defaults to Base Mounted Pump (as shown in the left side of Figure 10.6). You can specify a different subtype from the drop-down list or type in your own value.

10.Click the Layer Key browse button (…) select the M-MV-PUMP_BASEMTD layer key, and then click OK (see the right side of Figure 10.6).

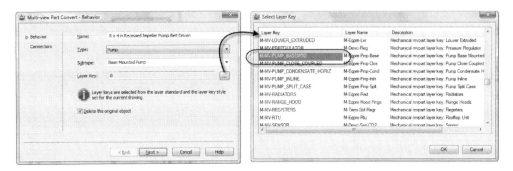

FIGURE 10.6

Setting the Layer Key

The "Delete original object" checkbox is selected by default; this will delete the item that is being converted. If you want to maintain the original block after creating the Multi-view Part, clear this checkbox. In this case we will leave it selected.

11.Click the Next button.

The "Multi-view Part Convert – Connectors" dialog appears (see Figure 10.7).

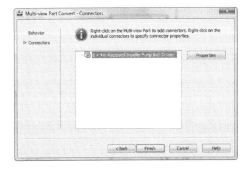

FIGURE 10.7

Multi-view Part Convert – Connectors dialog

ADD CONNECTORS

At this point you could simply click the Finish button if you do not want to add MEP connectors to the Multi-view Part. In some cases no connectors are needed. We will be adding connectors in the next series of steps.

Note: Defining a connector during convert is the same functionality utilized when you use the Multi-view Part wizard to create block-based equipment stored in the catalog.

12. Right click on 8 x 4 in Recessed Impeller Pump Belt Driven and choose **Add Pipe connector** (see Figure 10.8).

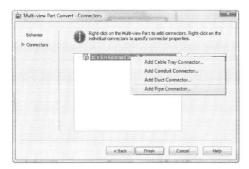

FIGURE 10.8

Adding a Pipe Connector to the Multi-view Part

CAD MANAGER NOTE: Pipe Connectors are used for both 2D Plumbing Connectors and Pipe Connectors. The determination of which command is started is controlled by the *AecbDomainConfig.xml* file located in the *Shared* folder. This file lists all content types and contains a variable IsPlumbingFixture="True" on a per type basis. Changing the True statement to False in the XML file will start the PipeAdd Command.

13. In the "Part Family Connector Properties" dialog that appears, configure the following settings (see Figure 10.9):

⇨ For Connection Name, type **Suction**.

⇨ From Flow Direction, choose **In** (Leave bidirectional for inline content such as valves).

⇨ For System Type, leave it set to **Undefined** (Allows the part to inherit the system from the connected object.)

⇨ Connection Domain: Pipe (Read-only due to selecting Pipe Connector above.)

⇨ Connection Shape: Round (Options appear for Duct Connectors.)

⇨ Engagement Length: 0.000 (This parameter is only for Pipe Connectors, this specifies if the connection is male or female. A zero value will indicate the connection is male. A value greater than zero indicates the pipe will connect inside this connector, thus defining it as female.)

⇨ Max Angle of Deflection: 0.00 (This parameter is only for Pipe Connectors, and specifies if the connection allows deflection when a pipe connects to it. This value is only used when the Engagement length is greater than zero—otherwise it is ignored. Angle of Deflection is only allowed on pipe connectors defined as female.)

⇨ For Unsized: leave it set to: **False** (This parameter determines if the connection can remain undefined in size.)

FIGURE 10.9

Part Family Connector Properties - Settings

14. Click OK to complete the connector settings.

   Back in the "Multi-view Part Convert – Connectors" dialog a new Suction connector appears under the part name (see Figure 10.10).

FIGURE 10.10

Suction connector defined for Part

15. Repeat the process to add another Pipe Connector.

⇨ Change the name of the connector to **Discharge**.

⇨ Change the Flow Direction to **Out**.

⇨ Leave the remaining settings as the defaults (see Figure 10.11).

FIGURE 10.11

Defining the Discharge Connector

Now that the parameters of both the Suction and Discharge connectors have been defined, we will define the location of each connector.

16. Select the Suction connector, right-click and choose **Edit Placement** (see Figure 10.12).

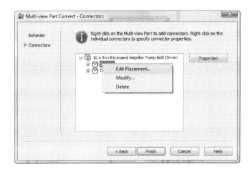

FIGURE 10.12

Accessing the Pipe Connector placement controls

17. The "MvPart Builder – Content Editor" palette will appear (see Figure 10.13).

This palette is used to define the connection type, location, and size.

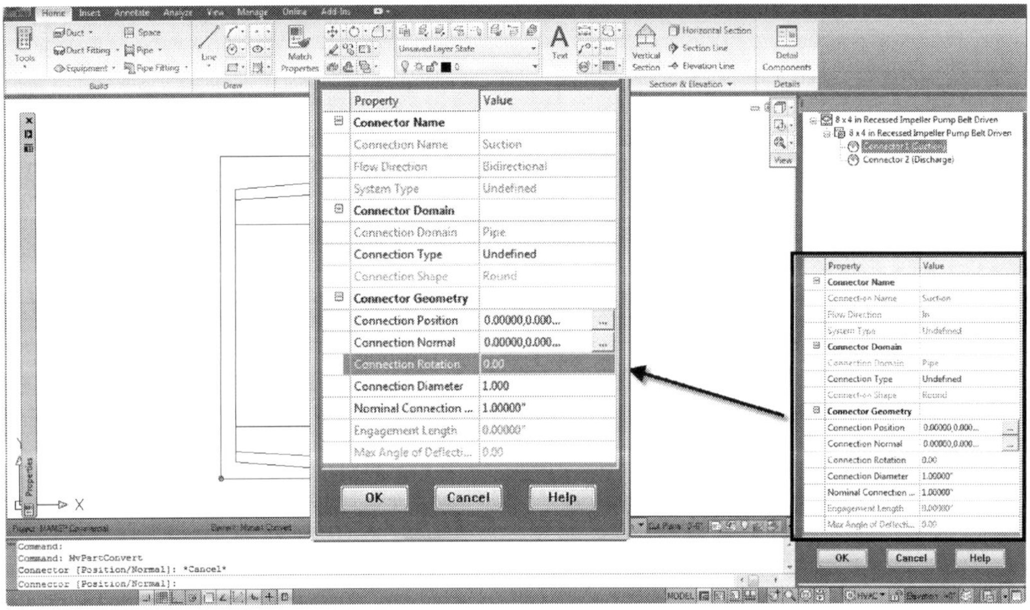

FIGURE 10.13

MvPartBuilder – Connector Editor

The Connector Editor is used to further define the properties of a connector. The Connection Type and Connector Geometry are specified using this palette. The connection type is used to indicate what connection should be used, such as flanged. The connection geometry uses position, normal, and rotation to define the three-dimensional location of the connector as well as the Diameter and Nominal diameter.

18. Select the Connection Type drop list and select Flange.

> **Note:** Pipe connections that are male and defined as "Undefined" will inherit the name of the connection from the connection assignments dialog in pipe preferences.

19. In the Connection Position row, click the browse button.

A Connector Location Graphic will appear at its current location (0,0,0 in this case). It will also appear on the cursor so that you can define the new position (see Figure 10.14).

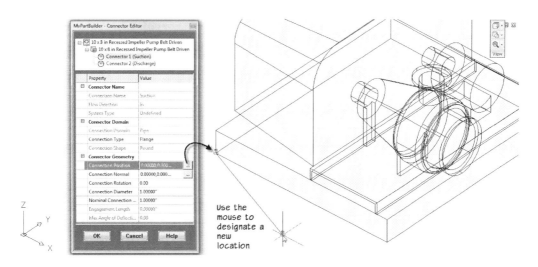

FIGURE 10.14

Defining the Connector Location

You are allowed to zoom, pan, and change the view direction using 3D orbit or your wheel mouse. To access 3D orbit, click the View tab, on the Navigate panel click the 3D Orbit icon. (You can also hold down the shift key and drag the wheel). Orientate the view in a southwest direction (see Figure 10.15).

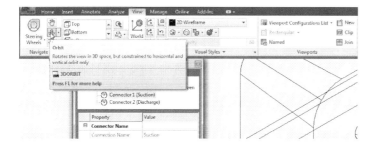

FIGURE 10.15

Use 3D orbit or your wheel mouse to orient the view

20. If you panned or zoomed, you may have inadvertently exited the Position command, and on the command line, the option for Position/Normal may become active. If this appears to be the case, type P for position so the Select Position prompt appears again. If the Select Position prompt is still active you can skip this step.

21. Use the AutoCAD Center snap to establish the suction connector location (see Figure 10.16).

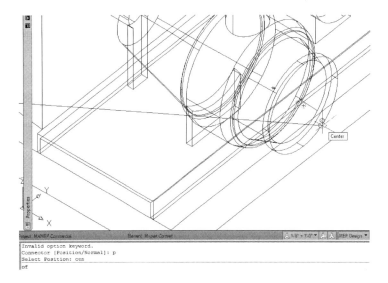

FIGURE 10.16

Locating the Connector position using the Center snap

The connector position is now located, but the connector still is not pointing in the correct direction; it is pointing up, and this is referred to as the Normal. (Normal is the orientation of a single point relative to the current UCS. AutoCAD MEP uses this value to determine draw direction/connection direction.)

22. To change the Normal orientation, type N at the command line.

23. Use the Center Osnap again to select the center point of the inside circle for the first point, then again for the outside circle for the second point (see Figure 10.17).

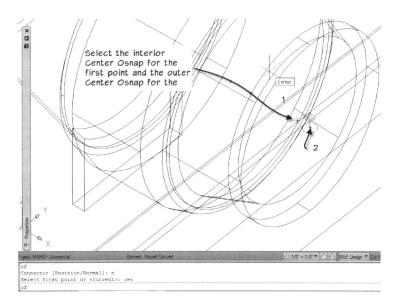

FIGURE 10.17

Defining the Connector's Normal Direction using the Center Osnap

Note: The specific points selected for defining the Normal are not important. In this case, as long as the second point is in the positive X direction from the first point you will be fine. You can verify on the palette that the Connection Normal is some positive X value, followed by 0 values for X and Y (i.e., 6.875, 0.0, 0.0).

The normal direction is now pointing away from the part; this is not the flow direction, this indicates the Draw Direction. The PipeAdd command uses this direction to determine which direction the pipe should go.

24. On the MvPart Builder – Connector Editor palette set the Nominal Connection Diameter to 8.00 (see Figure 10.18).

25. Set the Connection Diameter value to 13.25 (see Figure 10.18).

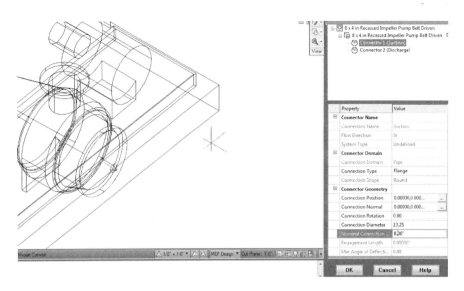

FIGURE 10.18

Configure the connection location and diameter values

The Suction Connector is now defined. We need to repeat the process for the Discharge Connector. We do this by selecting the next connector at the top of the palette.

26. At the top of the palette, select Connector 2 (Discharge).

27. Set the Connection Type to Flange.

28. Select the Connection Position on the palette, and then using the center osnap, click on the pump discharge location as indicated in Figure 10.19.

The normal direction is correct, we do not need to change it.

29. Set the "Connection Diameter" to **9.00** and the "Connection Nominal Diameter" to **4.00** (see Figure 10.19).

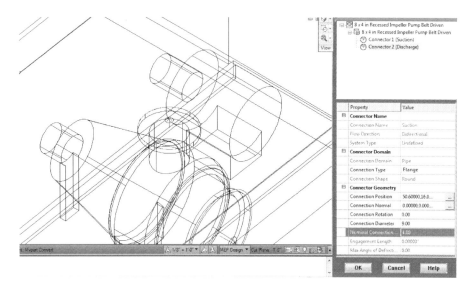

FIGURE 10.19

Locating the Discharge Connector and setting the Connection values

30. Click OK.

This will return you to the "Multi-view Part Convert – Connectors" dialog.

31. Click the Properties button in the upper right-hand corner.

32. In the "Property Editor" that appears, change the PrtSN Parameter Visible setting to True (see Figure 10.20).

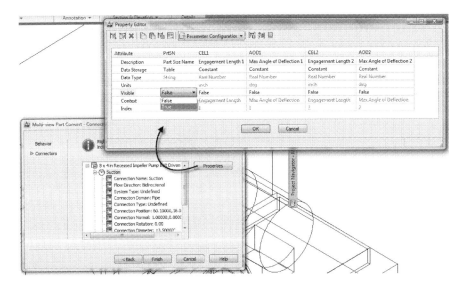

FIGURE 10.20
Set the Part Size Name Parameter Visibility to True

This allows the Part Size name to be visible inside AutoCAD MEP.

33. Click OK, and then click Finish.

You may want to have a simplified version of the part used for the top view block. The following steps will create that simplified version, and then associate it with the multi-view part style.

34. On the View panel, Appearance tab, click Top.

35. On the Insert ribbon, Block panel, click Insert Block.

36. From the Name list, select Recessed Impeller Pump Belt Driven.

37. Under Specify On-screen, specify X= 10', Y=0, Z=0.

38. Make sure Explode is unchecked.

39. Click OK.

40. Select the new block instance.

41. At the command line, type FLATTEN, and then press ENTER.

42. At the Remove Hidden Lines prompt, type Y, and then press ENTER.

43. Select all the newly flattened elements (See Figure 10.21)

FIGURE 10.21

Select the flattened elements

44. On the Insert ribbon, Block panel, click Create Block.

45. Enter the Name Recessed Impeller Pump Belt Driven Top.

46. Set the base point to X=10', Y=0, Z=0.

47. Click OK.

48. Select and erase the objects you just converted in the block.

49. Select the Multi-View part instance.

50. Right-click, and select Edit MvPart Style…

51. Select the Views tab.

52. Select #3 Plan.

53. In the View Block list, select Recessed Impeller Pump Belt Driven Top.

54. Uncheck all View Directions, except for Top.

55. Click OK.

The Multi-view Part conversion is complete. The block is now replaced with an AutoCAD MEP Multi-view Part version, and you have a simplified Top view.

We have just completed creating a unique piece of equipment that is not stored in the catalog. The typical use for this command is to create rudimentary versions of equipment during the conceptual design phase that will be replaced when the project is better defined. Adding connection locations and defining these locations is optional, since the part may be scrapped later in the project. However, many of the steps that we just covered are also incorporated when creating block-based content using Content Builder. Content Builder allows us to create the part and save it to the catalog. In the next exercise, we will create a part using Content Builder, New Block Based Part, then add additional sizes to the catalog and edit the appearance of the blocks created during the conversion.

> **Note:** There is no direct mechanism for copying a Multi-view Part created using MvPartConvert to the catalog. For a component to be in the catalog, you must use the Content Builder. Further, when using Content Builder, the wizard automatically creates flattened views of the block for each side.

CREATING A BLOCK-BASED MULTI-VIEW PART

In this exercise we will create a Multi-view Part using the Content Builder. The content will be Block-based as in the previous exercise. However, in this case we will define the part, store it in the catalog and add additional sizes. These features are not available when using the Multi-view Part Convert method. We will also look at how to revise the appearance of the defined blocks by editing the part.

Block-based parts are limited to multi-view parts. They require that each size defined in the part has a 3D solid within an AutoCAD block (in the current drawing) during creation. During creation you can also specify the two-dimensional blocks you want assigned for each view. If you don't specify 2D blocks, they will be created automatically. The creation of block-based equipment is intended in the following situations:

- You require a limited range of sizes.
- The 3D model for the equipment is overly complex or too difficult to define parametrically.
- You require the flexibility to customize the view blocks to suit your needs.

When defining a Block-Based part in Content Builder you can specify the representation for every view required. This exercise will create a new block-based part supporting multiple sizes and specify the blocks for the views. The drawing used in this exercise has AutoCAD blocks already defined with 3D solids which represent four sizes of a check valve. In addition, the drawing contains an AutoCAD block that can be assigned as the single line representation of each valve.

1. On the Project Navigator palette, click the Constructs tab.

2. Expand the *Elements\Equipment Creation* folder and then double-click to open the *MvPart Block Based* file.

3. On the Manage tab, on the MEP Content panel, click the Content Builder button.

4. In the "Getting Started – Catalog Screen" dialog, from the Part domain drop-down list, choose **Multi-view Part**.

   The *MAMEP Equipment* catalog should appear at the top of the list expanded to show a single *Valves* folder beneath it.

5. Select the *Valves* folder.

6. On the right side of the dialog, click the New Block Part icon (see the top left side of Figure 10.22).

7. In the "New Part" dialog, for the Name, type **Swing Check Valve**.

   The description is automatically updated to match the file name. The description appears in the catalog and in the Multi-view Part Add command dialog.

8. Click OK (see the bottom left side of Figure 10.22).

9. In the "MvPart Builder (New Part)" dialog, from the Type list, choose **Valve** (see the right side of Figure 10.22).

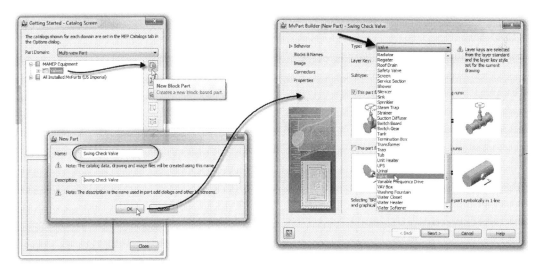

FIGURE 10.22
Creating the Block-based Part definition and assign the Type

10. Next to Layer Key click the browse button (…), select M-MV-VALVE-CHECK, and then click OK.

11. Set the Subtype to Check Valve.

12. Check the "This part family automatically BREAKS INTO existing runs" checkbox and then click Next (see Figure 10.23).

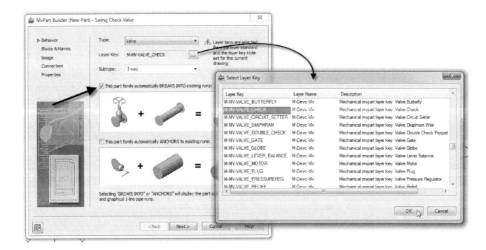

FIGURE 10.23
Setting the Layer Key

The Blocks & Names page appears.

13. Click the Add Part Size icon (see the left side of Figure 10.24).

The new line appears and the Model Block drop list automatically expands.

14. Select 04 check valve from the list, and then click in the gray area to update the row.

The row is automatically populated with the default / automatic block names for each view specified (see the right side of Figure 10.24).

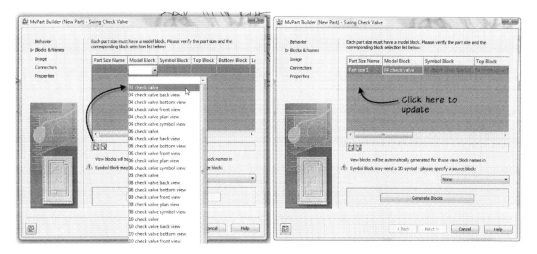

FIGURE 10.24

Select the first check valve block

15. Select **04 check valve symbol view** in the Symbol Block drop list located above the Generate blocks button.

16. Click the Add Part Size icon again; select the 06 check valve block this time and repeat the process.

17. Repeat the process to select the **08 check valve** block.

The catalog entry for this part will now support the three sizes we have selected.

> **Note:** When you want to add additional sizes after the catalog files are created, open the catalog drawing file, add the blocks, close the drawing, and then in Content Builder, select the part and click the modify button. This will start this process to add part sizes.

18. Select the first row, then select the 2D Symbol Drop list and select the **04 check valve symbol view** block (see Figure .10.25).

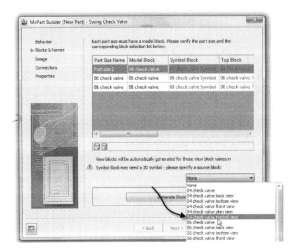

FIGURE .10.25
Selecting the Symbol block to be used with Graphical 1-Line display

The 2D symbol block is used for the pipe by size Graphical 1-Line display. The block is scaled to match the annotation scale of the drawing based on the settings assigned to the pipe system definition (for more information, refer to Chapter 6).

19. At the bottom of the dialog, click the Generate Blocks button (see left image of Figure 10.26).

A Multi-view Part requires a view block for each potential viewing angle. You can use different blocks for each of the orthographic views (top, bottom, left, right, front, and back) and you will also need a 3D view. A single 3D block can be used to generate all required views, or you can actually draw different blocks manually for each view. When you click the Generate Blocks button, AMEP generates a view block for each orthographic view from the 3D block you designated in the Model Block column. This can save you a tremendous amount of time that would otherwise be required if you had to draw each view block individually. You can specify blocks you have already defined, or modify the blocks that are automatically generated.

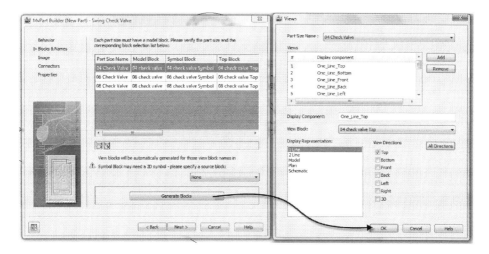

FIGURE 10.26

Generate the Blocks

20. In the Views list, select #15 Model_3D, click on the All Directions button to deselect all checkboxes, and then check 3D (see Figure .10.27).

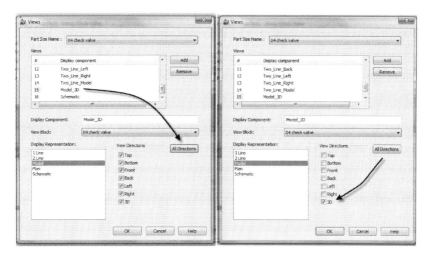

FIGURE .10.27

Defining the Model_3D block to only be used in a Model view

To help understand how AutoCAD Blocks are used to define block-based parts, during the next two steps we will replace the automatically generated representations from the 3D model with blocks that represent a standard check valve 2D symbol. The drawing contains a 2D block for the Top, Bottom, Front, and Back views. You can continue to replace the automatically generated blocks with the blocks ending in view as shown in Figure .10.28.

21. Select Views #8 Two_Line_Top, select the View Block drop list, and then select the 04 check valve plan view block (see Figure .10.28).

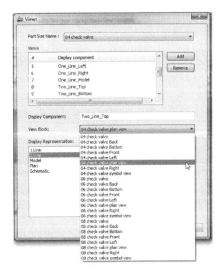

FIGURE .10.28
Replacing the automatically created block with a 2D AutoCAD block

22. Select View #16 Schematic and then select the 04 check valve symbol view.

23. Repeat for the 06 check valve and the 08 check valve by selecting the 06 check valve in the Part Size Name Drop list at the top of the dialog.

> Note: You can continue and define the Back, Bottom, and Front views with the supplied view blocks while you are in this dialog by selecting the appropriate view and selecting the block.

24. Click OK.

When the process is complete, notice that the previously red entries are now black. This indicates that the view blocks have been created for each size and view direction.

25. Click the Next button to continue.

On the "Image" screen, you can assign a preview image for your part that will be displayed in catalogs and the Multi-view Part Add dialog box. You can use an image you have drawn (in BMP format) or you can have the Content Builder generate an image automatically from one of your 3D blocks.

26. Select the "Generate image based on model block from the SW Isometric View" radio button.

27. Choose the 06 check valve block from the list, and then click the Generate button (see Figure 10.29).

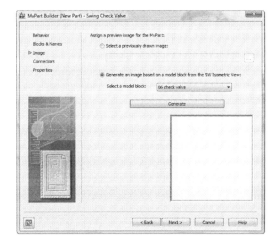

FIGURE 10.29

Creating the part image

> Note: We recommend that you replace the image by exporting a bitmap from a rendered view from the completed part and save over the file that was just created using AutoCAD's export command and selecting bmp as the file type.

28. Click Next to advance to the Connectors page.

When defining connectors, you first specify the types of connectors common to all sizes of the part family. Then, you specify the details of the connector for each size.

29. Right-click on Swing Check Valve (at the top of the list) and choose **Add Pipe Connector** (see Figure 10.30).

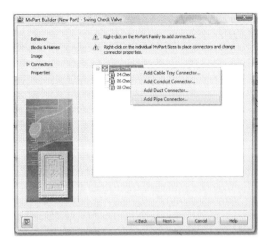

FIGURE 10.30

Creating a Pipe Connector

At this point, defining the connector positions and sizes is done the same way as during the Multi-view Part convert (see the "Add Connectors" topic above).

30. In the "Part Family Connector Properties" dialog box, for the Connector Name, type **Inlet** and then Click OK.

31. Repeat, creating a connector named **Outlet** (see Figure 10.31).

Notice that both connectors appear for the overall part at the top of the list and have been added to each individual size.

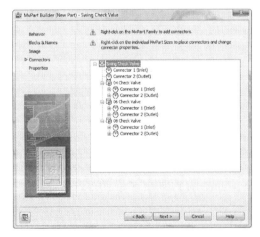

FIGURE 10.31

Creating connectors for each part size

In the next series of steps, we will locate the connector location, diameter, and flanges.

32. Select the Connector 1 beneath the 04 Check Valve, right-click and choose **Edit Placement** (see Figure 10.32).

FIGURE 10.32

Edit placement on the 4" part size

33. Configure the following settings:

- Set the Connection Type to: **Flange**.
- Set the Connection Diameter to: **9**.
- Set the Nominal Connection Diameter to: **4**.

Do not type the inch (quotes) key when specifying the size, otherwise the value will revert to the previous entry.

34. Click the browse button for Connection Position (see Figure 10.33).

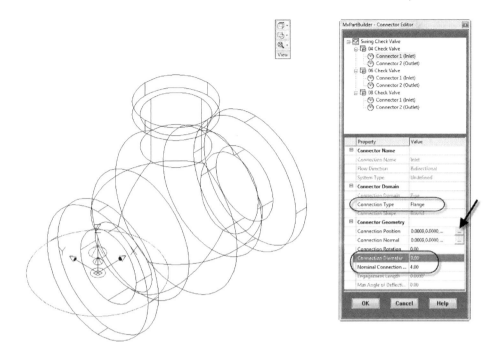

FIGURE 10.33

Define the Inlet Connector

35. Using the AutoCAD Center snap, select the lower left outer diameter. In this case, all the blocks in this exercise have the insertion point located at the same position as the default position (0,0,0) of Connector one. Thus, in this case, this step is redundant.

36. At the Command Line, type N.

37. Using the AutoCAD Center snap, Pick a point in space, and then click a point straight in the negative x direction. The result in the Connector Editor panel should be a Connection Normal of some value with a negative x component, followed by 0 for y and z (i.e., -0.93, 0, 0).

38. On the MvPartBuilder Connector Editor palette, select Connector 2 (Outlet) and repeat the steps to create similar settings (see Figure 10.34). In this case, the Connection Normal should have a positive x value, followed by 0 for y and z.

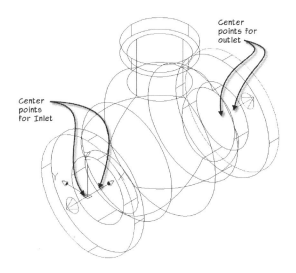

Center
points for
outlet

Center
points
for Inlet

FIGURE 10.34
Establish the Position and Normal locations for the Connectors

39. Repeat the process for the 6" and 8" part sizes (see Table 10.1):

TABLE 10.1—*SETTINGS FOR 6" AND 8" PART SIZES*

Size	Connection Type	Connection Diameter	Nominal Connection Diameter
6" Part Size	Flange	11	6
8" Part Size	Flange	13.5	8

40. On the MvPartBuilder Connector Editor palette, click OK.

41. Back in the "MvPart Builder," click Next to advance to the Properties page.

42. Click the Edit Properties button and change the Part Size Name visibility to True (see Figure 10.35). This parameter determines the visibility of all size names defined in the part.

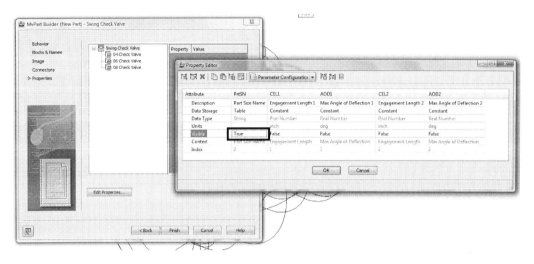

FIGURE 10.35

Configure the Properties to make the Connectors visible

43. At the top of the "Property Editor" dialog, select the Parameter Configuration drop-down list and choose **Values**.

44. Edit the Part Size Names to include the word "Swing" as shown in Figure 10.36 and then click OK.

FIGURE 10.36

Editing the Part Size Names

45. In the "MvPart Builder (New Part)" dialog, click Finish to complete the part.

The Swing Check Valve part is now defined. All we need to do now is test it out.

TEST OUT THE NEW PART

Our new part belongs to a catalog, which means we can use it in any drawing. The simplest way to test it out is in a new drawing.

46. On the QAT, click the New drawing icon.

47. On the Home tab, on the Build panel, click the Equipment button.

48. In the Multi-view Part Dialog, select the *MAMEP Equipment* folder, then the *Valves* folder and the Swing Check Valve.

49. The 04 swing check valve is selected in the part size name.

50. Click in the drawing window to place the valve.

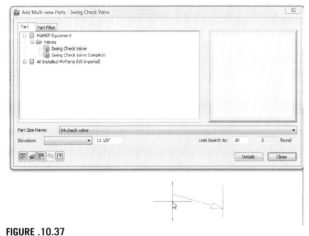

FIGURE .10.37

Testing the 04 check valve

51. On the view toolbar, switch to the SW ISO view (see Figure 10.38).

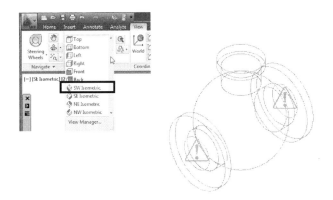

FIGURE 10.38

04 check valve in SWISO view

52. Select the Valve and select the Plus Grip to verify that the PipeAdd command begins and that flanges will be added. Change the Routing Preference on the Property Palette to **Slip on Flanged – 150 Lb. and Threaded,** and the system to **Chilled Water** (see Figure 10.39).

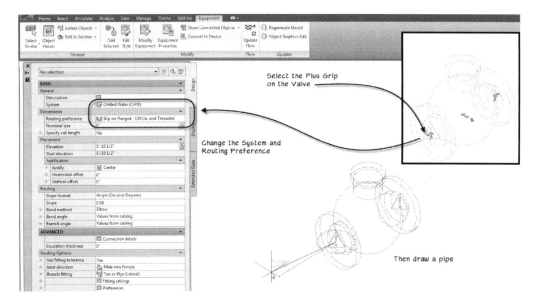

FIGURE 10.39

Testing the Valve

53. Repeat for the opposite connector.

54. Add the valve again, select the valve and then the Plus Grip, change the System to Standard to test the 2-Line representation of the block

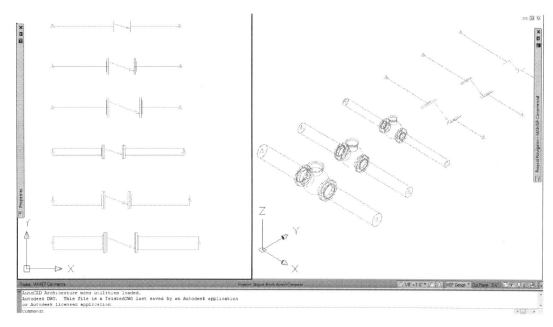

FIGURE 10.40

The tested valve (Mvpart – 04 Check Valve Complete.dwg)

We created a pump using MvpartConvert, which is a good way to create content that will only be used once in a project or is preliminary equipment that will change. Next we created a new Valve using Content Builder from AutoCAD Blocks. Inside Content Builder we created three sizes and redefined the automatically created blocks with 2D symbolic blocks in the orthographic views. This allows the plan and section representations to have the look and feel of true 2D AutoCAD drawings.

SUMMARY

✓ The MvpartConvert command creates a Multi-view Part from an existing AutoCAD block containing solid geometry.

✓ You can add converted parts to the catalog for later retrieval in the same or future projects.

✓ You can add custom catalogs to the ones provided with the default install by modifying the settings in the "Options" dialog on the MEP Catalogs tab.

✓ You can create block-based parts that contain multiple sizes.

✓ To create a piece of block-based content with multiple sizes, you must have a block for each size you wish to include.

✓ You can let MEP create all required view blocks for a piece of content, or you can replace any view block (for the orthographic view) with a symbolic block.

Chapter 11

Content Creation—Parametric Fittings

INTRODUCTION

Content Builder is a modeling program that runs on top of AutoCAD MEP (AMEP) that can create both block-based (see Chapter 10) and parametric content (this chapter). Content Builder contains unique parametric commands, independent of AutoCAD commands, which allows for parameters to be assigned to a model and for the model to be controlled by these parameters and accessed through the Work Planes.

In this chapter we will build a Duct Transition Elbow. The concepts we will be exploring apply to all fitting types, and additional tips are provided throughout to give you a better understanding on how Content Builder uses parametric commands. Best practice recommendations on fitting creation will be presented to ensure the best results with Auto Layout.

We will also learn how to add custom catalogs to AMEP, as well as create new catalogs to store equipment and fittings that you have created.

OBJECTIVES

After performing the exercises in this chapter, you will understand the power of Content Builder. Specifically, we will walk through the creation of a single fitting drawing. This fitting will have the capability of becoming thousands of fittings through the definition of parameters and constraints. In this chapter you will:

- Learn how to create parametric fittings
- Create workplanes
- Define Parameters
- Create Formula-based Parameters
- Edit Parameters
- Create Constraints

INSTALLING TUTORIAL CATALOGS

We need first to install the fitting and equipment part catalogs from the CD to store the new parametric fitting and access the completed version at the end of the chapter.

INSTALL THE DATASET FILES AND LOAD THE CATALOG

The lessons that follow require the dataset files provided for download with this book.

1. If you have not already done so, install the book's dataset files.

    Refer to "Book Dataset Files" in the Preface for information on installing the sample files included with this book.

2. Launch AutoCAD MEP 2012.

3. From the Application menu, choose **Options** (this is shown in Figure 5.1 in Chapter 5).

4. Click the MEP Catalogs Tab.

5. In the Catalogs area, select the *Duct* Folder and then click the Add button.

6. Browse to the *C:\MasterAME 2012\MAMEP Duct* folder, select the *MAMEP Duct.APC* file, and then click Open.

7. Back in the "Options" dialog, click the Move Up button to place the MAMEP catalog at the top of Duct list (see Figure 10.1).

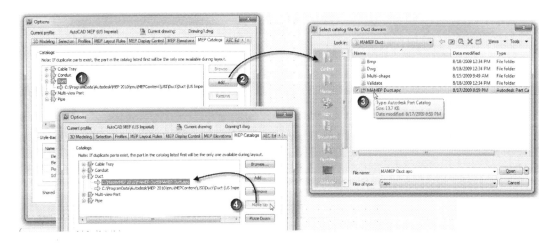

FIGURE 11.1

Adding the MAMEP Equipment Tutorial Catalog

8. Click OK to exit.

9. On the Manage tab, expand the MEP Content panel and then click the Regenerate Catalog button (see Figure 10.2).

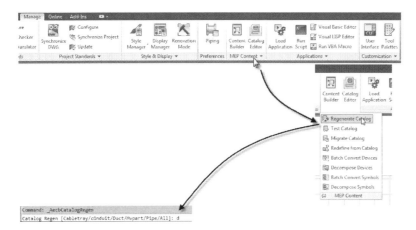

FIGURE 11.2
Regenerate Catalog

10. At the Command Line, type D (for the Duct Catalog) and then press ENTER.

11. In the dialogs that appear, click OK.

12. Press ENTER to complete the command.

The Tutorial Catalog is now accessible, and since it is now the first catalog, all new Multi-view Parts will be added to it.

CONTENT CREATION—FUNDAMENTALS

Content Builder is best described as an application within an application. Content Builder supplies its own commands and controls, it creates its own drawings and associated storage files and it indexes them into a catalog structure. The content built is based on parametric constraints, similar to Autodesk's Inventor product or other similar parametric modeling programs. The parametric constraints allow simple parts to supply hundreds or thousands of possible sizes from a single source file.

Parametric constraints are rules applied to the model that define what the model can do. Constraints can be simple rules, like parallel or perpendicular, or dimension-based rules that refer to a table or formula to get the correct value.

Dimension constraints are similar to AutoCAD dimensions. The dimensions can be restricted to vertical or horizontal, and can be defined as an angular dimension as well as a radius, parallel or perpendicular dimension. Dimensions are assigned parameters that require values to be specified and that in turn defines how the model is created.

Parameters can be described as formulas, like high school algebra: X+Y=Z. Parameters require values to complete the equation for a specific number. Parameters created in the parametric fitting definition must be defined with values or calculations to allow the model to be valid.

- Parameters are stored in spreadsheet form with rows and columns to store the values.
- Parameter values are defined as constants, tables, lists or calculations.

PARAMETER TYPES:

Constant—A single value applied for all instances of the parameter.

Table—A value assigned to each row of parameters and cannot exceed the associated rows. Tables also do not support adding of values within the Add dialog unless custom sizing is turned on. For more information on custom sizing refer to "Configure Bitmap Preview and Options" below.

Lists—Are values that exist either for a single row of parameters (multiple options for a single size) or multiple rows (multiple options for a multiple sizes) and are not bound to have equal number of rows to list values. Lists allow you to type in a value while adding or modifying the part and are not bound by the list stored inside the catalog. List values require you to add values through the catalog editor.

Calculations—Formulas created with other parameters to create a solution value. Calculations are not required to be number-based only. A calculation can be created for the part size description by assigning the description parameter with size parameters and addition text, as an example. Calculations can be done while editing the parameters or a formula can be defined inside the content drawing during modeling.

The values defined for the parameters are stored within the catalog folder structure as an XML file. The XML file is associated with the DWG file within the catalog APC file. If the XML is not found, the parameters have no value.

PARAMETRIC MODELING—FUNDAMENTALS

Parametric modeling requires one or more 2D shapes, defined on related Work Planes, with constraints and their associated parameters to define a 3D body Work Plane. Work Planes are defined as a surface or UCS plane in which the model is built. Default options for Work Planes are available to match the AutoCAD default UCS values such as Top, Front, Left, Right, etc. There is also a default option that creates all of these work planes by using the Default option.

Work Planes are the basis for defining the model. You must first define a Work Plane before the model commands will be available. Some key points to understand about modeling:

> Note: You cannot model a parametric part without defining a Work Plane.

- Depending on the type of part you are creating, your models can be created by offsetting extrusions, assigning a shape to a path or using primitives.
- Extrusions are created by extruding a shape or transforming from one shape to another, between two different Work Planes.
- Shapes are assigned to follow a defined path on a single Work Plane to create an extruded body, similar to extruding a shape along a polyline in AutoCAD.
- Primitives can be used in limited fashion, and are restricted to Multi-view Parts.
- Work Planes are determined by the complexity and orientation of the part when inserted inside AMEP. Parts can contain a single Work Plane, typically Top, several Work Planes, or offset Work Planes to define the distance between surfaces.

PARAMETRIC DIMENSION RULES FOR CREATING A FITTING

When building a Fitting for use during auto layout, it is required that the connection order matches certain defined rules and that all the dimensions shown in Figure 11.3 are defined inside the part. During validation of the Fitting, the dimensions are checked to ensure that the parameters meet the allowable values.

- **PathA1** must be = 180 degrees on all fittings except for an Elbow.
- **PathA2** must be = < 90 degrees and must be between Connectors C2 and C3.
- **Path A3** must be = < 90 degrees and must be between Connectors C3 and C4.
- **Elbows** must have a Path A1 defined and can be any angle = >1 to = <180.
- **Male x Female** fittings must have the Male end on connector 1.

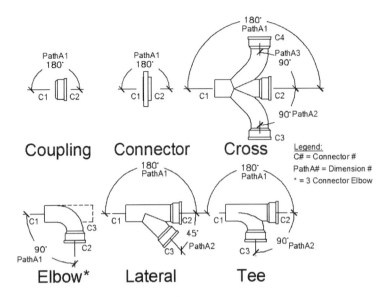

FIGURE 11.3

Required connector order and dimension locations for Auto layout fittings

CONTENT CREATION—BUILDING A DUCT TRANSITION ELBOW

This topic describes how to build Rectangular to Round Duct Elbow inside Content Builder.

GETTING STARTED

Content Builder is used to build and edit Fittings and Parts in the AMEP catalogs. Content Builder can be used to create Multi-view Parts as we explored in the previous chapter, or to create parametric fittings as we will explore here.

1. On the Manage tab of the ribbon, on the MEP Content panel, click the Content Builder button.

2. In the "Getting Started – Catalog Screen" dialog that appears, from the Part Domain list, choose Duct.

3. Beneath the *MAMEP Duct Catalog US Imperial* catalog, select the *Multi-shape* folder (see item 3 in Figure 11.4).

4. On the right side of the dialog, click the New Chapter icon.

5. In the "New Chapter" dialog, type **Elbow,** and then click OK (see items 4 and 5 in Figure 11.4).

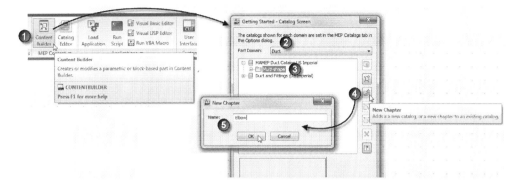

FIGURE 11.4

Creating a new chapter in a catalog (Steps 1 through 6)

6. With the Elbow Chapter selected, click the New Parametric Part Icon.

7. For the Name, type **Rectangular to Round Transition Elbow** and then press ENTER.

   The description will be added automatically (see Figure 11.5).

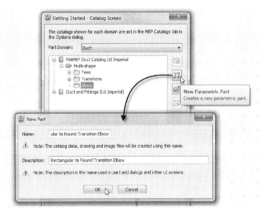

FIGURE 11.5

Creating the Rectangular to Round Transition Elbow

> Note: Name is the Drawing, XML and Bitmap name associated with this part and the Description that is displayed inside of AMEP.

8. Click OK to create the part and open Content Builder.

Content Builder is now active; the palette on the left contains all the parametric controls for Content Builder. All commands are accessible via the right-click menu when an item is selected within the Palette.

DEFINITIONS—PART CONFIGURATION:

Part File Name—The part file name shown at the top of the palette cannot be edited inside Content Builder. This is the file name that you entered in the "New Part" dialog box above. If required, you can "save as" to create a new part family from the current one with a different name.

Description—Describes the part family. You can edit the part description in the part browser. When you create and name a new part, you also enter a description in the "New Part" dialog. By default, the description is the same as the part name unless you change it.

Domain—Defines the family of parts, such as duct components, pipe components, cable tray components, conduit components, or MvPart components. You cannot edit the part domain in the part browser. The part domain is predefined based on the part catalog you selected in the "Getting Started" dialog of Content Builder. The part domain is selected from a list of predefined domains for AMEP.

Part Type—The actual type of part, such as elbow, tee, fan, damper, or tank. From the list of predefined part types in the part browser, you select the part type, which is based on the building system and loaded part catalogs. The part type is helpful during part selection.

Part Sub Type—Categorizes part types. In the part browser, you specify the part subtype from the list of predefined subtypes. You can also enter a custom part subtype if you wish. The part subtype is helpful during part selection to filter a large group of parts of similar type.

BUILDING YOUR FITTING

The process of building a fitting begins at the top and generally follows the top-to-bottom rule. We will start by defining your fitting.

9. Expand the Part Configuration control and then select the third item down (beneath Duct) currently listed as "Undefined."

10. Click on the same control again.

    A drop-down list will appear.

11. Choose Elbow from the list (see Figure 11.6) this defines the part type.

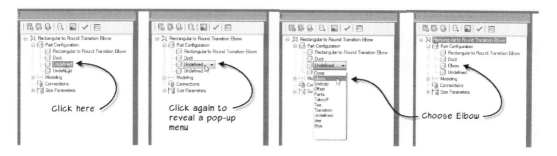

FIGURE 11.6

Configuring the Fitting

12. Select the fourth item down on the list (also currently listed as "Undefined") and then click again to access the pop-up menu.

13. Choose Transition from the pop-up menu (see Figure 11.7). This defines the part subtype.

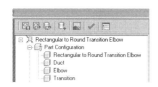

FIGURE 11.7

Configure the Part Type as Elbow and Sub Type to Transition

MODELING—ACCESS TO CONTROLS

Now that we have defined the fitting as a Transition Elbow, we need to begin to define the parameters required to model this elbow parametrically.

14. Expand the *Modeling* control below the part configuration in the Content Builder palette.

FIGURE 11.8

Expand the Modeling controls

The Modeling control is used for creating the Parametric Content, Constraints, and Dimensions. The commands for each of these tasks are only accessible via right-click beneath the associated subfolders.

Definition—Work Plane:

A **Work Plane** is an infinite surface upon which parametric parts are built. AMEP creates a rectangular graphic to indicate a portion of the Work Plane surface. You do not need to be within this boundary. Also, you can create multiple Work Planes within a part definition. You can think of a Work Plane as similar to a UCS.

CREATE A WORK PLANE

1. If you have not done so, expand the *Modeling* control.

2. Right-click on *Work Planes* and choose **Add Work Plane**.

3. In the "Create Work Plane" dialog, click the Top button.

4. Accept the default name of Top Plane and then click OK (see Figure 11.9).

FIGURE 11.9

Create a Top Work Plane

Once a Work Plane is created you will have access to the parametric drawing controls. The parametric drawing controls allow you to create the geometry, paths, and constraints necessary to build a parametric fitting. To access these commands, simply right-click on the Work Plane—Top Plane in this case (see Figure 11.10).

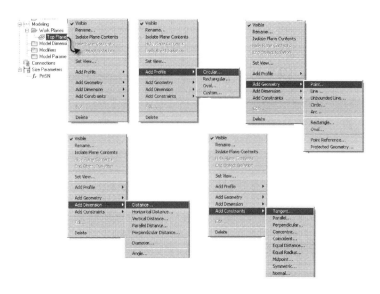

FIGURE 11.10
Work Plane modeling commands on the right-click menus

> Note: Content Builder does not recognize the UCS command for creating content. Instead, Work Planes define what view or orientation you are working in and are required to access Front, Side, Top, etc., views for the purpose of creating content.

Work Plane right-click commands:

Visible—This control allows you to turn on and off the visibility of the Work Plane. This is very helpful when creating multiple Work Plane parts.

Rename—Renames the Work Plane.

Isolate Plane Contents—This control only displays the objects on the current Work Plane and turns off the visibility of every other Work Plane and its objects.

Hide Plane Contents—Opposite of Isolate, hides the objects on the current Work Plane and turns off the visibility of every other Work Plane and its objects.

End Object Isolation—This control ends the object isolation and hides plane contents.

Set View—Allows you to zoom to the view of Work Plane. Since all Work Planes are in 3D space, this is an efficient way of orienting to a Work Plane

Add Profile—Profiles are shapes that get extruded along a path or from one Work Plane to another to generate the 3D body.

Add Geometry—This is the parametric geometry that defines the fitting—whether the parametric line or a parametric point—that is referenced by the defined parameters or constraints.

Add Dimension—These are parametric dimensions that control the geometry and shapes based on the values stored in the Size Parameters.

Add Constraints—Constraints are similar to dimensions, but limited to a single value like equal distance. Constraints are preferred when the condition only has a singular condition. For example, a perpendicular or parallel constraint will restrict the two constrained objects to the specified condition at all times.

Our next step is to create the path geometry required for the fittings. In the next sequence of steps, we are going to build a simple fitting like the one pictured here in Figure 11.11.

Note: The Fitting requires a straight throat for connection and a radius for the turn.

FIGURE 11.11

Rectangular x Round Transition Elbow

Note: Content Builder is an application that runs on top of AMEP and has its own commands. AutoCAD commands like Line, Arc, and Circle are not recognized by Content Builder in the context of defining the fitting. All geometry must be created using the Content Builder commands. In this case, the **Add Geometry > Line** command runs the Content Building Line command: AddColLine. Even though you cannot type normal AutoCAD commands, you can still use the Command Line to type commands in Content Builder if you wish.

5. On the Content Builder palette, right-click the Top Plane Work Plane.

6. Choose Add Geometry > Line.

7. In the Drawing window, you will begin by drawing a three point horizontal line and continue the line down with two additional points as shown in Figure 11.12.

> Note: You can use AutoCAD drawing aids like Ortho and Polar to keep the lines straight and use typical point entry methods like relative and absolute coordinate entry. Both Cartesian (X,Y,Z) and Polar (D<A) methods work. However, the constraints and dimensions ultimately control the behavior or the objects; therefore all items must be properly constrained and dimensioned.

8. Press ENTER to complete the command.

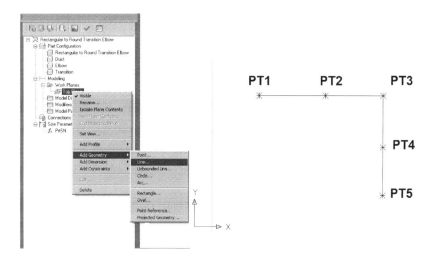

FIGURE 11.12
Adding the Col Lines to create the path of the Elbow

> Note: The Col Lines also automatically create Col Points which are located at the start and end points of the Col Lines, and coincident constraints are automatically added when two line endpoints meet. The points are used as markers and can be added via the AddColPoint command if needed. Col refers to the Cole Modeling engine inside Content Builder. Col commands are the parametric equivalent of the related AutoCAD commands.

9. Expand the Top Plane Work Plane and then expand the *Geometry* folder.

   You will notice that each point and line added in the drawing window is listed beneath the *Geometry* folder.

10. Select an object in the *Geometry* folder. Select a different item.

   As you select the items in the list, the associated objects in the drawing window will highlight.

11. Select the first Line in the *Geometry* folder, right-click and choose Fixed (see Figure 11.13).

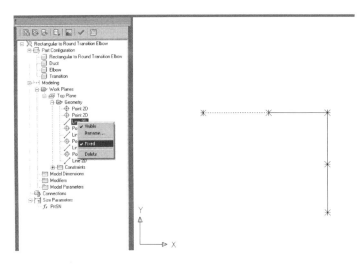

FIGURE 11.13
Mark the first line as Fixed

By selecting Fixed, the line will be restricted to its current position and will not move when constraints are added to the part. The line will turn green in the drawing window to indicate this is a fixed object.

12. Select the Top Plane, right-click and choose **Add Geometry > Arc**.

13. In the drawing window, pick the first point as indicated by PT1 in Figure 11.14.

   Content Builder will auto snap to the point.

14. Pick two more points as indicated. Remember to turn Ortho off (F8).

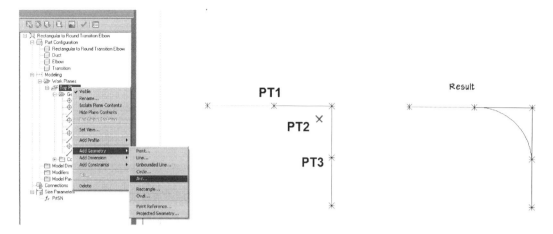

FIGURE 11.14

Draw a 3-point Arc

Note: The resulting Arc is shown on the right side of the figure. We will constrain it tangentially to the lines in the following steps.

This completes the basic path of the fitting. The next series of steps will add constraints to the paths so the fitting will react properly when dimensions are added.

Note: As a general rule of thumb, constraints are preferred when possible to minimize the amount of dimensions required for the fitting to be generated.

ADDING CONSTRAINTS TO THE PATH OBJECTS

1. On the Palette, select the Top Plane, right-click and choose the **Add Constraints >** **Tangent**.

2. At the "Select first geometry" prompt, pick the green line.

3. At the "Select second geometry" prompt, pick the arc (see Figure 11.15).

   The Arc and the Line are updated so the arc is tangential to the line. Remember, we set the line to Fixed, so the arc adjusts to it.

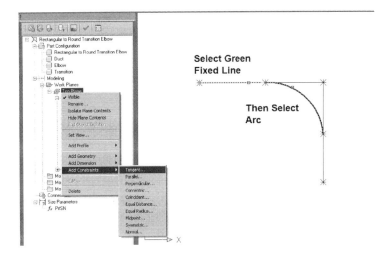

FIGURE 11.15

Adding a Tangent Constraint between the Fixed Line and the Arc

4. Repeat the process for the vertical line and the arc.

DEFINING THE ANGLE

Now that the lines and the arc are constrained tangentially, we will need to control the angle of the elbow. To do this we will add some additional Col Lines to allow for the dimension angle to be controlled.

5. Right-click on the Top Plane, and select **Add Geometry > Line**.

6. For the first point, click the upper left point, move the mouse straight down, and then click a point vertically below the first point.

    Do not worry that the line is not perfectly orthographic at this stage. We will use constraints to update this next.

7. Pick the next point on the lower right point on the existing vertical line (see Figure 11.16).

8. Press ENTER to complete the command.

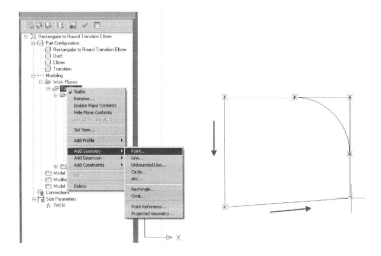

FIGURE 11.16

Draw Col Lines to define the angle of the elbow

9. Right-click on Top Plane and choose **Add Constraints > Perpendicular**.

10. At the "Select first geometry" prompt, pick the green horizontal line.

11. At the "Select second geometry" prompt, pick the adjacent vertical line.

12. Repeat for the lowest horizontal line and the lower right vertical line (see Figure 11.17).

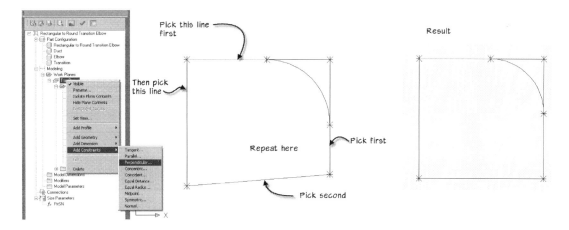

FIGURE 11.17

Define the Perpendicular Constraints

The Col Lines will adjust based on the new constraints and become perpendicular to each other.

13. Right-click on Top Plane and choose **Add Constraints > Parallel**.

14. At the "Select first geometry" prompt, pick the green horizontal line.

15. At the "Select second geometry" prompt, pick the adjacent horizontal line.

16. Repeat for the two vertical lines on the right (see Figure 11.18).

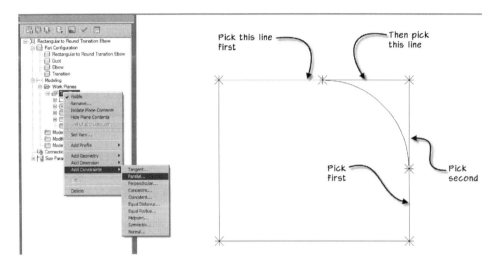

FIGURE 11.18
Define the Parallel Constraints

17. Select the Top Plane, right-click and choose **Add Dimension > Angle**.

18. At the "Select first line geometry" prompt, pick the left vertical line towards the top.

19. At the "Select second line geometry" prompt, pick the lowest horizontal line on the right side (see PT1 and PT2 in Figure 11.19).

20. At the "Pick dimension position" prompt, click above and to right of the drawing (see PT3 in Figure 11.19).

    This is where the dimension will be positioned.

21. At the "Enter angle" prompt, type **90,** and then press ENTER.

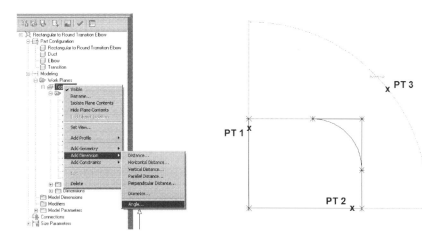

FIGURE 11.19

Add an Angle Dimension

> **Note:** A Col dimension (Path A1) will be added with an associated Size Parameter in the Size Parameter list. To see the list of parameters, expand the *Size Parameters* control.

The Path A1 Dimension allows you to specify all the angles that will be available for this fitting. We will add angles at the end of the chapter.

ADDING DIMENSIONS TO CONTROL THE LAYING LENGTH

First we need to add dimensions to control the Start and End Throats and the radius of the arc.

1. Right-click on Top Plane and choose **Add Dimension > Distance**.

2. At the "Select geometry" prompt, select the left Point associated with the green Line (PT1 in Figure 11.20).

3. At the "Select second geometry" prompt, select the right Point associated with the green Line (PT2 in Figure 11.20).

4. At the "Pick dimension position" prompt, pick a point above the green line.

> **Note:** If dimensions are appearing big relative to the model, change the Annotation Scale on the Drawing Status Bar.

5. Press ENTER to repeat the command. (If you change the drawing scale the last command has changed, you must Add Dimension > Distance again).

6. Following the prompts, pick the left Point associated with the green line and again point at the apparent centerline intersection (see PT3 and PT4 Figure 11.20).

7. Pick a point above the first dimension to place the dimension.

8. At the "Enter a dimension value" prompt, press **ENTER** to accept the value.

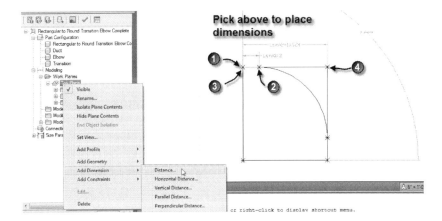

FIGURE 11.20

Add Distance Dimensions for the straight segments

Note: The dimension is placed automatically depending on the relative size. If the dimensions appear large, you can change your annotation scale on the drawing status bar to a smaller scale.

9. Right-click again on the Top Plane and choose **Add Dimension > Distance**.

10. Select the bottom point on the vertical line (PT5 in Figure 11.21) and then the point at the apparent centerline intersection (PT6 in Figure 11.21).

11. At the "Pick dimension position" prompt, pick a point to the right.

12. At the "Enter a dimension value" prompt, press **ENTER** to accept value.

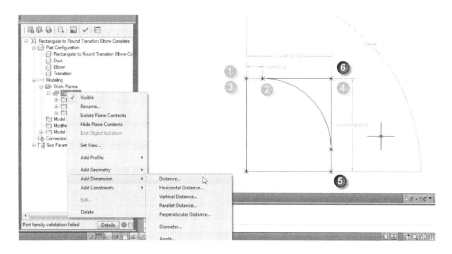

FIGURE 11.21
Distance Dimension to the Elbow Intersection

Now that we have the path and the angle dimension for the elbow created, we can begin to create the 3D Body for the Transition elbow. The command line should report "Right number of dimensions" to indicate the part is constrained properly.

> **Note:** Rather than defining a path angle and opposing dimensions to define the laying length of the fitting, an alternate strategy is to define a diameter dimension. This approach is useful for 90 Degree elbows only. Elbows with additional angles need opposing dimensions to accurately determine the laying length of the elbow at any angle.

ADDING PROFILES

Before we perform the next step, it will be helpful to see the Work Plane graphics. The boundary of the Plane is for reference only. The actual plane is infinite, and the graphics you draw can exist inside and outside this graphic.

1. At the Command Line, type z, press ENTER, and then type e and press ENTER.

> **Note:** As you can see, basic zoom and pan commands do work the same in Content Builder as in base AutoCAD.

2. Right-click on Top Plane and choose **Add Profile** > **Rectangular**.

3. At the "Select first corner" prompt, pick the upper left point as indicated by PT1 in Figure 11.22.

Note: The size of the rectangle should be drawn relative to the overall size of the items already in the drawing. This rectangle will be used for the 3D body.

---

4. At the "Select second corner" prompt, pick the lower right point of the profile as indicated by PT2 in Figure 11.22.

Note: We are drawing this profile next to our path as indicated in the figure. This will define the 3D body.

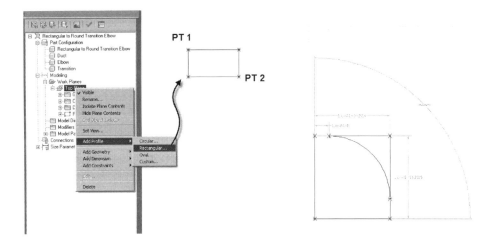

FIGURE 11.22

Create a Rectangular Profile

5. Right-click on Top Plane, and choose **Add Profile** > **Circular**.

6. At the "Select center point" prompt, click a point below the rectangle for the center (see PT1 in Figure 11.23).

7. At the "Select radius" prompt, pick a point to indicate the radius (see PT2 in Figure 11.23).

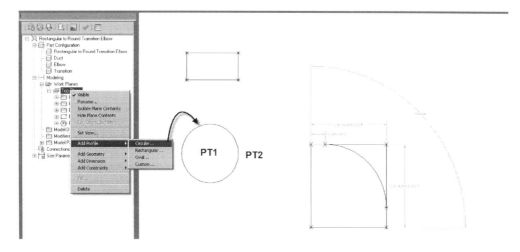

FIGURE 11.23
Create a Circular Profile

These two profiles will be our shapes. In order to make the transitional elbow, we will assign these shapes to the paths and generate a 3D body.

ASSIGNING SHAPES TO PATHS

1. Beneath the *Work Planes* folder, locate and select the *Modifiers* folder.

2. Right-click on the *Modifiers* folder and choose **Add Path**.

3. At the "Select path geometry" prompt, select the green Line.

4. At the "Select start profile" prompt, select the Rectangle.

5. At the "Select end profile" prompt, press ENTER to use the same profile at the end (see Figure 11.24).

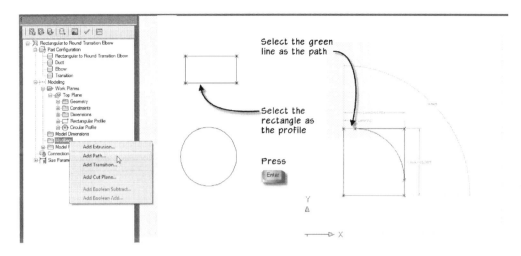

FIGURE 11.24

Assigning the Rectangle to a Path

Unlike most of the actions we perform in Content Builder, viewport navigation uses normal AutoCAD commands. In this sequence, we will switch to a 3D view. You can use any normal viewport navigation method to change to 3D. We will use the View ribbon tab here.

6. On the View tab of the ribbon, on the Appearance panel, scroll the view list and then click the SW Isometric view (see Figure 11.25).

FIGURE 11.25

Change to a 3D view

Notice the rectangular 3D body in light gray. This form is generated from the rectangular profile as it is applied on both ends of the green line.

7. Right-click on the *Modifiers* folder and choose **Add Path**.

8. At the "Select path geometry" prompt, select the arc near the lower right end.

9. At the "Select start profile" prompt, select the Circle.

10. At the "Select end profile" prompt, select the Rectangle (see the left side of Figure 11.26).

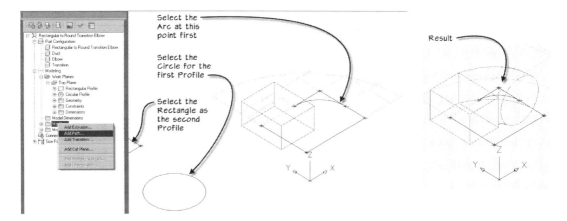

FIGURE 11.26
Create the Transitional Body

Notice that the Body has been generated in 3D to transition from the Circular to the Rectangular profile (see the right side of Figure 11.26).

11. Right-click on the *Modifiers* folder once more and choose **Add Path**.

12. At the "Select path geometry" prompt, pick the lower vertical Line.

13. At the "Select start profile" prompt, select the Circle.

14. At the "Select end profile" prompt, press ENTER to use the same profile at the end.

The basic 3D Body is now complete for the Transition elbow. The parameters are not yet complete, however. We still need to add the connectors to the body, assign the dimensions to the associated profiles, and add values for all the parameters.

CREATING THE CONNECTIONS

Before we begin to add the connectors on the Elbow, we need to switch back to the top view.

1. On the View tab, on the Appearance panel, click Top view.

2. On the Content Builder palette, right-click on the *Connections* control and choose **Add Connection**.

The next several steps refer to points shown in Figure 11.27.

3. At the "Select connector location" prompt, click the point on the left of the rectangular body (see PT1).

4. At the "Enter the connector number" prompt, press ENTER to accept the default of Connector 1.

On the Command Line, a message will appear above the normal prompt: "Please add the height dimension." In the next series of prompts, we will indicate both the height and width dimensions relative to the rectangular Profile.

5. At the "Select geometry" prompt, select the lower Line of the Profile (see PT2).

6. At the "Select second geometry" prompt, select the upper Line of the Profile (see PT3).

7. At the "Pick dimension position" prompt, pick a point to the left of the Profile to locate the dimension (see PT4).

The prompts will repeat for the width dimension.

8. At the "Select geometry" prompt, select the left Line (see PT5).

9. At the "Select second geometry" prompt, select the right Line (see PT6).

10. At the "Pick dimension position" prompt, pick a point above the Profile to locate the dimension (see PT7).

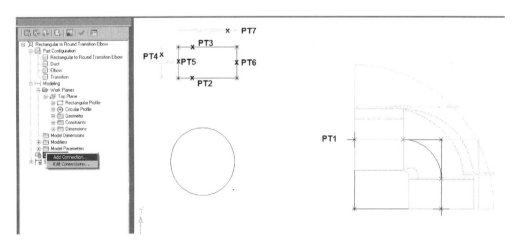

FIGURE 11.27
Create the Rectangular Connector

We will now repeat the process to create the circular connector.

11. Right-click on *Connections* and choose **Add Connection**.

12. At the "Select connector location" prompt, click the point on the bottom of the circular body (see PT1 Figure 11.28).

13. At the "Enter the connector number" prompt, press ENTER to accept the default of Connector 2.

14. At the "Pick dimension position" prompt, pick a point for the diameter dimension next to the circular Profile (see PT2 Figure 11.28).

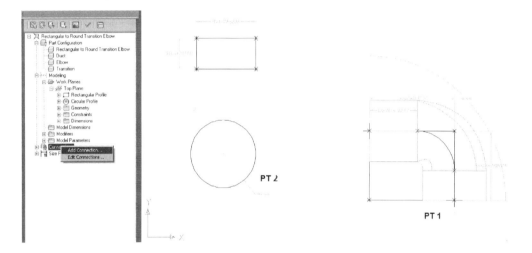

FIGURE 11.28
Create the Circular Connection

The connections are now assigned to the 3D Body. This allows you to control the height and width or the diameter dimensions on the connections based on the Size Parameters. This is a good time to save your progress.

15. At the top of the Content Builder palette, click the Save Part Family icon (see Figure 11.29).

FIGURE 11.29
Save the Part

When you save the fitting, the part will automatically have the validation utility run, and you will be prompted to allow the part to be added or keep the part hidden. The validation utility determines whether the part has the required elements to be a valid part family.

16. In the dialog that appears, choose YES (see Figure 11.30).

FIGURE 11.30
Part validation successful

PARAMETERS

One of the things that make parametric parts so powerful is their ability to be "parametrically controlled." In other words, they can use variables to help them conform to different circumstances. This is done using constraints and parameters. While each of these terms has several possible meanings, in the context of AMEP the following definitions found in an online dictionary are suitable to our discussion.

Essentially each of these is a rule applied to some part of the fitting's geometry or behavior. A constraint is a *fixed rule* that can only be manipulated by editing the Part Family. A parameter creates a rule or relationship that has *user-editable properties*. This next section covers how to create custom parameters to define how the dimensions added above will be controlled.

CREATE A CUSTOM PARAMETER

1. Beneath the Modeling node, select the *Model Parameters* folder.

2. Right-click on the *Model Parameters* folder and choose Edit.

The "Model Parameters" dialog will appear. Notice that each of the parameters we have defined so far is listed in this dialog.

3. In the "Model Parameters" dialog, click the New button (see Figure 11.31).

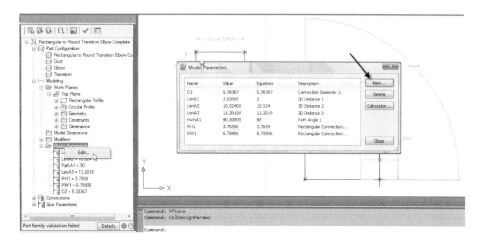

FIGURE 11.31

Use the "Model Parameters" dialog to edit, delete and create parameters

The "New Parameter" dialog will appear. We are going to define a formula for Laying Length (LL).

4. For the Name, type **LL**.

5. For the description, type **Laying Length**.

6. Next to the Equation field, click the Equation Assistant button (see the left side of Figure 11.32).

7. In the "Equation Assistant" dialog, use the buttons to write a formula:

⇨ Click the open parenthesis button: **(**.

⇨ Click the Variable button. From the pop-up that appears, choose **RW1**.

⇨ Use the buttons, or type ***2)+** (that is: multiplied by 2, close parenthesis, plus sign).

⇨ Click the Variable button again. From the pop-up that appears, choose **LenA1**.

The completed formula should read: (RW1*2)+LenA1

8. Click OK to close the "Equation Assistant" and return to the "New Parameter" dialog.

9. For Parameter Units, choose Inches and then click OK.

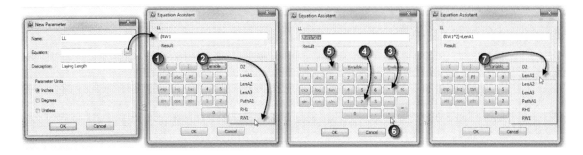

FIGURE 11.32

Create a Laying Length Parameter using the Equation Assistant

We will now assign the LenA2 and LenA3 dimensions to use the value calculated from LL to determine their value.

10. In the "Model Parameters" dialog, select the LenA2 parameter and then click the Calculator button.

11. Select the existing value, click the Variable button and replace it with the LL variable (see Figure 11.33).Then click OK.

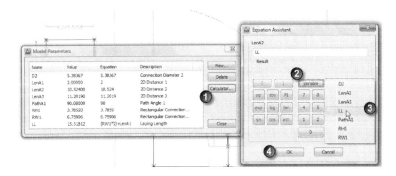

FIGURE 11.33

Configure the LenA2 parameter to equal LL (Laying Length)

12. Repeat the process to make LenA3 equal LL as well.

TIP: Another option at this point is to create additional model parameters, like the laying length parameter, and assign the dimensions so the parameter names match the cut sheet or have a name users would understand. For example, you can create a model parameter called Diameter instead of using the default parameter called BodyD1 by right-clicking on Model Parameters, selecting the New button, and typing Diameter for the name. Next select BodyD1 and assign Diameter as the value. The BodyD1 parameter now becomes a calculation equal to the values assigned to Diameter (this is done in the completed version).

Now that LenA2 and LenA3 are defined to equal the Laying Length (LL) parameter, we can define the rest of the parameters.

13. Close the "Model Parameters" dialog.

DEFINING THE PARAMETERS

In the next series of steps, you will edit the Size Parameters to utilize the new custom parameter above and define final values to the parameter dimensions.

DEFINITIONS—PARAMETERS:

Description—You can add a description to any parameter and it is recommended to add a description on complex parts.

Data Storage—There are four types available:

Calculation—A formula based on other parameters or an equation in the parametric model.

Constant—A single value that will be used for all the size ranges for the parameter.

List—A customizable value that allows user input to create the part size inside the drawing. The list can be edited through the Catalog Editor.

Table—A set of values that are determined inside the model parameters and will appear as part of the drop-down list inside the application.

Data Type—Determines if the data is a string (text) or a number.

Units—Units are the basis of measurement for the value specified. The value can be unitless or a specific unit based on the drawing units (Imperial or Metric). Unit values should always be specified for dimension parameters or associated parameters. Content placed in drawings with different units will be converted when the unit is specified. If the unit is not specified, the value is absolute. For example, if the value specified is 1 and the drawing units are set to Feet, then the 1 unit now becomes 1 foot instead of 1 inch.

Visible—Determines whether the parameter is visible or not inside the AMEP UI.

> **Note:** The FittingAdd command has a checkbox to expose the invisible parameters during selection.

Context—Context indicates how the parameter will be used. All default parameters are assigned a context.

Index—Indicates the index value of a parameter. For example, each length dimension is indexed with a progressive number each time it is added.

Custom parameter—All parameters defined by you. These can be the manufacturer's catalog dimensions, names, or parameters intended to be used inside formulas.

EDITING PARAMETERS

Our first editing task will be to assign the Data Storage value for several of our parameters. We will be using the Table data storage to define the values for specific part sizes. Refer to the Table definition above for more information.

1. Select the Size Parameters control, right-click and choose **Edit Configuration**.

2. In the "Size Parameters" dialog, click on the Data Storage value for RH1 (currently List) and change it to **Table**.

3. Repeat for RW1, D2, and PathA1 (see Figure 11.34).

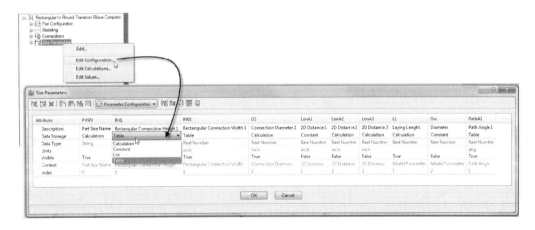

FIGURE 11.34

Accessing and editing the Parameter Configuration

> Note: When using the List data storage method, you can select via drop list or type values in the fitting add dialog inside AMEP. Please be aware that for List-based sizes, you need to edit the part after completion using the Catalog Editor to populate the list with more than one size.

Our next task is to determine which parameters will be visible inside AMEP when the fitting is inserted.

4. In the Visible row, change the value to True for each of the following parameters (see Figure 11.35):

➪ **PrtSN** (Part Size Name)

➪ **RH1** (Rectangular Height 1)

➪ **RW1** (Rectangular Width 1)

➪ **D2** (Circular Connection Diameter 2)

➪ **LL** (Laying Length)

➪ **Path A1** (Elbow Angle)

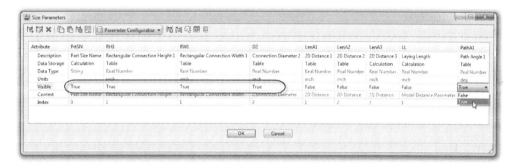

FIGURE 11.35
Setting Parameters to Visible

CREATING THE SIZE NAME CALCULATION

Our next task is to create a formula that will automatically generate the part name (Style Name) when the fitting is added into AMEP.

5. In the "Size Parameters" dialog at the top, click the Parameter Configuration drop-down list and choose **Calculations**.

6. Select the Part Size Name Column (PrtSN) and then click the Edit icon (see Figure 11.36).

FIGURE 11.36

Accessing the Calculation Assistant

The "Calculation Assistant" appears so you can create a calculation for the naming of the fitting instead of manually naming each line. Model parameter calculations should be edited through edit model parameters as indicated above. In the next few steps, we will concatenate additional parameters from the list of available parameters to create the part description that will be visible inside AMEP.

7. Place your cursor in front of PrtD in the Part Size Name field.

8. Scroll to the bottom of the list, select the RW1 variable, and then click the Insert button.

> **TIP:** To add parameters to the description, you need to use the arrow keys or your cursor to deselect the PrtD and select the location where you want the parameter to be added.

9. Type the letter **x** after the automatically added space.

10. From the variable list, select the RH1 and then click the Insert button.

11. Click the Evaluate button to verify that the spaces and parameters are correct.

The evaluation result should read: 10.00 x 10.00 Rectangular to Round Transition Elbow (see the left side of Figure 11.37).

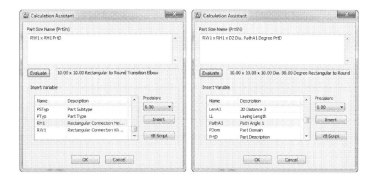

FIGURE 11.37

Define the Part Size Name formula and evaluate it

12. Type another **x** after the RH1 parameter and then insert the D2 variable.

13. Type **Dia.** after the D2 variable.

14. Insert the PathA1 variable next and type **Degree** after it.

The final result should read: RW1 x RH1 x D2 Dia. x PathA1 Degree PrtD.

15. Click the Evaluate button (see the right side of Figure 11.37).

The result should read: 10.00 x 10.00 x 10.00 Dia. x 90.00 Degree Rectangular to Round Transition Elbow.

Note: You can set the precision of the values by setting the Precision before selecting insert. For example, this allows you to change the value from 0.00 to 0. This is beneficial when creating Metric content.

---

16. Click OK when done, see Figure 11.38 for the Part Size name formula in row 1.

FIGURE 11.38
The PrtSN Calculation value

ADDING VALUES

We can now add values to the parameters to make the fitting sizes we need. You can manually add values here inside each parameter, or you can paste from a spreadsheet to populate values quickly. To edit the parameter you can double-click in the cell and type the value.

1. Back in the "Size Parameters" dialog, click on the Calculations drop-down list at the top and choose Values (see Figure 11.39).

FIGURE 11.39
Accessing Parameter Values for editing

2. Change the parameter values to the following by double-clicking in the field (see Figure 11.40).

⇨ RH1 = 4.0000

⇨ RW1 = 8.0000

⇨ D2 = 6.0000

⇨ LenA1 = 2.0000

⇨ Path A1 = 90.00 (90 is the default, so simply verify if the value exists).

Note: Calculations will automatically be updated based on the new values that are input for the parameters listed above.

FIGURE 11.40

Edit the parameter values for the first size

3. Select Row 1; then at the top of the "Size Parameters" dialog, click the New icon (see Figure 11.41).

FIGURE 11.41

Adding a new row of parameters

Note: The information in the first row will be copied to the new row; therefore you only need to change the values that must differ.

4. Change the parameter values in row 2 to the following by double-clicking in the field (see Figure 11.42):

⇨ RH1 = 6.0000

⇨ RW1 = 10.0000

⇨ D2 = 8.0000

⇨ LenA1 = 2.0000

⇨ Path A1 = 90.00

FIGURE 11.42

Editing the parameters for Row 2

5. To update the model based on the selected row, click the update icon (see Figure 11.43). This button pushes the size parameter values to the model parameters to depict the model based upon the row values.

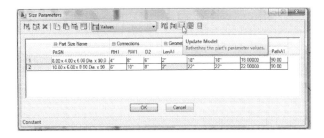

FIGURE 11.43

Updating the model to the dimensions specified in the selected row

6. Click OK to exit the "Size Parameters" dialog.

CONFIGURE BITMAP PREVIEW AND OPTIONS

We have a few finishing touches to complete our fitting. We will assign a bitmap preview image and verify that our other Options are set correctly.

> Note: We recommend replacing the default Bitmap with one created from a Visual Style once the part is added to a drawing. To do this, use the AutoCAD export command, export to a bitmap, and overwrite the file created in the catalog.

7. On Content Builder on the palette, click the Generate Bitmap icon (see Figure 11.44).

FIGURE 11.44

Generate a Bitmap image for the part

The "Bitmap Preview" dialog will appear.

8. Select a view, such as SW Isometric, to create the bitmap image and then click OK (see Figure 11.45).

> Note: Please be aware that Vista users may not see an image due to an AMEP bug. If this happens, you must replace the image file created manually by generating a Bitmap in AMEP using the Export command.

FIGURE 11.45

Set the Bitmap Preview to southwest view to create the image preview

> CAD MANAGER NOTE: you can replace the bitmap by creating your own bitmap inside AMEP and saving over the one that is created by Content Builder so you can add additional information like connections and dimensions.

9. On Content Builder on the palette, click the Options icon (see Figure 11.46).

FIGURE 11.46

Open the Content Builder Options Dialog

The options dialog will appear (see Figure 11.47).

FIGURE 11.47

Content Builder Options Dialog

In the Content Builder "Options" dialog you can decide whether the part can be used during auto-layout, can have custom sizes, and whether the part should be visible in the catalog.

> **TIP:** Use the "Hide Part Flag" feature to hide incomplete parts that you do not want the team to use until the part is ready.

Please verify that the Hide Part Flag is unchecked and Custom Sizing Flag is checked (this allows the part to support custom sizes). We will not be checking the AutoLayout flag on this part, since auto layout does not support multi-shape elbows.

10. Click OK.

11. In the Content Builder palette, right click on the Connections and select Edit Connections

12. Select the Browse button in the Connector 1 column next to Type: Undefined

13. Check Banded, Clipped, Flange, Slip Joint, Undefined, and Vanstone.

14. Click OK twice.

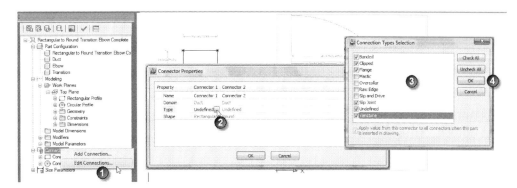

FIGURE .11.48

Assigning the Connection Type

VALIDATE THE MODEL

At this point we need to validate that the model and the fitting will work and finish editing the part so it can be inserted into AMEP and a Green Icon will appear (check). If the part fails, you will see a red icon and a message indicating that it failed at the bottom of the Content Builder Palette (see Figure 11.50).

15. On Content Builder on the palette, click the Validate icon (see Figure 11.49).

FIGURE 11.49

Validate the fitting

In the lower section of the palette the validation results are listed. If a serious problem exists, an error will appear indicating problems at the bottom of the Content Builder palette with a Red icon (see Figure 11.50).

Part family validation successful   Details

FIGURE 11.50

Results from Validation on bottom of Content Builder palette

16. Click the Save icon, and then click the Close box at the top right corner of the Content Builder palette.

Test your Fitting

Now that you have completed building your part and the validation was successful, you are ready to test it out in AMEP.

1. On the QAT, click the New icon.

2. Switch to the HVAC Workspace.

3. On the Home tab, on the Build panel, click the Duct Fitting button.

    If necessary, expand the MAMEP Duct Catalog, then the *Multi-shape* folder and then the *Elbow* folder.

4. Select the Rectangular to Round Transition Elbow.

5. Verify that the sizes are listed on the Part Size Name list (see Figure 11.51).

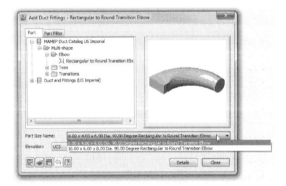

FIGURE 11.51

Results from Validation on bottom of Content Builder Palette

6. Insert one of each size in the drawing. View them in plan and 3D.

BONUS STEPS

Now that you have successfully inserted the fitting and it behaves correctly, you can use this fitting as is or edit the part and add 45-degree angles for each elbow size.

1. Start Content Builder again.

2. Double-click the Round to Rectangular Transition Elbow to edit it.

3. Right-click on Size Parameters and select **Edit Values**.

4. In the "Size Parameters" dialog, select the first line by selecting number 1.

5. Click the New icon.

    The first row will be copied to Row 3.

6. Repeat for row 2.

7. In Rows 3 and 4, change the PathA1 value from 90 to 45 and then click OK.

8. Click the Save part Family icon and then Close the Content Builder palette.

    We have now added two additional angle options for this fitting (see Figure 11.52).

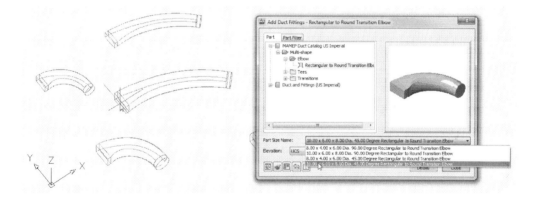

FIGURE 11.52

After adding two new sizes, try inserting them in a drawing

9. Return to your drawing and start the DuctFittingAdd command again.

    Notice that both 90- and 45- degree angles are now available for each size.

CONTENT BUILDER TIPS

Here are some additional tips to keep in mind when creating parametric fittings in Content Builder.

▪ Sketch out the skeleton of the fitting and all associated dimensions before trying to define the paths. This will minimize errors when creating the fitting.

- Once the sketch is created with dimensions, review it to determine if any constraints are required (parallel, perpendicular, equal distance, etc.).

- With the dimensions and constraints identified, you next want to determine the order in which the paths will be defined. The order of creation is important because if there are conflicting dimensions/constraints, the first one created "wins."

- Once the creation order is identified, determine which path should be defined as Fixed. This will reduce the possibility of the model's "blowing up" in Content Builder (the model looking completely different). This is also helpful when experimenting with new types of fittings, since applying values to the parameters will be restricted by the fixed object. Usually this is the first item drawn.

- Construction lines/paths can be created to control the model. The elbow created above uses two lines that are considered construction lines to control the angle of the elbow. You can create as many as needed.

- Paths can be on more than one Work Plane. The fitting we just built uses only the top Work Plane, but you can define paths on additional Work Planes if needed. However, this is considered a complex approach to creating content. AMEP has a pipe fitting that is built with this approach. See the "Brazed - DWV 90 Deg Elbow with Side Inlet" fitting in the Copper Pipe Catalog.

- Editing Parameters supports copy/paste from Excel and other programs on a per column basis.

- Creating unique Model Parameters to assign as the value for default parameters is recommended. For example, we replaced LenA3 with a custom parameter called Laying Length (LL). This allows you to understand what the parameter controls. The common approach is to create parameters to match cut sheets and define dimensions to match the cut sheet parameter names.

- When adding Constraints or Dimensions always read the command line; this will indicate if the part is under-constrained or dimensioned, contains the right amount of constraints or dimensions, or if it is over-constrained or dimensioned. This message appears each time a constraint or dimension is added.

- Catalog Editor can be used to edit values in content, create new chapters, organize catalogs, and create new catalogs. Catalog Editor can be run outside of AMEP by starting the *AecbCatalogEditor.exe* located in the AMEP program folder.

- Creating a new catalog with Catalog Editor requires at least a single valid fitting in the catalog to allow the catalog to be used in AMEP. Simply copy a fitting from the existing catalog to the new one; once you create your custom fittings you can delete the initial copied fitting.

Congratulations! You have successfully completed your first parametric part in the AMEP Content Builder.

SUMMARY

- ✓ Content Builder creates Parametric Parts using paths, dimensions, and constraints to define the part.

- ✓ Using Content Builder, we can create a parametric part which can contain potentially thousands of variations and sizes.

- ✓ In this chapter we created a new parametric Duct Elbow Fitting which transitions from a rectangular connection to a round connection. However, the principles employed apply to any parametric part.

- ✓ Add parametric lines and arcs to define the path of the geometry.

- ✓ Assign constraints and dimensions to the parametric lines and arcs to control their behavior.

- ✓ You can control different aspects of the parts using constraints or dimensions depending on the need

- ✓ You can add profiles to a path to define the body.

- ✓ Add custom parameters to define the overall distances or other important values such as the Laying Length parameter created here.

- ✓ Create a custom parameter (Laying Length) to define the overall distance to the apparent intersection and make it visible inside AMEP to better understand the purpose of the value.

- ✓ You can create Formula-based parameters to calculate the size.

- ✓ You can edit a part to add additional sizes.

- ✓ Remember to validate your new parts to ensure that they are built correctly.

Chapter 12

Display System

INTRODUCTION

The Display System in AutoCAD MEP is powerful and complex. Without taking the time to understand the concepts, fundamentals, and how to navigate the controls, you will surely be lost. This chapter will explain the Display System fundamentals and how to exploit the power of Display Manager. In Chapter 2, we explored the fundamentals of the Display System with a hands-on look at its three main components: Display Representations, Sets, and Display Configurations. In this chapter we will explore these items in more detail so you can begin to leverage the potential of this powerful tool.

OBJECTIVES

In this chapter, we will explore Display Representations, Sets, and Configurations in greater detail than we have so far. Furthermore, we will take a look at the display by elevation functionality, system definitions, overrides, XREFs and how all these features relate to one another in AMEP. In this chapter you will:

- Explore the Display Manager and its functions.

- Learn how how to control Display Settings.

- Look at display overrides in many circumstances.

- Understand Display by Elevation.

- Take control of XREFs through the Display Manager.

DISPLAY SYSTEM DEFINITIONS

Each AMEP object has a collection of display settings. These settings are referred to as the object's Display Properties. These include Visibility mode (On or Off), Layer, Color, Linetype, Lineweight, LTScale, and Plot Style. They also include settings for hatching where appropriate and other object-specific settings unique to each kind of object, such as the center line. Display Properties can be applied at several levels, and are applied in a hierarchy, so that if an override is not applied, for example, the next higher level on the hierarchy will apply. The levels of display in order of hierarchy are:

Drawing Default—In the hierarchy of display settings, the Drawing Default settings come first and establish the baseline for a particular type of object.

Domain Drawing Default—Controls the default for all objects within a Domain. So All Segments, Custom Fittings, Fittings and Flex share common display settings and can be changed in one action.

System Definition Style Override—A System Definition represents Pipe, Fittings, Custom Fittings and Flex Pipe assigned to do a specific task (for example, Chilled Water). This includes all shapes and sizes of these objects. Controlling display at this level allows objects and parts of the same type or style to display differently if they are in different systems. For example, the Demo System displays the hatch pattern while the other systems have the hatch display component off.

Style Override—Style Override affects all instances of a single object, such as a 6" Round Duct Elbow. MEP objects typically have a style for each Part Size Name from the catalog, so any override only affects all instances of a single size of the object. The other sizes of the Duct Round Elbow are not affected by this change. Style-level settings override the System Definition Style Override and Domain Drawing Default settings.

Object Override—Object Override affects only a single object selected in the drawing. Object-level settings override those of the Drawing Default and, if present, any other overrides. In general, frequent application of object-level overrides should be avoided. This is similar in concept to overriding the color of a line in-lieu of having it's color display bylayer.

UNDERSTANDING DISPLAY HIERARCHY

The Display System has three main parts: Display Representations, Sets, and Configurations. The Display Manager is used to access, configure and manage these items. Each AMEP object has one or more Display Representations used to draw the graphics of the object in one or more common drawing situations (i.e. Plan, 1 Line, 2 Line, Model, etc.). A Display Set is akin to a large switchboard wherein each object is configured to display in one or more of its available Display Representations. If no Representations are selected for an object, that kind of object will be invisible in that Set. Finally, a Display Configuration enables the display of Sets in the drawing based on viewing direction. In other words, a Configuration can display a Plan Set when the drawing is viewed from top and a Model Set when it is viewed in 3D.

The default AutoCAD MEP template files provide several premade Display Configurations offering many different display possibilities and allow you to customize the display settings for your company.

The best way to understand the definitions presented above is to work through a quick exercise. The Display System follows a hierarchical structure. Drawing Default settings apply if there are no overrides. Overrides can be applied at three levels: System, Style, and Object. A System override will apply to all members of the System unless they have another override. A style override will apply to all members of the style unless there is an object override. Finally, an override at the object level overrides everything else. However, managing such an edit is the most labor intensive because you must select objects that have the override and edit them directly. You cannot use Style Manager to make global edits to object overrides.

INSTALL THE DATASET FILES AND OPEN A PROJECT

In this exercise, we use the Project Browser to load a project from which to perform the steps that follow.

1. If you have not already done so, install the book's dataset files.

   Refer to "Book Dataset Files" in the Preface for information on installing the sample files included with this book.

2. Launch AutoCAD MEP 2012 and then on the Quick Access Toolbar (QAT), click the Project Browser icon.

3. In the "Project Browser" dialog, be sure that the Project Folder icon is depressed, open the folder list and choose your *C:* drive.

4. Double-click on the *MasterAME 2012* folder.

5. Double-click *MAMEP Commercial* to load the project. (You can also right-click on it and choose **Set Project Current**.).

IMPORTANT: If a message appears asking you to repath the project, click Yes. Refer to the "Repathing Projects" heading in the Preface for more information.

Note: You should only see a repathing message if you installed the dataset to a different location than the one recommended in the installation instructions in the preface.

6. Click the Close button in the Project Browser.

When you close the Project Browser, the Project Navigator palette should appear automatically. If you already had this palette open, it will now change to reflect the contents of the newly loaded project. If for some reason, the Project Navigator did not appear, click the icon on the QAT.

DEFINING OVERRIDES

Let's get started in a simple file designed to give us a better understanding of the Display System definitions just presented.

1. On the Project Navigator, click the Constructs tab.

2. Expand the *Elements\Display System* folder and then double-click the *Display System Definitions* file.

All the objects in this file are controlled by the Systems to which they are assigned. The Systems in turn are controlled by the Domain Drawing Default. We are going to add the different levels of overrides to gain a better understanding of the hierarchy.

Four duct runs are included in this file. Each has a text label. Use the text labels to assist you in making selections for the following steps.

Let's begin in the area labeled: Controlled by Domain Drawing Default.

3. Select a Duct segment.

4. On the Duct tab, on the General panel, click the Edit System Style button.

5. In the "Duct System Definitions – Supply" dialog, click the Display Properties tab.

> Note: In this case, we did not check a Style Override box, thus, we are modifying the Drawing Default settings.

6. On the right side, click the Edit Display Properties icon.

7. On the Layer/Color/Linetype tab, select Contour and change the color to Red and the Linetype to HIDDEN2 (see Figure 12.1).

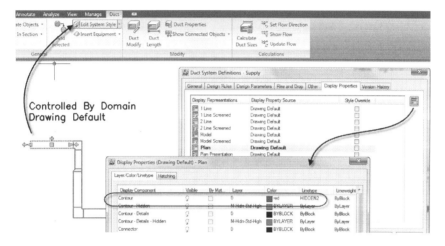

FIGURE 12.1

Changing the Domain Drawing Default through the System Definition

8. Click OK twice to accept the change and return to the drawing.

The Contour display component on all Ducts and Fittings has been changed to display Red and with line type HIDDEN2. Since both the segment and fitting styles have changed, this indicates Domain Drawing Default has been changed (see Figure 12.2).

9. Deselect the Duct.

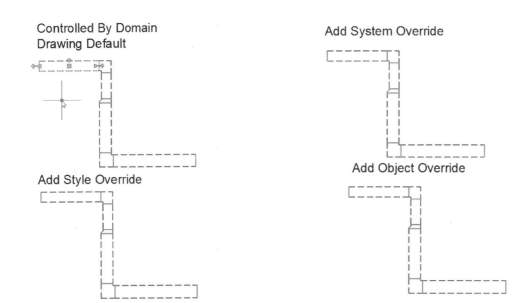

FIGURE 12.2

Domain Drawing Default Contour display component is now red and Hidden 2

Let's move on to the area labeled: Add System Override.

10. Select a Duct segment.

11. On the Duct tab, on the General panel, click the Edit System Style button.

12. In the "Duct System Definitions – Supply (2)" dialog, click the Display Properties tab.

13. In the Plan row, click the Style Override check box.

14. Select Contour and change the color to Green and the Linetype to HIDDEN (see Figure 12.3).

Controlled By Domain
Drawing Default

Add System Override

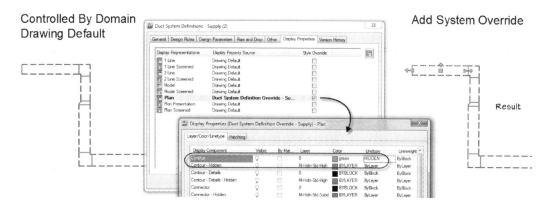

Result

FIGURE 12.3

A System-Level Override only applies to members of that System

15. Click OK twice to accept the change and return to the drawing.

The Contour display component on all Ducts and Fittings that are in the Supply (2) system have been changed to color Green and linetype HIDDEN (see Figure 12.4).

16. Deselect the Duct.

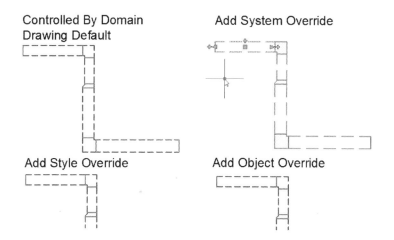

FIGURE 12.4

Only the Supply (2) System Definition Contour display component is now green and Hidden

Our next edit will be in the area labeled: Add Style Override.

17. Select a Duct segment.

18. On the Duct tab, on the General panel, click the Edit System Style drop-down button this time and choose Edit Duct Style.

19. In the "Duct Style Properties" dialog, click the Display Properties tab.

20. Next to the Plan Display Representation, click the Style Override check box.

21. Select Contour and change the color to Blue and the Linetype to PHANTOM2 (see Figure 12.5).

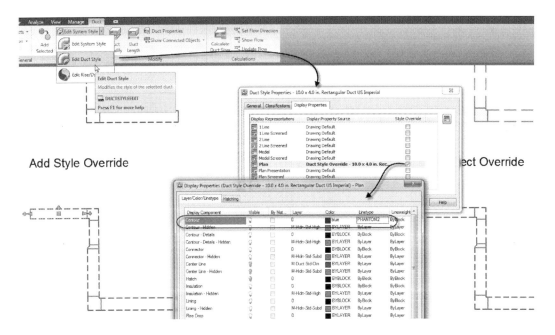

FIGURE 12.5

A Style-Level Override only applies to members of that Style

22. Click OK twice to accept the change and return to the drawing.

The contour display component on all Ducts that use the style: 10.0 x 4.0 Rectangular Duct US Imperial have been changed to Blue and PHANTOM2 (see Figure 12.6).

23. Deselect the Duct.

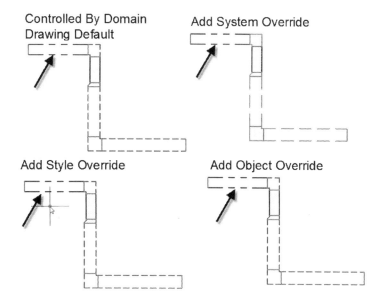

Controlled By Domain
Drawing Default

Add System Override

Add Style Override

Add Object Override

FIGURE 12.6

Any Duct using the 10.0 x 4.0 Rectangular Duct US Imperial Duct style is now Blue and Phantom2

Our final edit in this sequence will be in the area labeled: Add Object Override.

24. Select a Duct segment.

25. Right-click and choose **Edit Object Display**.

26. In the "Duct Style Properties" dialog, click the Display Properties tab.

27. In the Plan row, click the Object Override check box.

28. Select Contour and change the color to Magenta and the Linetype to DASHED2 (see Figure 12.7).

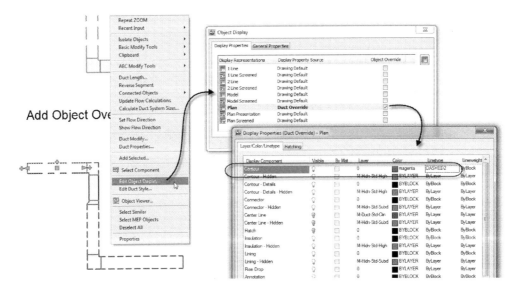

FIGURE 12.7
An Object-Level Override applies only to the selected object

29. Click OK twice to accept the change and return to the drawing.

The contour display component on only the selected Duct has been changed to Magenta and
DASHED2. No other Duct will reflect this change (see Figure 12.8).

30. Deselect the Duct.

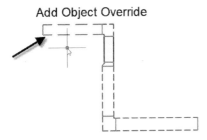

FIGURE 12.8
Only the selected Duct is Magenta and Dashed2

As you study the final results in the figure, you should be able to see the hierarchy more clearly.
An object-level override will override a style-level override, which will override the system
override. All of these will override the drawing default.

UNDERSTANDING DISPLAY SYSTEM DEFAULTS

Now that we have a better understanding of the hierarchy, let's begin to dig a little deeper into the overall Display System settings. A good way to understand it is to explore the out-of-the-box defaults.

EXPLORE THE DISPLAY MANAGER

1. On the Project Navigator, in the *Elements\Display System* folder, double-click the *Display Drawing Defaults* file.

2. On the Manage tab, on the Style & Display panel, click the Display Manager button.

3. Select the *Representations by Object* folder (see Figure 12.9).

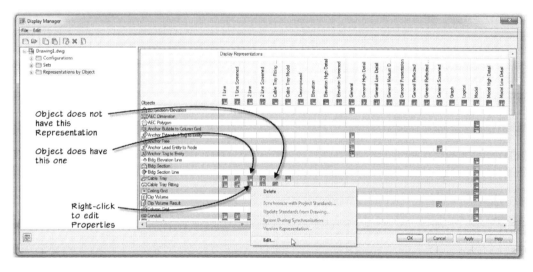

FIGURE 12.9
Display Manager showing the Representations by Object folder

The grid shown in the right pane indicates at a glance all AutoCAD MEP (and AutoCAD Architecture) objects and their associated Display Representations. You can see the complete list of MEP and Architectural objects in the column at the right. Along the top row, all Display Representations that exist in the drawing are listed. (This list is compiled from all objects in the left column). In the grid matrix itself, an icon indicates that the object has that particular Display Representation. A quick study of this list will tell you a particular Representation is available for an object. You can also right-click the icons to edit the properties of any object and Display Representation combination.

4. In the left pane, expand the *Representations by Object* folder (see Figure 12.10).

Notice that all of the objects are listed here. You can access the specific Display Representations (Reps) here. Simply click the plus (+) sign next to the object to expand it and reveal its Display Representations.

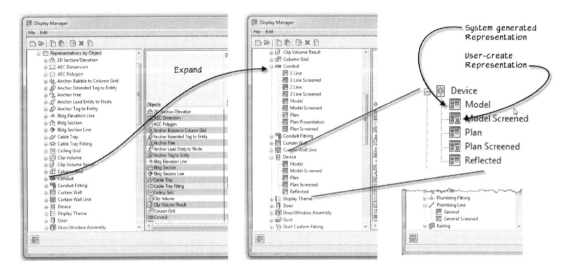

FIGURE 12.10
Using the left pane, you can access the Display Representations for objects directly

The icons that appear without a little person are system-generated Display Representations that can be edited, but not renamed or deleted. The names of the Display Representations usually indicate their use. Each object will have different Representations available depending on the needs for that object. For example, Conduit, Devices, and Plumbing Line (shown in the figure) share some Reps in common, but do have some unique ones as well.

AutoCAD MEP objects have the ability to be displayed in multiple ways, such as: 1-Line, 2-Line, Model (3D), Plan, Presentation, and Screened. In most cases, the names indicate their use, but sometimes, as with Plumbing Line, the same Representation is used for all views so it is named General. (General Screened is a screened version of General and is user-created, as the small person icon indicates).

UNDERSTANDING DISPLAY REPRESENTATIONS

Object Display Representations are as important as the objects themselves; each one defines how the object will be displayed throughout AutoCAD MEP. There are several dozen Display Representations in total (as seen above in Figure 12.9). There are too many to cover them all in detail. Several common ones are available for most MEP objects. Following is a list of the most common Display Representations:

- **1 Line**—will display MEP objects as single-line objects in any view.
- **2 Line**—will display MEP objects as 2-line and will not use the hidden line functionality. Also, 2-Line will display Pipe and Equipment objects with their 2-line bodies in any view (see Model definition).
- **General**—is used for 2D components that do not require different Representations (Schedules, Schematic Lines, etc.).
- **Model**—will display the object with its true 3D body regardless of view direction. This will also honor the Pipe by Size functionality that can display piping as 1-line, graphical 1-line or 2-line based on size.
- **Plan**—will display objects using their 2D/plan graphics and hidden lines will be generated for overlapping objects. This display only works for Plan (Top) views.
- **Screened** and **Presentation**—these are copies of the system controlled displays and provide different settings while honoring the rules listed above. For example, the Plan Screened display representation has all objects assigned to a gray color, and Plan Presentation uses solid hatching to display different systems as filled objects.

The settings modified in the Display Manager under Representations by Object are the drawing default settings. As mentioned in the display properties and definitions section above, AutoCAD MEP uses the term Domain Drawing Default. Domain drawing default slightly differs from drawing default to any single item within a domain (Cabletray, Conduit, Duct, etc.); other domain items will automatically update (Custom Fitting, Fitting, and Flex). AutoCAD MEP Display Representations use layers and the Layer Key defined in the System Definition to control each sub-component. Domain drawing default and drawing default control the display of any object that does not have an override assigned to it.

> **IMPORTANT:**
> Drawings that are XREFed into another drawing and contain the same Display Representation names as the host will be controlled by their own settings *not* the host drawing's. For example, Architectural is XREFed into the MEP drawing, and both drawings have the Plan Screened Display Representation for Wall objects. The Architectural drawing's settings for Plan Screened will control the display of the XREF in the MEP drawing. Refer to the "External Reference Control" topic below for more information.

Overrides are not defined or edited inside Display Manager; these are only available on the objects and styles themselves.

5. Still in Display Manager, in the left pane, expand Conduit.

6. Select the 1-Line Representation.

   Take note of its components in the right-hand pane.

7. Select the 2-Line Representation.

Comparing the two Representations. You will notice that the display components are somewhat different for each display (see Figure 12.11).

FIGURE 12.11

Compare the difference with 1-Line and 2-Line Display Representations

The list of components is based on the definitions listed above. For example, the 2-line Display Representation includes a Hatch component, whereas 1-line does not have a Hatch component (since you cannot hatch a single line).

You will notice that in addition to the varying list of components within each Display Representation, you have a great deal of control over each individual component. As you explore different objects, you will find that most of the components are assigned to Layer 0 with the other properties set to ByBlock. Think of the "ByBlock" setting as "ByObject". In other words, the component is deferring to the parent object for its color, linetype, and lineweight. So in the case of the Conduit depicted in the figure, instead of the Contour component's determining its own color, linetype, and lineweight, ByBlock ensures that the Conduit object itself will determine these settings. Furthermore, since most company standards work "ByLayer," you can expect your parent objects (Conduit in this case) to be drawn ByLayer. This means that your layers will ultimately determine the color, linetype, and lineweight of your objects, just as you'd expect. There are exceptions, of course. For example, you can see that the Center Line component shown in Figure 12.11 above has been assigned to a specific layer. In that case, the other properties are assigned to ByLayer rather than ByBlock to ensure that the E-Cndt-Std-Clin layer will determine color, linetype, and lineweight of the Center Line component. This is a very powerful aspect of the Display System.

As we saw briefly above, AutoCAD MEP objects have an additional level of control, that being the System Definitions. System Definitions assign layers to objects within the System via a Layer Key assignment. (This has been discussed in several previous chapters). Because of this additional level of control, you will rarely find the need to modify the drawing default level layer, color, and linetype settings. When modifications are necessary, System overrides usually provide the best means to achieve the required modification.

If you decide to edit the drawing default settings here, you should change the layer of the component and then set the color, linetype, and lineweight to ByLayer. Keep in mind that here in the Display Manager and at the drawing Default level, you cannot assign a Layer Key and instead must hard code an actual layer. This limitation is one of the primary reasons System Definitions are typically used instead. A System Definition as we have previously seen assigns a Layer Key rather than a hard coded layer, making them more flexible. For some components, such as Center Line, Insulation, Lining, and Connector, the Drawing Default settings are set to a specific layer. As such, these settings are common to all systems by default, but could be overridden if necessary.

SETS AND CONFIGURATIONS

There are two other folders in the Display Manager. We will discuss them briefly since they have already been detailed in Chapter 2. The purpose of a Set is to decide which objects should display onscreen and how (using what Display Representation) they should display.

8. In the left-hand pane, expand the *Sets* folder and then select the MEP Design – Plan Set.

9. Select the Display Representation Control tab.

Notice that the same grid matrix appears again. Objects still appear in a column along the left and in the row along the top are all Display Representations. Notice, however, that this time instead of icons, there are checkboxes. A checkbox indicates that a particular Display Representation is available for display in the Set. If the box is checked, that Representation is displayed (or turned on) and if it is unchecked, it is turned off.

10. Scroll to the right and locate the Plan column.

11. Scroll vertically and study the checkboxes.

Notice that most AutoCAD MEP objects have Plan checked. For this Set, all MEP objects will display in using their Plan Representation. For an object to appear in the drawing, a Display Representation must be checked; if you do not want certain objects to appear (you want them invisible) uncheck all the boxes for that object. This is most common in ceiling plans to make floor plan information disappear.

Sets have additional controls available on the Display Options tab. The top section contains the classification filter which allows you to determine whether a MEP object defined with a specific classification will be displayed when the set is active or not. Refer to Chapter 2 for more information on using Classifications with Display Sets.

The bottom portion of the Display Options tab allows you to control Live Sections and Materials. This control will indicate if a Live Section is currently enabled. You can control whether the display of AEC Objects are controlled by the cut plane defined in the Display Configuration. This setting is especially useful in complex layouts when a live section is enabled in a 3D view. The live section will be restricted to the display range settings defined in the cut plane, and will update to changes in the cut plane settings.

The final folder in the Display Manager is the *Configurations* folder. These were also explored in Chapter 2. Configurations decide which Set to display based on the viewing direction in the drawing. For example, a single Configuration is capable of switching from a 2D Set to a 3D Set when the viewing direction changes from top to isometric.

Configurations also allow you to define what the cut plane is set at for all objects. AutoCAD Architecture objects that are being displayed in a plan view will always respect the cut plane setting. AutoCAD MEP objects have the ability to ignore or respect the cut plane by the display by elevation setting.

12.Close the file. It is not necessary to save it.

DISPLAY BY ELEVATION

The Display by elevation functionality allows most MEP objects to respect the cut plane (this has no effect on Devices, Panels, Plumbing Line, and Schematic Line). This is extremely useful for creating sectional plan views. You enable the feature in the "Options" dialog on the MEP Display control tab.

When Display by elevation is enabled, the display components of AutoCAD MEP objects are redefined to expose unique components for each range of the display in a Plan view (Plan based Display Representation). The components that can be configured within the Display Representations will be duplicated the first time display by elevation is turned on. Each component becomes four unique components based on the cut plane controls. You can turn on Display By Elevation in the template to have this control in all new drawings or turn it on only in drawings you want to control by the cut plane.

Below—Below the "display below range"

Low—Above "display below range" and below the "cut plane"

High—Above the "cut plane" but below the "display above range"

Above—Above the "display above range"

Display by Elevation and the cut plane values are drawing specific settings. XREFed drawings respect the host drawing's cut plane values and if it has Display by Elevation turned on or off.

However as noted above, when both XREF and host have the same Display Representation names, the settings for the Display Representations will be controlled by the XREF file, not the host.

USING DISPLAY BY ELEVATION

1. On the Project Navigator, in the *Constructs\Display System* folder, double-click the *Display System - Display By Elevation* file.

2. From the Application menu, choose **Options**.

3. Scroll as required and click the MEP Display Control Display tab.

4. At the bottom left corner, click the "Enable Display By Elevation" checkbox and then click OK.

    AutoCAD MEP objects now respect the defined Display Configuration cut plane.

> Note: The Display by Elevation setting can be toggled on and off. As such, the Display System stores both the on and off conditions and will remember the settings for both.

> CAD MANAGER NOTE: You may need to update the visibility settings of some display components the first time the Display by Elevation feature is turned on. Override visibility settings applied to display components that were placed in the drawing when Display by Elevation was off may need to be configured again. Template display settings should be updated with Display by Elevation turned on to assure that the display is correct when the feature is enabled.

The sprinkler system exists above the ceiling at 9'-6" and the fire pump room exists at the floor level. With Display by Elevation enabled, select the cut plane control.

5. On the Drawing Status bar, click the value next to Cut Plane.

6. In the "Global Cut Plane" dialog, change the "Display Above Range" value to **8'-0"** (see the left side of Figure 12.12).

The drawing is updated to show only the items that exist below 8'-0" above 0'-0". This range is defined by the Display Above Range and Display Below Range values. You may be required to perform a Regenerate Model (OBJRELUPDATE on the Command Line) on the drawing to update the graphics. The command is located on the View tab, on the Appearance panel (see the right side of Figure 12.12).

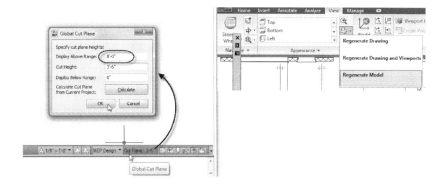

FIGURE 12.12
Update the Display Above Range and then Regenerate the Model

Now let's view the MEP elements that exist above 8'-0".

7. On the Drawing Status bar, click the Cut Plane control again.

8. In the "Global Cut Plane" dialog, make the following changes:

   ⇨ Set the "Display Above Range" value to **14'-0"**.

   ⇨ Set the "Cut Height" value to **9'-0"**.

   ⇨ Set the "Display Below Range" value to **8'-0"**.

9. Click OK to view the results. Regenerate the Model if necessary.

   The fire pump room equipment is no longer visible.

10. Save and close the file.

SHEET SETUP WITH DISPLAY BY ELEVATION

Now that you understand how to display MEP objects by elevation in your constructs, let's learn how to create section plans in the Sheets using Display Configurations. The Cut Plane is a global drawing setting, which also affects all viewports. However, different Display Configurations can be assigned to each viewport independently in Paper Space. This allows the same model objects to display differently in two or more viewports. In this sequence, we will create two new Display Configurations in a Sheet drawing and apply them to the two different viewports.

11. On the Project Navigator, click the Sheets tab.

12. Expand the *Display System* Subset and then double-click to open the *F-1 Display By Elevation* file (see Figure 12.13).

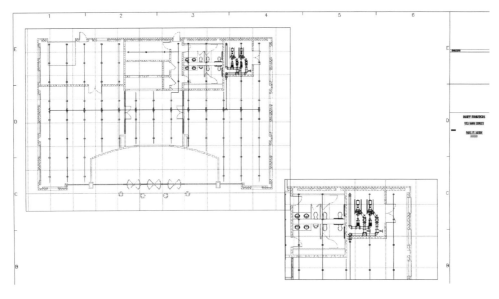

FIGURE 12.13
Fire Protection Plot Sheet

The drawing already has Display by Elevation enabled. The *Display System - Display By Elevation* Construct is already XREFed into this file and two viewports have been created. One viewport is displayed at 3/16" scale to show the sprinkler system and the other is at ¼" scale to show the fire pump room.

13. Double-click within the sprinkler system (larger) viewport.

14. On the Manage tab, on the Style & Display panel, click the Display Manager button.

You may have several drawings listed in the left pane. Be sure that you are performing the next steps in the *F-1 Display By Elevation* file.

15. Expand the *Configurations* folder, select the MEP Design configuration, and then click the Copy icon at the top of Display Manager (see Figure 12.14).

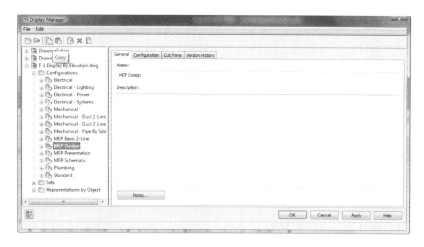

FIGURE 12.14

Copy a Display Configuration

16. Click the Paste icon (next to the Copy icon).

A new Configuration named MEP Design (2) will be created at the bottom of the Configurations list.

17. Select MEP Design (2), and in the right pane on the General tab, change the name to **MEP Design 0 Ft to 8 Ft** (see Figure 12.15).

FIGURE 12.15

Rename the new Display Configuration

18. Click the Cut Plane tab.

19. Change the "Display Above Range" to **8'-0",** and then click Apply.

20. Repeat the process to copy MEP Design again and create **MEP Design 8 Ft to 13 Ft**.

21. On the Cut Plane tab, change the "Cut Plane" to **9'-6"**, and the "Display Below Range" to: **8'-0"** (see Figure 12.16).

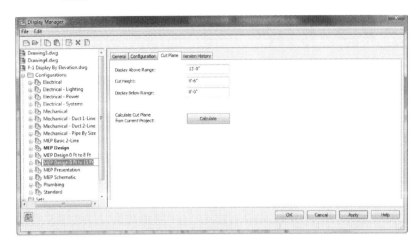

FIGURE 12.16

Setting the Cut Plane values for a Display Configuration

22. Click OK to complete the changes and dismiss the Display Manager.

23. While still in the sprinkler viewport on the left side of the drawing, on the Drawing Status Bar, choose MEP Design 8 Ft to 13 Ft from the Display Configuration list.

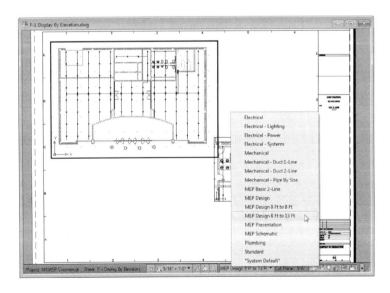

FIGURE 12.17

Apply the MEP Design 8 Ft to 13 Ft Display Configuration to the left viewport

24. Click inside the right viewport for the pump room and set the Display Configuration to MEP Design 0 Ft to 8 Ft.

25. Double-click outside the viewport to deactivate it.

The result is that all MEP objects located below 8 Ft are not visible in the left viewport and all objects above 8 Ft are not visible in the right viewport (see Figure 12.18).

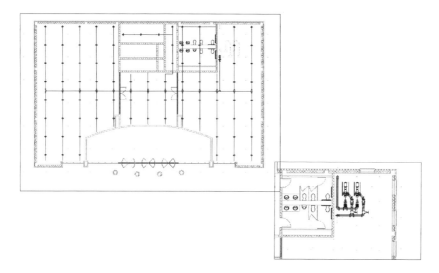

FIGURE 12.18
Results of using Display Configuration Cut Planes to control the display in viewports

With Display By Elevation you can create sectional plans within a single model and display these plans on a single plot sheet by using Display Configurations to control the cut plane.

EXTERNAL REFERENCE CONTROL

Controlling XREFs inside the Display System provides some unique challenges. If both drawings contain Display Representations by the same name, the XREF file's (base file) settings will be used, not the host file (receiving the XREF). Furthermore, the most common default Display Representations—Plan, Plan Screened, Model, and Model Screened—exist in both AutoCAD Architecture and AutoCAD MEP. This limits the control you will have over AutoCAD Architecture objects. If there is a conflict in the settings between the AMEP (host) drawing and the architectural background drawing, the architectural file's settings will prevail.

To eliminate this issue you need to perform a simple work-around procedure. It involves renaming the Display Representation that is used for all architectural objects.

1. On the Project Navigator, click the Constructs tab.

2. Expand the *Constructs\Display System* folder, and then double-click to open the *Display System - Rename Arch* file.

The file is currently displaying with the default AutoCAD Architecture screened display. Perhaps you want remove some of the hatching from the walls or adjust the display settings of some of the architectural objects without opening and editing the architectural file.

3. On the Manage tab, on the Style & Display panel, click the Display Manager button.

4. Expand the *Representations by Object* list and then expand the Wall object.

5. Select the Plan Screened Representation.

6. Select the Shrinkwrap Display Component and then change the color to Red (see Figure 12.19).

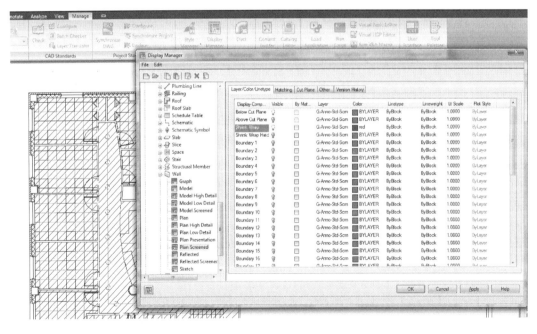

FIGURE 12.19

Change Wall Shrinkwrap to Red

7. Click Apply, then click OK.

   Notice the color of the wall did not change (see Figure 12.20).

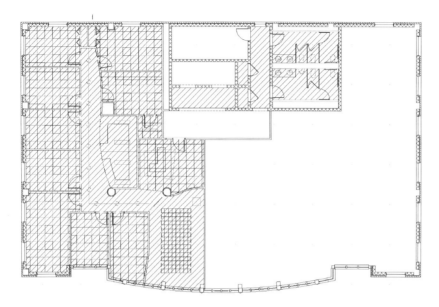

FIGURE 12.20
Changing Display Properties in the host file has no affect on the XREF

8. Return to Display Manager, expand *Representations by Object* again and then select Wall.

9. In the grid view on the right, right-click Plan Screened and choose **Rename**.

10. Change the name to **Arch Plan Screened,** and then click OK (see Figure 12.21).

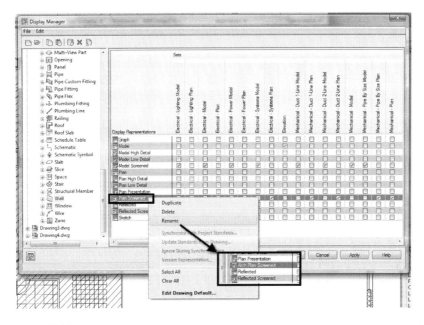

FIGURE 12.21

Rename Plan Screened to Arch Plan Screened

Notice the Hatch pattern for the walls disappeared and the outline of the walls is now red. The hatch pattern is off since all Hatch display components are turned off by default in the AMEP template files. Since the background architectural file does not contain a Display Representation called "Arch Plan Screened," the one we have here in the host drawing is now in control. This makes the Walls red with no hatching (see Figure 12.22).

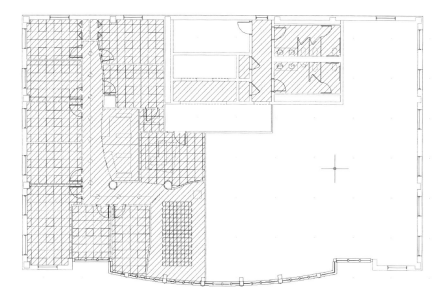

FIGURE 12.22
XREF Wall display is now controlled by Arch Plan Screened Display Representation in the host Drawing

You will need to repeat the process for any other Display Representations that share the same name in both drawings. The *Display System - Rename Arch Complete* Construct contains renamed Display Representations for all Architectural objects.

CAD MANAGER NOTE: If you always want to control the display of XREFed Architectural drawings from your host file, drawing, make this change in your Construct, View and Sheet templates. Please note that this technique only works for user-created Representations (the ones with the little person on the icon). If you are using Screened display for architectural backgrounds, this is no problem since Plan – Screened and Model – Screened are both user-created. But if you use another Representation like Plan or Model, this technique will not be possible.

When architectural drawings are XREFed in, their display will now be controlled by the new Arch Plan Screened Display Representation inside your drawing. You can also duplicate different Display Representations per object; for example you can create a low-detail copy for Structural elements, and these elements in the XREF will adhere you your settings as well. This is also useful for other MEP disciplines. Once the rename process is completed, the Sets and

Configurations are automatically updated to use the renamed Plan Screened Display Representation. It is best to make such a display change inside the office standard template.

> Note: Piping display by size is stored inside the System Definition and always controlled by the base drawing. There is no way to override this setting through an XREF on a per System basis.

SUMMARY

- ✓ Every AutoCAD MEP and AutoCAD Architectural object allows overrides to be assigned to it, either at the object level, style level or for AutoCAD MEP at the System level.

- ✓ Object overrides affect only the object(s) selected when the override is applied.

- ✓ AutoCAD MEP supplies several Configurations. Each configuration is designed to control the display for a specific task. For example, Mechanical Duct – 1-Line sets the duct objects to 1-line and all other MEP and architectural objects to 1-Line – Screened.

- ✓ Display Manager is used to configure drawing default/domain drawing default settings.

- ✓ Display Manager allows you to define which Display Representations are assigned to a Set.

- ✓ A Set determines which objects are visible and how they display.

- ✓ Display Manager allows you to define which display Sets are assigned in Display Configurations, and what the default cut plane values are.

- ✓ Any XREFed drawing that contains the same Display Representation name is actually controlled by the original drawing.

- ✓ Using the Display by Elevation feature, AutoCAD MEP objects can honor or ignore the Cut Plane.

- ✓ When Display Representation names are the same in the host and base drawings, the base drawing settings will be in control.

- ✓ Rename the common Display Representations for objects in the XREF that you wish to control from the host.

Documentation and Coordination

WHAT'S IN THIS SECTION?

In this section, we move beyond modeling and begin taking our building model into construction documentation. The next several chapters explore the topics of extracting sections from the building model, managing updates, running interference detection, generating schedules, and adding annotation. When you are ready to print your documents, all of this data can be placed onto sheets and then printed or exported to electronic formats such as PDF and DWF.

SECTION IV IS ORGANIZED AS FOLLOWS:

Section IV

Section IV

Chapter 13

Sections

INTRODUCTION

There are two basic approaches to generating sections and elevations in AutoCAD MEP (AMEP): sections and elevations can be generated from the AMEP model as a linked graphical "report" of the data contained within it, or the AMEP model can be viewed "live" in an appropriate Display Configuration. When generating sections in the "linked report" approach, AMEP offers a three-dimensional Section/Elevation object that is suitable for presentation drawings and "cut away" perspectives, as well as a two-dimensional Section/Elevation object that is useful for inclusion in design development and construction documents. Live sections are also very useful for design and presentation purposes. The 2D Section/Elevation object is style-based and robust. It will be the main focus of this chapter.

OBJECTIVES

In this chapter, we will look at the 2D Section/Elevation object in detail. We will cover ways to make this tool produce top-quality sections and elevations from your AMEP model. We will work in the MAMEP Commercial Project in the process of creating Sections and Elevations. At the end of the chapter, we will also look briefly at Live Sections. In this chapter, we will explore the following topics:

- Learning to Add Section/Elevation lines.

- Working with Callouts.

- Generating a 2D Section/Elevation object.

- Working with 2D Section/Elevation Styles.

- Updating 2D Section/Elevation objects.

- Understanding Edit Linework and Merging Linework commands.

WORKING WITH 2D SECTION/ELEVATION OBJECTS

2D Section/Elevation objects are useful for generating section drawings from an AMEP model suitable for design development and construction documents. 2D Section/Elevation objects have a broad scope; use them for full building sections or as an underlay for wall sections and even to get started with details. To create a 2D Section/Elevation, you must first add a Section/Elevation Line object to indicate where you wish the section cut to occur and in which direction it ought to "look." Section/Elevation lines are added when you run the tool from the Home tab, on the Section & Elevation panel or through the Callout tool on the Annotation tool palette. Callouts are robust routines that add required annotation and cross-link it throughout the project. The tools on the ribbon only add the Section/Elevation Line object with no cross-referenced annotation. You configure and fine-tune the appearance of the 2D Section/Elevation object in much the same way as with other objects. You can edit its style, change its display properties, and/or edit the actual component linework within the object.

THE BLDG SECTION LINE

The Bldg Section Line object is actually a three-dimensional "box." The purpose of this three-dimensional box is to determine what portion of the Building Model will be included in the Section (see Figure 13.1).

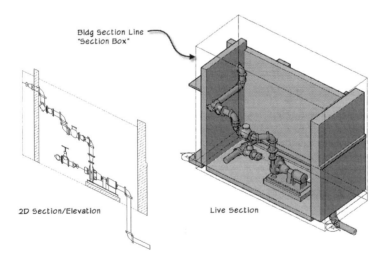

FIGURE 13.1
Bldg Section Line at right with a Live Section, middle showing subdivisions, and left showing 2D Section/Elevation

The Display Representations of the Bldg Section Line object contains three sub-components (see Figure 13.2). Two of the three subcomponents—the Boundary and the Subdivisions—are used merely for purposes of configuring the extents of 2D Section/Elevation object. The third component, the defining line, is the only component that would potentially have value when printed. The Defining Line can be used in conjunction with a Section Bubble in your drawings, or you can simply use the bubbles by themselves. Depending on the section style used, the boundary and subdivision display components are often assigned to a non-plotting layer. The Defining line is the "cut line," drawn through the building in plan view. It forms a plane three-dimensionally that determines where the bold cut line in the Section will be. The Boundary, illustrated above, *is* the "box". Nothing outside the boundary is included in the Section or Elevation. The subdivisions are drawn as lines parallel to the back edge of the Section box and determine where the Lineweight zones occur. Every section has at least Subdivision 1. Additional subdivisions must be added for each section individually. For more information on subdivisions, refer to the online help.

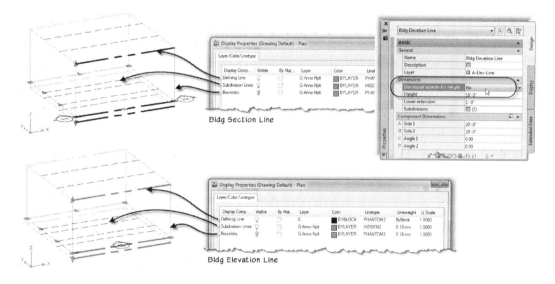

FIGURE 13.2
The components of the Bldg Section Line and Bldg Elevation Line objects

Another way to think of the Bldg Section Line object is like the field of view for a camera. When you look through your camera lens, you can only see so much of the scene both side-to-side and moving back. If you imagined mapping the field of view of your camera on the ground, you would have a pretty good approximation of the Section/Elevation Line object in plan. However, this boundary is actually three-dimensional; by default AMEP will automatically include the entire height of the model in the 2D Section/Elevation object. This default can be changed if required, but, ironically, when this option is active "Use model extents for height" is set to Yes on the Properties palette, and no height will be shown graphically in 3D views since it includes *all* of the height. When it is set to "No" you see a 3D box when you view the model (see the right side of Figure 13.2).

SECTIONING METHODS

There are two ways to create a section: Callout tools and basic Section line creation from the ribbon.

The Callout tools utilize Project Navigator's drawing management to allow the creation of a section in a View drawing (refer to Chapter 3 for more on View files). The basic section line tool (on the ribbon) creates a section in the current drawing. Each section is linked to the model and can be updated when the model changes (see Figure 13.3).

FIGURE 13.3
Create Sections from the Ribbon or the Tool Palettes

Both section tools create either a 2D section or a live section. A 2D section is "linked" to the model. This means that rather than being a live view of the model, it is like a snapshot of the section. Since the sections are linked, you can update the snapshots any time. Workflows vary from firm to firm as we discussed in Chapter 3, but often the section is stored in a separate drawing file and these separate drawings function as "reports" of the model that maintain a link to the live model data. Refer to Chapter 3 for more information on workflows with AutoCAD MEP. Again, these Views must be periodically refreshed to capture changes made to the model.

On the Home tab of the ribbon, on the Section & Elevation panel, several section tools are available. There is a Vertical Section tool, a Horizontal Section tool and a Section Line tool. Each works in similar fashion. Click the tool and follow the prompts to draw a Bldg Section Line object. The Vertical Section and Section Line tool can draw multi-segment section lines. This is helpful if you wish to have your section line "jog" around an item in your plan. The horizontal section creates a section line object rotated with respect to the UCS so that it is actually "looking down." This creates a section that is more like a plan. This can be useful in some circumstances if the normal plan graphics are not producing the desired results. However, with the "Display by Elevation" feature in AMEP (refer to Chapter 12) this is rare. Each of the tools on this ribbon panel shares the same basic procedure. After you draw the section line, you will need to select it and generate a section from it (or enable a Live Section). When generating a section, follow additional command prompts to complete the task.

INSTALL THE DATASET FILES AND LOAD THE CURRENT PROJECT

In this exercise, we use the Project Browser to load a project from which to perform the steps that follow.

1. If you have not already done so, install the book's dataset files.

    Refer to "Book Dataset Files" in the Preface for information on installing the sample files included with this book.

2. Launch AutoCAD MEP 2012 and then on the Quick Access Toolbar (QAT), click the Project Browser icon.

3. In the "Project Browser" dialog, be sure that the Project Folder icon is depressed, open the folder list and choose your *C:* drive.

4. Double-click on the *MasterAME 2012* folder.

5. Double-click *MAMEP Commercial* to load the project. (You can also right-click on it and choose **Set Project Current**.).

> IMPORTANT: If a message appears asking you to repath the project, click Yes. Refer to the "Repathing Projects" heading in the Preface for more information.

Note: You should only see a repathing message if you installed the dataset to a different location than the one recommended in the installation instructions in the preface.

6. Click the Close button in the Project Browser.

When you close the Project Browser, the Project Navigator palette should appear automatically. If you already had this palette open, it will now change to reflect the contents of the newly loaded project. If for some reason, the Project Navigator did not appear, click the icon on the QAT.

7. Make the Piping Workspace active (refer to the "Workspaces" topic in Chapter 1 for more information).

CREATING A 2D SECTION FROM THE RIBBON

In this exercise, we will create a basic 2D Section of the first floor Fire Pump Room using the tool on the ribbon.

8. On the Project Navigator, click the Constructs tab, expand the *Sections* folder and then double-click the 01 Fire Protection Section drawing to open it.

The drawing is based on the Sprinkler System created in Chapter 6.

9. Zoom to the upper right-hand corner of the building (Fire Pump Room). (See Figure 13.4.)

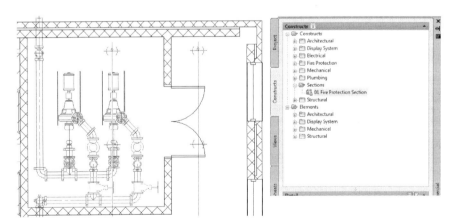

FIGURE 13.4
Zoom in on the Fire Pump Room

10. On the Home tab, on the Section & Elevation panel, click on the Section Line button.

Note: The orientation of the Section line is based on the points you pick: Top to Bottom—Section points Right; Bottom to Top—Section points Left; Left to Right—Section points Up; Right to Left—Section points Down. If you draw it the wrong way, you can flip it with the tool on the ribbon.

Picking the next two points will be easier if you toggle Object Snaps off (f3).

11. At the "Specify start point" prompt, click outside the fire pump room below the discharge pipe.

12. At the "Specify next point" prompt, move the mouse horizontally to the right and click outside the fire pump room (see Figure 13.5).

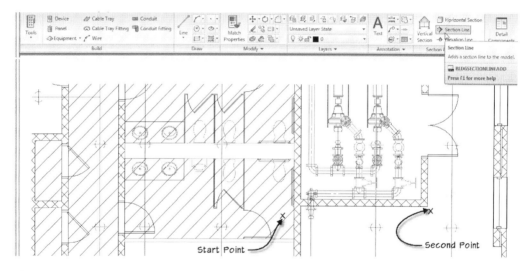

FIGURE 13.5

Click two points to indicate the width of the section

The prompt will repeat. You can add additional points to the section line to avoid obstructions. In this case, two points is fine.

13. Press enter to complete the section line without adding additional points.

14. At the "Enter length" prompt, press enter again to accept the default distance of 20'-0".

The "length" of 20'-0" can be better thought of as "depth" or, using our camera analogy, this is the "depth of field" or how far into the model our section will look. If there are objects in the background that you want to add or exclude, you can select the Bldg Section Line object and use the triangular grip at the back edge to adjust the depth (see the right side of Figure 13.6). In this case, it is not necessary to adjust it.

15. Select the Bldg Section Line (section box) object.

16. On the Building Section Line tab, click on the Generate Section button (see Figure 13.6).

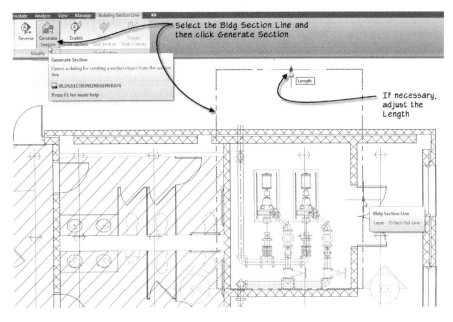

FIGURE 13.6
Generate the Section from the tool on the ribbon

The "Generate Section/Elevation" dialog appears. There are actually two types of section that can be generated from this dialog. One type is the 3D Section which is rarely used. It has some application as a presentation tool, but typically you will want to use the default option of 2D Section/Elevation Object with Hidden Line Removal. As this name implies, a 2D object will be generated and all lines that would be concealed from our vantage point, as if standing where we drew the section line, will be removed in the result.

2D Section/Elevation objects are style-based objects. You can find them listed under *Documentation Objects* in the Style Manager. As such, you are able to choose the style from which to generate the result. We will modify a style below.

In the Selection Set area, it will report that currently "0 items selected." In order to generate the section, we must make a selection of objects in the drawing. You can modify the selection set at any time by using the Regenerate tool. If you generate a section from XREFs, maintaining the selection set is easy. Simply select the XREF and it will maintain itself. When you create a section in the same file as we are doing here, you must maintain the selection set manually as new objects are added to the model. Using a crossing selection box is recommended to ensure you select all objects in the region of the Bldg Section Line.

In addition to selection set, sections also use a Display Set. Recall from Chapter 12 that a Set determines which objects will display in a view and how they will display based on the Display Representation. This leverages the power of the Display System with your section object creation.

Finally, on the right side are the Placement options. Usually the Pick Point icon is the easiest. With it you simply use your mouse on screen to indicate the location to generate the 2D Section object.

17. In the "Generate Section/Elevation" dialog, verify that MEP Standard is chosen for "Style to Generate."

18. Beneath "Selection Set," click the Select Objects icon (see Figure 13.7).

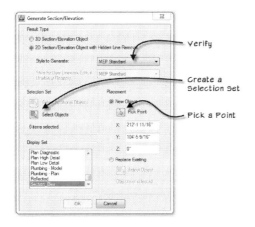

FIGURE 13.7

The "Generate Section\Elevation" dialog

19. Using a crossing window selection box, select the entire fire pump room (see Figure 13.8).

20. Press ENTER to complete the section and return to the "Generate Section/Elevation" dialog.

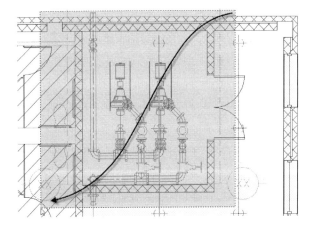

FIGURE 13.8

Select the objects for the section using a crossing window

21. Click the Pick Point icon.

22. In the drawing, pan to the left of the model and pick a point.

The "Generate Section/Elevation" dialog appears again.

23. For the Display Set, choose the MEP Section & Elevation Display Set (see Figure 13.9).

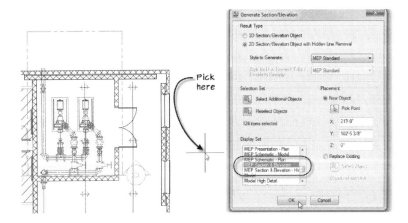

FIGURE 13.9

Pick a point to locate the section result and then choose a Display Set

24. Verify all settings and then click OK to generate the section.

A progress bar will appear and then the 2D section will appear with its lower left corner at the location you indicated above. Notice the two walls of the room that are being cut are shaded in purple, and all of the fire pump equipment appear in the room beyond the cut (see Figure 13.10).

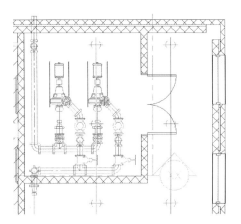
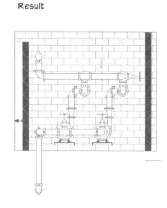

Result

FIGURE 13.10
The 2D Section accurately depicts the geometry drawn in the model

It took a few steps to create this section. We used the manual method to illustrate the entire process. When you use the tools on the ribbon, creating the section line and generating the section are separate steps. Later, in the "Using a Callout to Generate a Section" topic below, we will look at performing both steps in a single operation. Either way, as you can see, these steps are still much quicker than the time it would take to draw such a section manually, even with a well-stocked library of pre-drawn CAD components. Generating such drawings quickly is certainly one important benefit of this tool. However, the real power of the tool comes when you make modifications to your design.

EDITING THE MODEL AND SECTION

Now, let's explore what happens when a change occurs to your model. Frequent design changes cause havoc on a production schedule and budget if you are making the changes manually. Using a 2D Section/Elevation object, you can edit your model and then simply regenerate or refresh your section to capture the changes. Not only is this much quicker, but you can also be confident that the change is properly reflected in all project drawings!

To illustrate this, we will perform a simple edit. Look carefully at our pumps—they are floating! Let's assume that our Structural Engineer has just sent us the equipment pads in a separate Construct file. All we need to do is drag them into our drawing and then add them to the section.

1. On the Project Navigator, on the Constructs tab, expand the *Structural* folder.

2. Drag and drop the *01 Structural Fire Pump Pads* file into the drawing window.

   You will see two equipment pads appear beneath the fire pumps.

3. Select the 2D Section object.

4. On the 2D Section/Elevation tab, click the Regenerate button.

This displays the "Generate Section/Elevation" dialog again. Since we have added a new object to the drawing, the XREF of the structural file in this case, we need to add that to the section object's selection set.

5. In the Selection Set area, click the Select Additional Objects icon.

6. At the "Select objects" prompt, click on one of the equipment pads (both will highlight because it is an XREF) and then press ENTER.

7. In the "Generate Section/Elevation" dialog, click OK to complete the change and refresh the section.

The pads now appear beneath the two pumps in the section. It is even more common that changes will occur on geometry already in the model. For example, suppose that the elevation of our discharge piping needs to change.

8. Select the two small pieces of angled Pipe attached to the pumps.

9. On the Properties palette, change the Elevation to **5'-0"**.

10. Press ESC to deselect the Pipe.

11. Select the horizontal Pipe in the lower left corner of the room.

12. On the Properties palette, change the Elevation to **8'-0"**.

13. Deselect the Pipe.

14. Select the 2D Section object.

This time we do not need to reopen the dialog because we are only updating geometry that is already part of the section. Therefore, Regenerate is not necessary. We can use Refresh instead.

15. On the 2D Section/Elevation tab, click the Refresh button (see Figure 13.11).

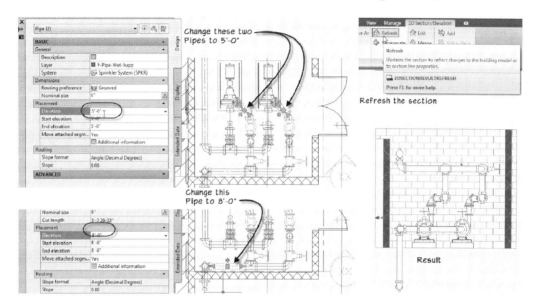

FIGURE 13.11
Modify the model and then refresh the section

Notice that the piping we modified and the other components connected to it has moved up to the new elevation.

> **Note:** 2D Section/Elevation objects are not live data. They can be edited for their own purposes as we will see below, but manual update is required to keep them current with the state of the model. There is also a batch process routine available that will update all elevations and sections within an entire project. On the Home tab, expand the Section & Elevation panel and then click the Batch Refresh tool.

16. Save the file.

EDITING THE 2D SECTION STYLE

Let's now shift our attention to making the look of the section more acceptable. In this topic, we will edit the section style.

17. Select the section object.

18. On the 2D Section\Elevation tab, click the Edit Style button.

19. In the "2D Section\Elevation Style Properties" dialog, click Display Properties tab.

20. Click the Edit Display Properties icon.

21. Turn off the visibility of the three Hatch components (see Figure 13.12):

- Shrinkwrap Hatch
- Surface Hatch Linework
- Section Hatch Linework

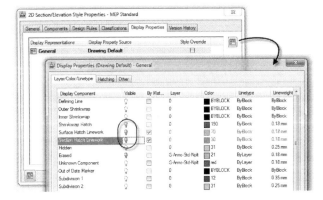

FIGURE 13.12
Edit the 2D Section Style and turn off hatching

22. Click OK twice to the return to the drawing.

The section object should update immediately. This change removes all the hatching from the architectural components (see Figure 13.13).

Note: The changes made will effect all sections. To limit the change only to this section style, select the Style Override check box in the 2D Section/Elevation Style Properties – MEP Standard dialog noted in Figure 13.12.

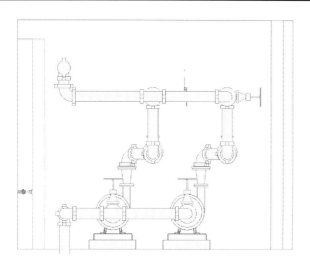

FIGURE 13.13
After editing the style, the architectural hatching no longer displays

2D Section styles have some unique display components like the hatch display components turned off here. For objects that support hatching, like architectural objects, these components help give you control of the section. To see other display components on the various MEP objects, you can update the object's display components. When generating the section above, we chose the MEP Section & Elevation Display Set. If you open Display Manager you will see that most MEP objects use the Model Display Representation in this Set (see Figure 13.14).

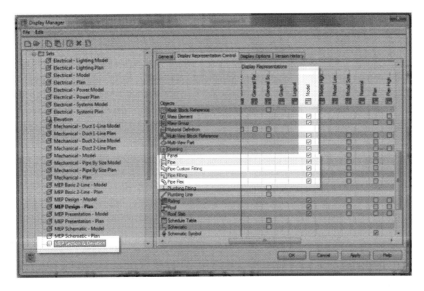

FIGURE 13.14
MEP Section & Elevation Display Set uses the Model Display Representation for most MEP objects

You can edit this Set to change which Display Representations you want to use when creating a section, such as using Model Screened for architectural components. For more information refer to Chapter 12

USING A CALLOUT TO GENERATE A SECTION

Having used the ribbon tool to create a section, you have an understanding of how the Bldg Section Line object is used to generate sections, the selection set, style, Display Set and the generated 2D Section/Elevation object itself. The alternative to the ribbon tools is a Callout tool on the Annotation Tool Palette. A Callout tool is like a "wizard" for generating sections. It first prompts you to create a Bldg Section Line object like the ribbon tools. It then presents a worksheet with several options including style, scale, and the ability to create a View file in Project Navigator for the section.

ADD A CALLOUT

Using the tool on the Annotation palette, let's create another section in this file to learn about the callout method.

1. On the Annotation palette, click the Section Mark A2T tool (see Figure 13.15).

If you do not see the Tool palettes on your screen, on the Home tab, on the Build panel, click the Tools button.

2. At the "Specify first point of section line" prompt, click above the room just to the left of the right-hand pump.

3. At the "Specify next point of line" prompt, move the mouse straight down and click outside the room (see Figure 13.15).

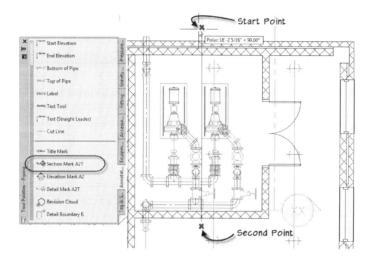

FIGURE 13.15

Use the Callout routine to draw a second Bldg Section Line

4. Press ENTER to complete the section line points.

5. At the "Specify section extents" prompt, drag to the left and click just outside the pump room wall (see the left side of Figure 13.15).

6. In the "Place Callout" dialog, for the New Model Space View Name, type **Fire Protection Section 1**.

Note: This Model Space View Name must be unique in your project. Project Navigator will not allow you to create a New Model Space View using a name that already exists in the Project, even if it is within a different drawing.

7. From the Scale list at the bottom, choose 1/4" = 1'-0".

Be sure that both "Generate Section/Elevation" and "Place Titlemark" are checked.

When you instruct the Callout routine to "Generate Section/Elevation" it will create the 2D Section/Elevation object for you. Place Titlemark will add a Title Mark Callout to the drawing you specify in the Create in area above, and that Title Mark will be scaled to whatever you indicate in the Scale list. This scale will also be assigned to the New Model Space View named at the top of the worksheet.

8. Click the New View Drawing icon in the center of the worksheet (see the right side of Figure 13.16).

> **Note:** When adding to an existing drawing, a worksheet will appear listing all the View files within the current project.

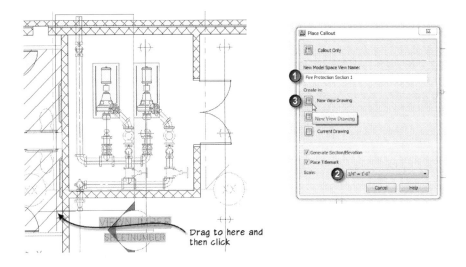

FIGURE 13.16
Indicate the extents of the section and then configure the callout options

Since we chose to create a new View file, the Add View wizard will begin and the "add Section/Elevation View" dialog will appear.

9. On the "General" page, for the Name, type **F-SC01**.

10. Add a description if you like and then click Next.

11. On the "Content" page, right-click and choose **Clear All**.

12. Check only the First Floor and then click Next.

The "Content" page appears next with a complete list of all Constructs in the project that belong to the first floor. If you simply click Finish, the Constructs from all disciplines will be added to the section. In any project you will want to consider the list of presented files carefully. Often, there will be files that you will want to exclude. You can uncheck individual drawings, or uncheck a folder to remove an entire discipline. In this case, since we have some redundant files from the previous chapters in the book, we will deselect most of the categories presented.

13. Uncheck all folders except *Architectural, Sections,* and *Structural,* and then click Finish (see Figure 13.17).

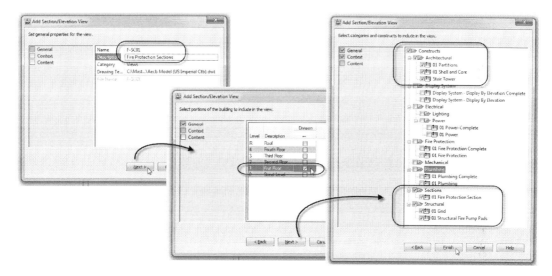

FIGURE 13.17
Add a new View using the Callout wizard

> CAUTION:
> Do not press ENTER or the ESC key; you are not done with the Callout routine yet!

Look at the command line and notice the message that has appeared. It will read:

```
** You are being prompted for a point in a different view
drawing **
```

When you create a section using the Callout routine to an existing or new View file, you still must indicate within the current drawing where you would like the section to be created. This prompt serves to inform you of this. It is usually best to pick a point off to the side of the plan.

14. At the "Specify insertion point for the 2D elevation result" prompt, click a point in the drawing to the right of the section generated above (see Figure 13.18).

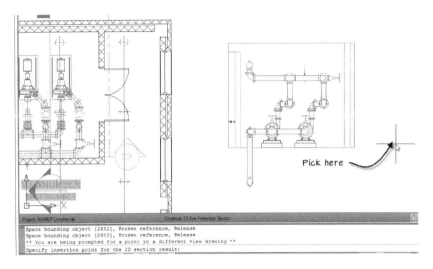

FIGURE 13.18
Indicate the location for the section within the new View drawing

Here in the current drawing, a Callout has appeared where we indicated. Nothing else appears to have changed. The 2D Section/Elevation object was created in a new View file named *F-SC01*. We will need to open that file to see it.

15. On the Project Navigator palette, click the Views tab.

16. Double-click to open the F-SC01 file.

17. Zoom in on the newly created Section.

Notice that a titlebar appears beneath it reading "Fire Protection Section 1." The scale also appears on this titlebar (see Figure 13.19).

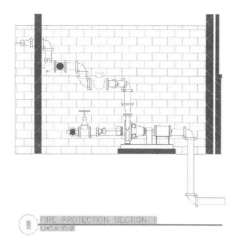

FIRE PROTECTION SECTION

FIGURE 13.19
The section has been generated in the F-SC01 file complete with titlebar

As you can see, this section is very similar to the one previously created. If you wish, you can repeat the process above to edit the style and turn off the hatching.

MODIFY THE MODEL

Suppose a design change caused the position of the pump in the section to move. Since your section is now in a View file, the process will vary slightly from the steps you performed above in the "Editing the Model and Section" topic.

18. On the Project Navigator, click the Constructs tab and then double-click the 01 Fire Protection Section file to open it.

19. Select the left Pump.

20. Using the Location grip, the square one in the middle, move the Pump up **2' 0"** (see Figure 13.20).

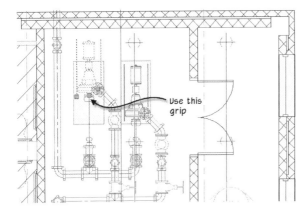

FIGURE 13.20
Move the left pump up in the Construct

> 21. Save and close the file.

Notice that the equipment pad also needs to shift. Since this would be the responsibility of the Structural Engineer, we would inform them of the change and await their updated file. In the meantime, we can proceed updating our section in the View file.

REFRESH THE SECTION

Back in the *01-Fire Protection Section* file, you should receive an alert that the *01 Fire Protection Section - Callout* XREF has changed at the lower right corner of your screen (see Figure 13.21).

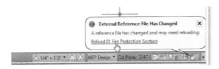

FIGURE 13.21
You must reload the External Reference file before refreshing the section

> 22. Click the link in the balloon to reload the XREF.

TIP: You can also right-click the XREF quick-pick icon for a menu to open the XREF Manager. Use this technique if the balloon does not appear.

> 23. Select the Section and on the 2D Section/Elevation ribbon tab click the Refresh button.

Keep your eyes on the Pump as the section refreshes. Notice the update to the Section. The Pump has moved 2'-0" to the right (see Figure 13.22). To enhance the clarity of the figure, the Shrinkwrap Hatch and Surface Hatch components have been turned off. Also, the MEP Section & Elevation Set has been edited in the Display Manager. Both the Wall and Slab objects have been changed to Model Screened instead of Model.

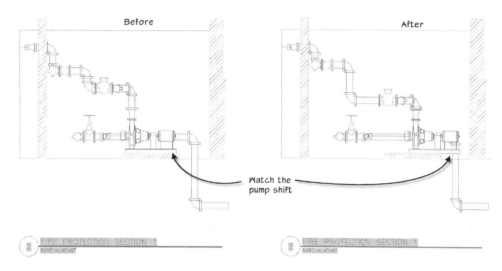

FIGURE 13.22
Reload the XREFs and then refresh the Section to see the change

IMPORTANT:

Make sure that you first update the XREFs and then refresh the sections. If you refresh the section first, you will see no change since the change took place in a remote file. We have to load those remote changes before the section can "see" them.

You have seen how to create a section both in the current drawing and using the callout tools. You have also learned how to edit the section style to control display components and update the section when changes have occurred in the model. Sometimes you will need to perform a little "fine-tuning" of the section result. In the next sequence, you will learn how to edit the linework in the section.

MODIFYING SECTION/ELEVATION LINEWORK

We are going to edit two areas of the section. The first is where the pipe penetrates the wall in the upper left-hand corner and the second is to remove the linework for the pipe behind the pump suction.

1. To add the linework to indicate the pipe passes through the wall, zoom into the upper left-hand corner where the pipe goes through the wall.

2. Draw two lines to indicate the pipe.

3. Select the lines and, on the Properties palette, change the following:

   ⇨ Color: 183.

   ⇨ Layer: F-Pipe-Wet-Supp.

   ⇨ Linetype: Hidden2.

   ⇨ Linetype scale: **0.500**.

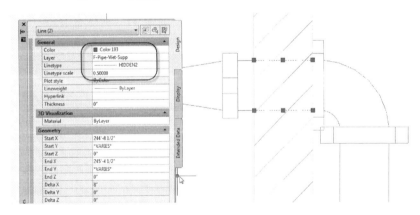

FIGURE 13.23
Creating the Wall penetration graphics for the pipe

4. Deselect the lines.

5. Select the section and, on the 2D Section/Elevation tab, on the Linework panel, click the Merge button.

6. At the "Select objects to merge" prompt, select the two lines and then press ENTER.

7. In the "Select Linework Component" dialog, click the Match existing linework icon.

8. Click on the linework for the Pipes already in the section, and then click OK (see Figure 13.24).

The linework has now been added to the Section and resides on the Subdivision 1 component.

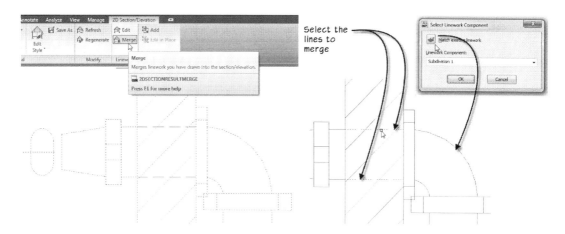

FIGURE 13.24

Merging lines into the 2D Section

You can also edit the linework already existing in the elevation. Use this technique to erase extraneous lines or change the lineweight of existing lines.

9. Select the section.

10. On the 2D Section/Elevation tab, on the Linework panel, click the Edit button.

    The Edit in Place: Linework tab appears on the ribbon.

11. Zoom in on the Suction pipe going to the Pump and erase the top and bottom lines as indicated in Figure 13.25.

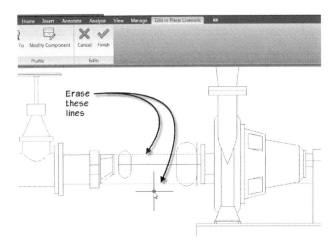

FIGURE 13.25

Edit the 2D Section to remove linework

12. On the ribbon, click the Finish button.

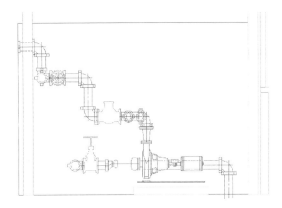

FIGURE 13.26

The completed 2D Section

13. Save the file.

UPDATE 2D SECTION/ELEVATION OBJECTS WITH USER EDITS

Inevitably, your design will change and your section will need to be updated. When this occurs, the 2D Section/Elevation object will automatically reapply all user edits. However, sometimes the nature of the change is such that some of the edits will no longer be in the same physical location as the newly regenerated model geometry. When this occurs, user edits can be saved to a second 2D Section/Elevation object and merged back into the section after the update. To do this effectively, you need to build another 2D Section/Elevation style. The basic process is as follows:

14. Create a New 2D Section/Elevation style from the Manage tab, Styles & Display panel, Style Manager button and name it User Edits.

15. In this new style, On the Display Properties tab, turn on all components, especially Erased. Set the Erased component to an alert color such as Magenta.

This is very important. Most user edits are erasures, however, erased in elevation and section user edits really mean that the linework is simply invisible. This is the desired effect in the actual elevation, but if it is invisible, you will not be able to reapply your user edits because you will not be able to see them during the refresh process.

16. Regenerate (not Refresh) the Section and choose the User Edits style from the "Style for User Linework Edits if Unable to Reapply" list (see Figure 13.27).

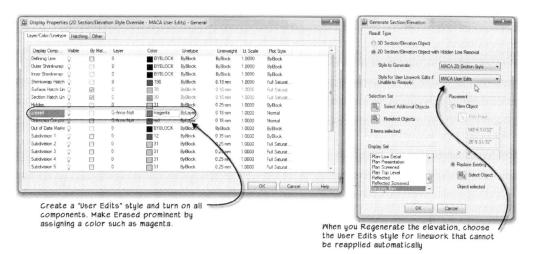

Create a "User Edits" style and turn on all components. Make Erased prominent by assigning a color such as magenta.

When you Regenerate the elevation, choose the User Edits style for linework that cannot be reapplied automatically

FIGURE 13.27
User Edits can be saved to another style if AMEP is unable to reapply them automatically

17. After the update, explode the User Edits Section, move the linework into place on the Section, and then reapply it with the Merge command.

CAD MANAGER NOTE: Although "off" is the proper setting for Erased Vectors in styles used in production, it is recommended that one style be maintained for saving user edits that has the Erased component turned on and set to color 6-Magenta. You can name the style "User Edits." This will make it possible to easily see and merge the second section after an update. You can even do this as the Drawing Default, because all your production styles will have style-level overrides. The choice is up to you. All 2D Section/Elevation styles, including this one, should be saved as part of the office standard template files.

LIVE SECTIONS

Live Section operates on the actual building model data and can be edited directly, unlike the 2D Section/Elevation object. A Live Section *is* the model. Live Sections make great design tools. Adding one is simple—just add a Bldg Section Line where you would like the Section to be cut, select it and on the ribbon, click the Enable Live Section button. Use the Section Line tool on the ribbon as we did at the beginning of this chapter. Switch to 3D View and have a look. It is that simple!

Live Sections are applied to the active Display Configuration. So if you wish to have a Live Section applied to a configuration other than the currently active one, be sure to choose the appropriate Display Configuration prior to enabling the Live Section.

SUMMARY

✓ The Section Line boundary determines how much of the model will be considered by the Section.

✓ The Section must be updated to reflect changes to the floor plan.

✓ Callouts completely automate the process of adding a Bldg Section Line and then generating the Section or Elevation, including placement of a linked Callout.

✓ 2D Section/Elevation styles can be assigned to the Section and Elevation to give them more graphical definition.

✓ When cutting a Section or Elevation from XREF models, remember first to reload the XREFs and then update the Section/Elevation.

✓ 2D Section/Elevation styles offer a great way to manage Layer, Color and Linetype settings on 2D Section/Elevation objects.

✓ Edit Linework allows the 2D Section/Elevation object to be directly edited.

✓ Merge Linework allows custom drawn linework and previously edited linework to be incorporated into the 2D Section/Elevation object after it has been updated.

Chapter 14

Managing Updates and Interference Detection

INTRODUCTION

Change occurs frequently in architectural and engineering projects. If not communicated effectively, such change can place team members at odds with one another. Having tools available to help the team manage change and coordinate concurrent efforts is critical to overall project success. AutoCAD MEP (AMEP) provides tools to assist in two critical areas: tracking changes and detecting clashes.

The Drawing Compare tool allows the engineering design professional to review newer versions of architectural drawings and compare them to the originals. This helps to determine what is new in the drawings, what has been modified, and even what has been deleted. This tool is an extremely effective means to help manage project-related changes and minimize potential tensions with your architectural partner.

Not only is identifying changes critically important, but early detection of conflicts between building elements, structure, and other systems is equally important. Interferences or clashes between building components, if not identified in a timely fashion, can be devastating to the project schedule and budget. At a minimum, costly redesign can result. On the other end of the spectrum, change orders, litigation, or worse consequences can ensue. Catching such conflicts early and resolving them efficiently will have a direct impact on a project's schedule and budget.

The Interference Detection tool allows the coordination of MEP systems with other MEP systems, structural building components, or even other building components like walls. This tool is an extremely effective means to help manage interferences and clashes between disciplines and give you peace of mind when delivering your projects.

OBJECTIVES

In this chapter we will focus on the Drawing Compare and Interference Detection tools to help you understand and manage project changes and help identify potential conflicts before they are identified in the field. In this chapter, we will explore the following topics:

- Using the Drawing Compare tool to quickly identify changes in the project drawing files
- Using the Interference Detection tool to quickly identify conflicts between project components
- Tagging interferences in the drawing to share with other project members
- Using the Interference Detection tool to identify locations where wall openings might be required

DRAWING COMPARE

The Drawing Compare tool allows you to compare the changes in one file to an existing version of the file. Consider the common scenario where the Architect provides a background file and the Engineer bases their design on this file. Project changes occur frequently. Although the Architect and Engineer endeavor to communicate project-related changes to each other effectively, it is virtually impossible to identify *all* the important changes. This is particularly so when you consider that changes can occur daily or even hourly, yet updated files are typically sent and loaded with much less frequency. This is where Drawing Compare can help. Using it will help you understand what changes have occurred and make good decisions about how to address them.

INSTALL THE DATASET FILES AND LAUNCH PROJECT BROWSER

In this exercise, we use the Project Browser to load a project from which to perform the steps that follow.

1. If you have not already done so, install the book's dataset files.

   Refer to "Book Dataset Files" in the Preface for information on installing the sample files included with this book.

2. Launch AutoCAD MEP 2012 and then on the Quick Access Toolbar (QAT), click the Project Browser icon.

3. In the "Project Browser" dialog, be sure that the Project Folder icon is depressed, open the folder list and choose your *C:* drive.

4. Double-click on the *MasterAME 2012* folder.

5. Double-click *MAMEP Commercial* to load the project. (You can also right-click on it and choose **Set Project Current**.).

IMPORTANT: If a message appears asking you to repath the project, click Yes. Refer to the "Repathing Projects" heading in the Preface for more information.

Note: You should only see a repathing message if you installed the dataset to a different location than the one recommended in the installation instructions in the preface.

6. Click the Close button in the Project Browser.

When you close the Project Browser, the Project Navigator palette should appear automatically. If you already had this palette open, it will now change to reflect the contents of the newly loaded project. If for some reason, the Project Navigator did not appear, click the icon on the QAT.

FILE STRUCTURE IS IMPORTANT FOR DRAWING COMPARE

The exercises in this book have centered on the use of the Project Navigator tool. The Project Navigator is a very useful tool that helps you manage your project and your project files (refer to Chapter 3). The Drawing Compare tool works best when your project is organized in a structured environment like the one imposed by Project Navigator. This doesn't mean that Drawing Compare cannot be used outside of Project Navigator, just that a good structured file organization is important to use the tool successfully.

If you are not working in Project Navigator, you will need to arrange your folder structure in a manner that the Drawing Comparison tool will recognize. Place your project drawings in a *Project* folder, and the drawings that you wish to compare to your project drawings will need to be placed in a *Review* folder. Within the *Project* and the *Review* folders you must maintain the same folder structure (see Figure 14.1).

For the exercise below, we will continue to work in Project Navigator as we have in several other chapters. This will give us the structured environment that Drawing Compare requires.

FIGURE 14.1

Mirror the project folder structure in the review folder

UNDERSTANDING DRAWING COMPARE

Before we begin the exercise it is best that we gain a better understanding of the Drawing Compare tool and how it behaves. Here are a few key points.

- Drawing Compare works only in model space. If a drawing was saved in paper space, Drawing Compare will automatically change to model space to perform the Compare.

- Drawing Compare will not list Sheets in the Drawing Compare dialog box. This is because Sheets, and the content of the Sheets, are exclusive to paper space.

- The result of the Drawing Comparison is not saved with the drawing. When you close the Drawing Comparison tool, the results are completely discarded.

PROJECT COMPARISON SETUP

Most engineering projects rely on background files generated from an Architect. While the Architect often begins their work before the Engineer, the Engineer typically comes on board early in the project. This means that both parties are working concurrently and actively making design decisions, producing drawings and most importantly—making changes! Therefore, regular communication about changes that affect both parties and frequent drawing updates are common. Managing the flow of information and coordinating updates can become a daunting task. Drawing Compare can help.

The first step to using the Drawing Compare tool is proper setup. As noted above, folder structure is very important, and for the tool to function properly, the structure must be mirrored between the files being compared and the newly received updates. Project setup strategies vary from firm to firm and even from project to project. Several common strategies were discussed in Chapter 3. However, it is likely that you already maintain two copies of the Architectural files. One copy is saved with your project files and becomes the background files, and the other is stored as a record or backup of what was received. This approach allows you more control by allowing you to implement architectural changes when the time is appropriate in your own workflow.

The only potentially new requirement of adding Drawing Compare to this existing workflow is that the two separate folder structures must be parallel. In other words, the "backup" or "holding" location for newly received updates must have a folder structure that mirrors that of the live project. This is the only way Drawing Compare can correctly make associations between the two versions of any particular file. If you use Project Navigator, creating such a folder structure is simple. From the QAT, open the Project Browser. Right-click the current project and choose Copy Project Structure. In the dialog that appears, input a name, number and location to which you will copy the files. The result will be a duplication of all the folders in the project, but none of the files. Open this folder structure in Windows Explorer and copy any newly received architectural files to the appropriate folder. Once this is complete, you can proceed with the Drawing Compare setup. To save some effort, we have provided such a copy in the *Chapter14* folder with the files you installed from the book dataset.

USING DRAWING COMPARE

Before you can begin a comparison, you must first tell AMEP where the two file structures are located. When you have a Project Navigator project active, it will assume that the Project Folder is the current project. This means that you will only need to configure the location of the comparison files (the updated files you received from the Architect). If you are not using Project Navigator, you will need to configure both locations.

1. From the Application Menu choose **Drawing Utilities > Drawing Compare** (see Figure 14.2).

FIGURE 14.2

The Drawing Compare tool can be accessed from the Application Menu through Utilities

On the Drawing Compare palette, at the top, notice that the Project Folder is already set to *C:\MasterAME 2012\MAMEP Commercial*.

2. Next to the Review Folder, click where it says Browse.

3. Navigate to the *C:\MasterAME 2012\Chapter14* folder and then click OK.

   A confirmation dialog will appear asking if you "Would like to automatically match and analyze drawings now?"

4. Click Yes to confirm and begin the process.

A progress bar will appear as AMEP analyzes the drawings in the two folders (see Figure 14.3).

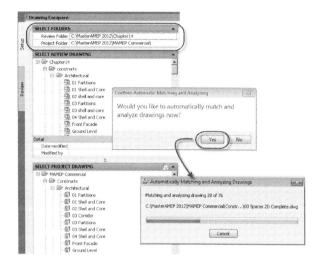

FIGURE 14.3

After the Review and Project Folders have been selected, AMEP will begin analyzing the drawings

---

Note: If you use this tool outside of Project Navigator, configure the Project Folder path first, answer No when prompted, configure the Review Folder next, and then click Yes to begin the process.

Review drawings, the updates received from the Architect, and their folder will appear at the top of the palette in the "Select Review Drawings" list. Project drawings (your firm's live project files) will appear in the "Select Project Drawings" list at the bottom of the palette.

5. Scroll down in the "Select Review Drawing" list and select the *03 Drawing Compare Project* file.

When you select the drawing in either list, the other list will scroll to the same file and you should notice two green arrows pointing to each other appear on the files. This indicates that AMEP has matched the two drawing files and is ready to do a comparison (see Figure 14.4).

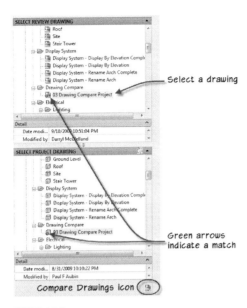

FIGURE 14.4
AMEP will note which drawings are ready for comparison after it has analyzed the files in the directory

---

Note: When comparing project drawing files, the file's names must be an exact match. If they are not an exact match, Drawing Compare will not be able to compare the files.

---

6. In the "Select Project Drawing" list, select the 03 Drawing Compare Project file, right-click and choose **Begin Comparison from this drawing** (see Figure 14.5).

You can also click the Compare Drawings icon in the lower right-hand corner of the Drawing Compare palette.

FIGURE 14.5
Begin the comparison with the right-click or the icon in the corner of the palette

> Note: You will get the best results with the Drawing Compare tool if you open the old file (Project) first and compare it to the new file (Review). The Project drawing that is opened on your screen is overlaid with the Review drawing matched to it.

AMEP will begin analyzing all the objects between the two drawings in an effort to determine which items are new to the project, have been modified, or even removed from the project (see Figure 14.6).

FIGURE 14.6
AMEP will search all the items in a drawing to determine what is new, has been modifed or removed from the project.

Once the comparison is complete the Drawing Compare palette will switch to the Review tab and display the comparison's findings (see Figure 14.7).

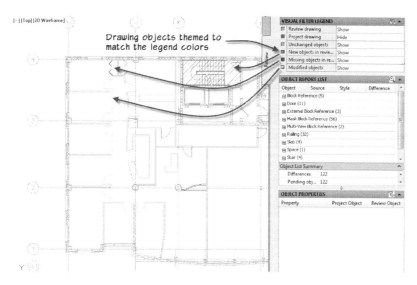

FIGURE 14.7
When the comparison is complete, the results of AMEP's findings will be displayed in the Review tab of the Drawing Compare dialog box. The onscreen display of the drawing will be themed to match

Let's break down the components of the Review tab of the Drawing Compare dialog box.

Setup—This information is the same as on the Setup tab.

Visual Filter Legend—The drawing is themed using the colors in this legend. You can show or hide different aspects of the comparision. For example, you can choose to show only the new items in the project by hiding the Missing and Modified items. To get more control, click the Visual Filter icon on the "Visual Filter Legend" titlebar. This will allow you to go one step further and begin to limit the comparison to specific objects (see Figure 14.8). Initally, all objects are being compared.

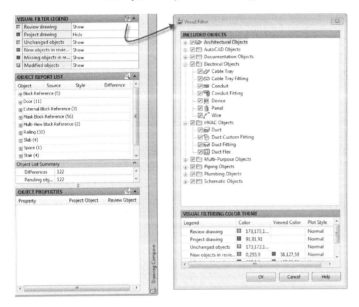

FIGURE 14.8
The Visual Filter allows you to limit comparison to selected objects

TIP: One of the nicest features in the Drawing Compare tool is the Visual Filter. With the Visual filter you have the ability to isolate new items from modified and/or missing items. You can even go further to isolate changes in particular objects. You also have the ability to plot those specific snapshots to your plotter for others to review. If your company has a color plotter, the green, red, and yellow color scheme can be an effective means of conveying project-related changes to other team members.

Object Report List—The Object Report List gives the overall results of changes grouped by object type. Simply expand an object category to see additional information on the differences found (New, Modified, or Missing) (see Figure 14.9). You can even select an item in the list and AMEP will automatically zoom to the object. You can toggle the zoom behavior on and off with the icon on the "Object Report List" titlebar.

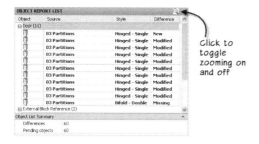

FIGURE 14.9

The Object Report List allows you to review detailed information on the changes to the individual objects in the project

In Figure 14.9 there were 11 modifications to the door objects. Within the 11 modifications, one new door was added to the project, one was deleted, and nine others were modified. In addition to this there was one new external reference file added to the project as well.

> **Object Properties**—Select an object in the "Object Report List" to see detailed Object Properties displayed in the "Object Properties" panel. Here you can see specifically what has changed for the selected object (see Figure 14.10). Click the Object Properties Filter icon on the "Object Properties" titlebar to see only the properties that vary between the two compared drawings.

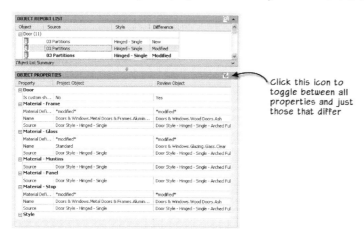

FIGURE 14.10

Object Properties provides more detail on how the individual object was modifed

With the exception of the Home ribbon, whenever the Drawing Compare tool is active you should note that the Drawing Compare panel will appear in your current ribbon tab (see right side Figure 14.11). To access the Drawing Compare panel from the Home ribbon click the Drawing Compare pull down (see left side Figure 14.11). The Drawing Compare panel allows you to move to the next or previous object based the current selection in the Object Report List, zoom to objects, select objects to display the properties of the changed object, show or hide the Drawing Compare palette, plot the results of the Drawing Compare session or close the Drawing Compare session.

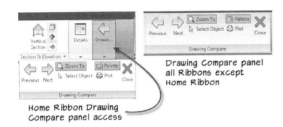

FIGURE 14.11

Whenever the Drawing Compare tool is active the Drawing Compare panel will appear in your current tab

Following is a description of the tools of the Drawing Compare panel:

Previous—After picking an object in the palette you can use Previous to zoom to the previous object identifed in your drawing.

Next—After picking an object in the palette you can use Next to zoom to the next object identifed in your drawing.

Zoom To—Toggle this button to turn on or off the automatic zooming to the selected interference in your drawing. You can also locate this button on the "Object Report List" titlebar on the palette.

Palette—Click this button to hide or unhide the Drawing Compare palette.

Select Object—Click this button to select an object in the drawing and reveal it in the "Object Report List" in the palette.

Plot—Plot out the results of the Drawing Compare session with this button.

Close— After you finish comparing the modified objects between the selected project files, you can click Close in the Drawing Compare panel to end the Drawing Compare session.

Continue exploring the results if you wish.

7. When you are finished, on the Drawing Compare ribbon panel, click the Close button.

The "Exit Drawing Compare Mode" dialog will appear.

> **Note:** Currently there is no way to save the results of the Drawing Comparison with your drawing file. The only possible way to save the results of the Drawing Comparison is through printing. You can print a hard copy or create a DWF. A DWF of the Drawing Compare Session has been included in the *Chapter 14* dataset directory.

8. After reading the prompt, click Yes to confirm exit (see Figure 14.12).

FIGURE 14.12

The results of the Drawing Comparison will be discarded after you exit the Drawing Comparison mode

9. Close the Drawing Compare tool palette

INTERFERENCE DETECTION

When two building elements end up in the same spot, a clash or interference occurs. Traditionally, discovering such interferences involves the labor-intensive process of manually comparing the drawings produced by the multiple parties working on the project. Such a task is further complicated by discrepancies that often occur between plan, section, and elevation. One of the benefits of using AutoCAD Architecture and MEP is that we are simultaneously creating a three-dimensional model even as we create plans, sections, and elevations. Having a 3D model and the powerful AMEP Interference Detection tool allows us to discover clashes between building elements long before they appear in the field as costly change orders.

Let's look at how the Interference Detection tool can help identify potential interferences before they are discovered in the field.

> **Note:** To have the most success with the Interference Detection tool, the project will need to be modeled as accurately as possible. This means that all objects placed in AMEP will need to be shown at their correct elevation and location. The importance of this point cannot be understated.

1. On the Project Navigator, on the Constructs tab, expand the *Mechanical* folder and then double-click *03 Interference Detection* to open it.

Now let's bring in the structural framing and see what mischief lies above the ceiling.

2. Expand the *Structural* folder.

3. Drag *04 Framing* and drop it in the drawing window.

> **TIP:** When you are finished performing your Interference Detection, you can either detach the structural XREF or unload it. To perform either action, right-click the *03 Interference Detection* file on Project Navigator and choose External References. Right-click *04 Framing* to unload or detach it.

4. On the Analyze tab, on the Inquiry panel, click the Interference Detection button.

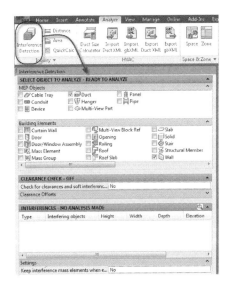

FIGURE 14.13

Select the objects to include in the Interference Check

There are four areas on the Interference Detection palette (see Figure 14.13):

Select Object to Analyze—Objects eligible for analysis are grouped as MEP Objects and Building Elements. You can select the objects that you want to include in your interference detection analysis.

Clearance Check—This is an on/off toggle for the "Check for clearances and soft interferences" option. A soft interference is an additional amount of space applied to an object to ensure that there are no interferences within the additional space. For example, a duct is placed in the drawing at a certain size. However, when the duct is installed in the field, that duct might have insulation around it (effectively making it larger than drawn). You can add 1-1/2" for the insulation in the Clearance Offsets box to represent the ductwork insulation. This will apply to all duct objects. Selecting Yes for this option turns it on and allows you to set soft interferences for several engineering systems. Input the desired clearance next to each item to which you want to add clearance. Selecting No turns this option off.

Interferences—After running an Interference Detection on your project, the detected interferences will be listed in this box. You can select one of the items in the list and then zoom to the selected item in the drawing. The Zooming feature can be toggled on or off with the icon on the "Interferences" group titlebar.

Settings—Several settings are available.

Keep interference mass elements when ending analysis mode—will leave a mass element in the drawing for each clash found once you exit.

Tag Location and Name—Browse to a location and choose a tag for labeling interference conditions.

Opening Symbol Location and Name—Browse to a location and choose an Opening Symbol for indicating required penetrations.

5. On the Interference Detection palette, select only Duct (deselect the others).

6. In the Building Elements area, select Structural Member (see Figure 14.14).

FIGURE 14.14

The Interference Detection tool allows you to limit analysis to selected objects

7. Leave the other settings as-is and then click the Start Interference Analysis icon at the lower left-hand corner of the palette (see Figure 14.15).

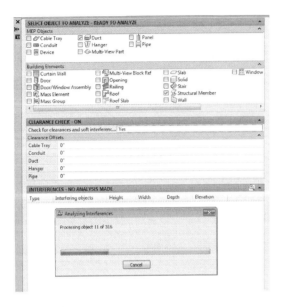

FIGURE 14.15

Configure your settings and then run the analysis

After the analysis is complete AMEP will show the conflicts in the Interferences area of the palette. The total number of interferences found will be indicated in the Interferences title bar. Each interfering object will also include the Height, Width, Depth, and Elevation (see Figure 14.16).

> **TIP:** The objects in the Interferences box can be sorted in ascending or descending order simply by picking on the columns (Type, Interfering objects, Height, Width, Depth, or Elevation).

INTERFERENCES - 97 INTERFERENCES FOUND					
Type	Interfering objects	Height	Width	Depth	Elevation
Hard	Beam & Duct	3 11/16"	3 1/2"	3'-10 13/32"	10'-3 1/2"
Hard	Beam & Duct	8 1/16"	2'-2"	6"	9'-11 3/32"
Hard	Beam & Duct	3 11/16"	3 1/2"	15'-0 11/32"	10'-3 1/2"
Hard	Beam & Duct	1 11/16"	3 1/2"	1'-3"	10'-3 1/2"
Hard	Beam & Duct	1 11/16"	3 1/2"	1'-3"	10'-3 1/2"
Hard	Beam & Duct	1 11/16"	3 1/2"	1'-3"	10'-3 1/2"
Hard	Beam & Duct	1 11/16"	3 1/2"	1'-8"	10'-3 1/2"
Hard	Beam & Duct	1 11/16"	3 1/2"	1'-8"	10'-3 1/2"
Hard	Beam & Duct	1 11/16"	3 1/2"	1'-8"	10'-3 1/2"
Hard	Beam & Duct	1 11/16"	3 1/2"	1'-8"	10'-3 1/2"
Hard	Beam & Duct	6 1/16"	6 13/16"	1'-8"	9'-11 3/32"
Hard	Beam & Duct	1 11/16"	3 1/2"	1'-8"	10'-3 1/2"
Hard	Beam & Duct	1 11/16"	3 1/2"	1'-8"	10'-3 1/2"
Hard	Beam & Duct	1 11/16"	3 1/2"	1'-8"	10'-3 1/2"
Hard	Beam & Duct	1 11/16"	3 1/2"	1'-8"	10'-3 1/2"
Hard	Beam & Duct	1 11/16"	3 1/2"	1'-8"	10'-3 1/2"

FIGURE 14.16

Any interferences that AMEP discovers will be indicated in the Interference box, including size and elevation

8. To zoom to the exact location of the interference, simply select one of the interferences listed on the palette.

The Interference Detection tool is a contextual tool. This means that when this tool is selected and active an Interference Detection panel will appear on the Ribbon (see Figure 14.17). You can use the Previous or Next buttons on the Interference Detection ribbon panel to cycle between the objects in the Interferences box. You can even close the palette if necessary. This will not exit the interference mode. Use the Palette button on the ribbon to reopen it when needed.

FIGURE 14.17

The Interference Detection panel is a contextual panel that appears when the Interference Detection tool is active

Following is a description of the tools of the Interference Detection panel:

Previous—After picking an interference in the palette, you can use Previous to zoom to the previous interference identifed in your drawing.

Next—After picking an interference in the palette, you can use Next to zoom to the next interference identifed in your drawing.

Zoom To—Toggle this button to turn on or off the automatic zooming to the selected interference in your drawing. You can also locate this button on the Interference titlebar on the palette.

Select Interferences—Choosing this button allows you to select one or more interference objects on screen (select the interference mass element—not the tag). This will highlight the corresponding item(s) on the palette.

Palette—Click this button to hide or unhide the Interference Detection palette.

Tags—Choosing this button allows you to associate a tag to the interference mass element. You can specify the tag you wish to use in the Settings area of the Interference Detection palette.

> **Note:** If you add Tags and/or interference mass elements in the drawing, they will need to be erased manually once the interference has been resolved or addressed.

Opening Symbols—Choosing this button allows you to place an opening symbol at an interference between an AMEP object and a wall. This can help better coordinate your project with the Architect in regards to where wall openings are required.

Close—Choosing this button will end the interference detection analysis mode and discard all interference mass elements and tags placed in the drawing unless the "Keep interference mass elements when ending analysis mode" in the Settings area of the palette has been set to Yes.

9. With the Interference Detection tool still active, zoom into our return air ductwork system that we placed in Chapter 5 (see Figure 14.18).

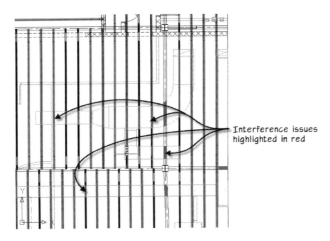

Interference issues highlighted in red

FIGURE 14.18
Interferences are reflected graphically on the screen in red to draw attention to the conflicts

The return air ductwork system that we placed in Chapter 5 is in conflict with quite a few of the structural members. Let's resolve this conflict.

10. Select the 40"x12" horizontal main.

11. On the Duct tab, on the Modify panel, click the Modify Run button.

12. Select Elevation: and change the elevation to 9'-9" and click OK (see Figure 14.19).

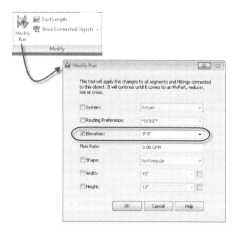

FIGURE 14.19

Eliminating conflicts might mean adjusting the overall elevation of objects

13. On the Interference Detection palette, click the Start Interference Analysis icon again to rerun the Interference Detection.

This will allow AMEP to review the changes made to the ductwork system to see if there are still any interferences between this duct system and your structural members. Once the analysis is completed, you can review the changes to see if the two objects are still in conflict with each other (see Figure 14.20).

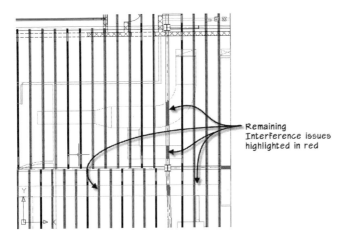

Remaining
Interference issues
highlighted in red

FIGURE 14.20
*After changes are made to the objects in conflict, Interference Detection can be
rerun to determine whether those items are still in conflict*

After rerunning Interference Detection, AMEP indicates that most of the conflicts between
this ductwork system and the structural members have been resolved. However, not all the
conflicts have been resolved, so further adjustments will need to be made to the ductwork
system and/or the structural member system. Feel free to experiment further with such
adjustments.

Before we close the Interference Detection tool, let's look at conflicts between our ductwork
system and the project walls. This will helps us coordinate wall opening locations with the
Architect.

14. On the Interference Detection palette, in the Building Elements area, deselect the Structural
    Member box and select the Wall box.

15. Click the Start Interference Analysis icon once more.

Since we are coordinating opening locations with the Architect, let's tag those locations in our
drawing and then we will elect to keep the interference mass elements as well.

16. Select the first interference listed on the palette.

    The drawing should zoom to the location.

17. On the ribbon click the Tags button (see Figure 14.21).

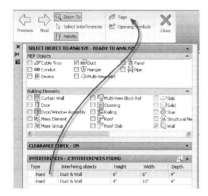

FIGURE 14.21

Tagging interferences gives you the ability to coordinate your drawing with others

18. Repeat the process to tag the other opening.

    If you wish, you can click the Opening Symbols button on the ribbon to add these symbols to the opening as well.

19. In the Settings area at the bottom of the palette, for the "Keep interference mass elements when ending analysis mode" option, choose **Yes** (see Figure 14.22).

FIGURE 14.22

Tags and interference mass elements can be maintained with the drawing after ending the Interference Detection analysis

20. Choose Close in the Interference Detection panel in the Ribbon.

21. Open the XREF Manager and Unload the 04 Framing plan.

You have now coordinated wall openings, and you are ready to share this information with the Architect (see Figure 14.23).

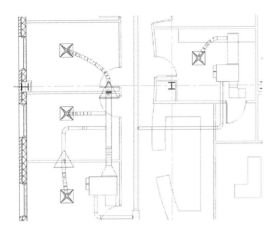

FIGURE 14.23

Add Tags to Conflicts to assist communication

Applying tags and keeping the interference mass elements in the drawing can be a great visual representation of where conflicts are in your project or where wall openings might need to be located. Finally, since the interference tags contain property set information regarding the interference, a schedule could be created to reflect this information. See Chapter 14 for information on Schedules and Property Set Data.

22. Save and close all files.

SUMMARY

The Drawing Compare and Interference Detection tools are simple to use and provide huge benefits. Using them, you can realize significant time savings in regards to project management and help identify potentially costly project conflicts before they are discovered in the field. In this chapter we learned:

✓ Drawing Compare offers a way to help manage project-related changes.

✓ Drawing Compare interactively overlays two drawings, and themes change elements using color codes for new, edited and deleted items.

✓ Results of Drawing Compare cannot be saved to the DWG, but can be printed to DWF files (or printed to hard copy).

✓ Interference Detection offers a way to identify project conflicts before the project leaves your office.

✓ Interference Detection allows you to better coordinate your wall openings for your MEP systems.

✓ Results of Interference Detection can be saved as mass elements in the DWG file.

Chapter 15

Annotation, Property Sets, and Schedules

INTRODUCTION

AutoCAD MEP offers several forms of automated annotation. These include labels and tags that are associated with MEP elements, such as duct and pipe labels, lighting fixture tags, circuit tags, diffuser tags, and schedules. AMEP elements have certain properties that are commonly used for tagging and scheduling, and AMEP facilitates automated annotation by providing options such as placing labels as ducts or pipes are placed. These labels typically show the size of the related component and adjust automatically if the size is modified. Other annotations such as circuit tags are closely related to the circuiting operations needed to tabulate connected and demand load. In other cases, such as with diffuser tags or lighting fixture tags, the annotation is based on a property that defines the type of diffuser, or the type of fixture. In these instances, the annotation information is derived from information inherent to the object itself; you don't need to specify the values for the annotation—it comes automatically from the object, and will update automatically if the object itself changes.

OBJECTIVES

To automatically annotate effectively, you need a fundamental understanding of Property Sets, Tags, Labels, and Schedules. This chapter will walk you through these items so that you may effectively define them on your own to suit your company standards or requirements. In this chapter you will:

- Learn how to create and use Property Sets
- Work with Tags, Schedules, and Labels
- Learn how to centralize annotation content
- Learn how to create Tool Palette Tools

PROPERTY SETS

Any information that is displayed in a tag or a schedule must be defined in a Property Set. A Property Set is a grouping of data related to a particular object or object style. At first, a Property Set can seem similar to attributes in AutoCAD Blocks. A Block can have attributes (which are simple text fields) that are displayed as text when the block is inserted. Some attributes within a block definition may be invisible and not displayed with the Block, but the associated value may still be edited and extracted. However, the similarities end here, and Property Sets are so much more. Property Sets are not part of a Block, but rather are defined in Style Manager like other AMEP objects (see Figure 7.1).

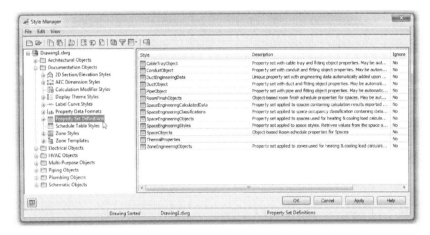

FIGURE 15.1

Property Sets are styles, not block attributes

> Note: It is important to give it a name that you can live with, as changing it will break any tag definitions that refer to the Property Set. It is possible to fix the tag definitions, but you probably have better things to do.

You will find Property Sets in the *Documentation Objects* folder in Style Manager. Like other styles, you give them a name and description on the General tab. However, unlike other objects, they have no graphical representation; they are just groupings of properties. As such, when a Property Set is defined, the most important setting is determining to what types of objects the Property Set applies. In other words, you must specify if the Property Set applies to objects or styles and to which object(s) it should apply (see Figure 15.2). The first choice is for

"Applies To," which can either be Objects or Styles. Depending on your choice, the list of available object types will adjust. You can select one or more object types in the list. Check all that apply. Finally, on the right side, you can filter the list further using Classifications.

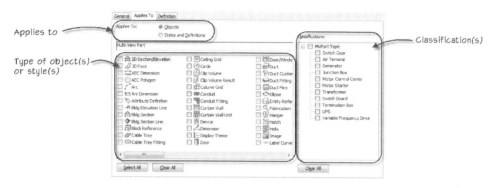

FIGURE 15.2

You must configure how the Property Set is applied to objects before adding properties

A Property Set can be applied to nearly any element in the drawing, including AutoCAD objects such as Blocks, Lines, and Circles, and also any MEP object such as Ducts, Devices, etc. AutoCAD entities are available only if the Property Set applies to Objects. If you choose Styles and Definitions, choices will include any MEP object style such as Duct Styles, Device Styles, etc. A Property Set cannot apply to both objects and styles simultaneously. Therefore, when defining a Property Set, you must first determine if the Property Set will be associated with Objects or Styles and Definitions (see Figure 15.3).

FIGURE 15.3

Each Property Set can be either Object-based or based on Styles and Definitions (Style-based)

Object properties can contain unique values for each instance of a particular object, such as the flow at a specific diffuser or the length of a particular line. For example, you may have a 24 x 24 Inch Square Louver Face Ceiling Diffuser with a 10-inch neck in your project. Each instance of this diffuser may contain different airflows; for example, in one room, the diffuser may have 110 CFM per terminal, while in another room, you may specify 125 CFM. Since the airflow *could* be different for each instance, you would specify that a Property Set containing such values would Apply To: Objects. Similarly, the circuit property of a lighting fixture would be an Object (or instance) property; that is, each fixture instance may be connected to a different circuit. Since an air terminal is a Multi-view Part, when defining the Property Set, you would first be sure that it applied to Objects and then select the Multi-view Part in the list of object types (see the top of Figure 15.4). For a lighting fixture, you would select Device instead.

Note: The Light object listed in the object type list in Figure 15.4 refers to an AutoCAD Light object, used for rendering. Lights of this type cannot be circuited.

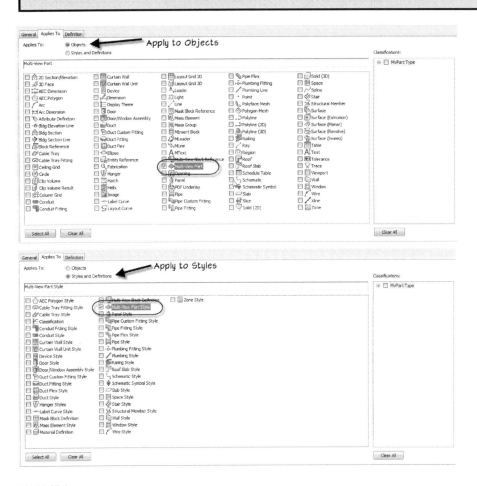

FIGURE 15.4

The list of Object types varies when switching from object-based to style-based

Style and Definition properties contain the same value for all instances of a particular object style. Using the diffuser example again, all instances of the 24 x 24 Inch Square Louver Face Ceiling Diffuser with a 10-inch neck would have the same type designation, such as SD-1. As such, a Property Set containing these values would Apply To: Styles and Definitions. Similarly, the fixture type property of a lighting fixture would be constant for all instances of a particular fixture type (or Style). As such, the fixture type would belong to a Property Set that Applies To: Styles and Definitions. For the air terminal and lighting fixture, you would select Part Style (see the bottom of Figure 15.4) or Device Style, respectively.

Further, Objects can be defined with a particular Classification. For example, Air Terminals are a specific Classification of Multi-view Part and Lighting fixtures are a specific Classification of Device (Figure 15.5). Thus, when creating a Property Set, you can be very specific about what type and classification of object (or style) the properties will apply to. Specifying a Classification facilitates associating Property Sets with objects. For example, it is not possible to associate a Property Set that applies to a Fire Safety Device Style with a Lighting Device Style. However, if no Classification is selected, the Property Set will apply to all Classifications of the particular object type. For example, a Device Property Set with no Classification selected will apply to Lighting, Other Power, Receptacles, Devices, etc.

FIGURE 15.5
Classifications of Multi-view Parts and Devices

The Definition tab is used to define the specific Properties contained within the Property Set. There are various types of properties that can be defined for use in scheduling or tagging. Use the icons on the Definitions tab to create them (see Figure 15.6). Examples of several types of properties will be demonstrated in the various exercises in this chapter.

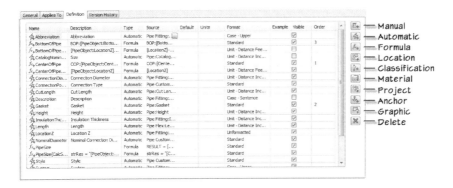

FIGURE 15.6

The Definition tab contains a table of the Properties within the Property Set. Icons on the right are used to create new Properties.

Each row in the table represents an individual Property within the Property Set Definition.

Manual properties—are simple text fields. Values are entered manually by the user, such as the ID of a fan.

Automatic properties—are obtained automatically from the object geometry, such as the length of a pipe, or the area of a room.

Formula properties—are calculated dynamically by evaluating an expression written in the property. Formulas can perform simple arithmetic or robust programmatic expressions. Examples include the volume of a pipe or a total allowable lighting power for a room.

Location properties—are obtained automatically based on the object's physical location in the model. These properties provide the means for a light to be able to report the room it is in, or for a fan to report the space it serves.

Classification properties—provide information on how an object is classified, such as the Air Changes per Hour, or Lighting Power Density.

Material properties—are values are obtained from an object's component materials or from property data specified for an object's component materials. This is more commonly used in AutoCAD Architecture, and not commonly used in AutoCAD MEP.

Project properties—queries information from Project Navigator, such as the Level on which an object resides or the Project Number.

Anchor properties—allow one object to get data from another object to which it is anchored. For example, an anchor property of a hot water coil in a VAV unit may display the GPM of the coil as a property of the VAV unit.

Graphic properties—are values that can be blocks in the current drawing or image files, such as BMP, GIF, JPG, PNG, or TIF. These may be used in a light fixture schedule to show the fixture symbols.

Delete—is used to delete the selected property in the list.

There are several columns of metadata about the properties on the Definition tab. The Name column has an icon indicating the type of property selected, corresponding to the icon used to create it. Additionally, the Type column indicates the property type. Note that there are no spaces in the names of any of the properties—this is standard practice. It is done because tags use AutoCAD Blocks with Attributes to display Property data. AutoCAD Block Attributes do not support spaces. You can rename a property by "slow double-clicking" the property name (click, pause for brief moment, then click again). Double-clicking on the name quickly will allow you to edit certain property definitions (Automatic, Formula, etc.)

For manual properties, the Type can be specified as one of the following:

Auto Increment – Character—a character value that automatically increments as you tag objects, such as A, B, C, etc.

Auto Increment – Integer—a numeric value that automatically increments as you tag objects, such as 1, 2, 3, etc.

Integer—any integral number, such as 123, or -44.

List—limits the user to a list of values based on list definitions. List Definitions are defined in Style Manager, under Multi-Purpose Objects.

Real—any non-integral number, such as 1.24, or -0.44

Text—any free-form text.

True/False—limits the user to a selection of True or False when specifying the property value. Values can be customized to Yes/No, In/Out, Up/Down, etc.

The Description column allows you to provide an explanatory description of the property. This is displayed as a tool tip when you hover over the property name on the Properties palette. The Source column displays a button to edit the property when it is selected. This is similar to quick double-clicking on the property name.

The Default column displays the default value if one is specified, and the Units column displays a selectable list of input units based on the Format column if the property is a manual property with the Type set to Real (as in numerical real value).

The Format column allows you to select a Property Data Format to apply to the value; for example, to display 60 as 5'-0" AFF (instead of just 60). The Example column provides a sample of the data with the Property Data Format applied if the value is a manual property. Property Data Formats are defined in Style Manager as well.

The final two columns control if and where the property displays in the Properties palette. Unchecking Visible will hide the property from the palette. This is handy for hiding non-manual properties that are only used in formulas or schedules; it reduces the clutter for the user when entering or viewing essential properties. The Order column allows you to specify the order in which properties appear in the Properties palette. By default, properties are displayed alphabetically; however, you can force certain properties to the top by specifying the order.

Property Sets associated with an object or its style are displayed and edited on the Properties palette Extended Data tab (see Figure 15.7).

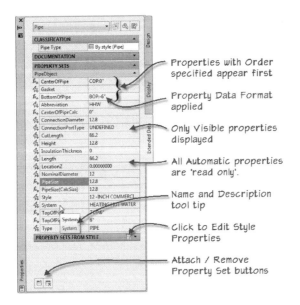

FIGURE 15.7

Various aspects of how a Property Set Definition relates to the display of the data on the Extended Data tab of the Properties palette

INSTALL THE DATASET FILES AND OPEN A PROJECT

In this exercise, we use the Project Browser to load a project from which to perform the steps that follow.

1. If you have not already done so, install the book's dataset files.

   Refer to "Book Dataset Files" in the Preface for information on installing the sample files included with this book.

2. Launch AutoCAD MEP 2012 and then on the Quick Access Toolbar (QAT), click the Project Browser icon.

3. In the "Project Browser" dialog, be sure that the Project Folder icon is depressed, open the folder list and choose your *C:* drive.

4. Double-click on the *MasterAME 2012* folder.

5. Double-click *MAMEP Commercial* to load the project. (You can also right-click on it and choose **Set Project Current**.).

IMPORTANT: If a message appears asking you to repath the project, click Yes. Refer to the "Repathing Projects" heading in the Preface for more information.

Note: You should only see a repathing message if you installed the dataset to a different location than the one recommended in the installation instructions in the preface.

6. Click the Close button in the Project Browser.

When you close the Project Browser, the Project Navigator palette should appear automatically. If you already had this palette open, it will now change to reflect the contents of the newly loaded project. If for some reason, the Project Navigator did not appear, click the icon on the QAT.

DEFINE OBJECT AND STYLE-BASED PROPERTY SETS FOR AIR TERMINALS

1. On the QAT, click the Open icon.

2. Browse to the *C:\MasterAME 2012\Chapter15* folder and open the *03 HVAC.dwg* file.

3. On the Manage tab, on the Style & Display panel, click the Style Manager button.

4. Expand the *Documentation Objects* folder.

5. Right-click on Property Set Definitions, and choose **New**.

6. For the name, type **MvPartAirTerminalObjects,** and then press ENTER.

Note: When defining Property Set and Property Names, it is best not to use spaces. The properties can ultimately be used to define an attribute in a tag block, and AutoCAD doesn't allow spaces in attribute definition names.

---

7. On the right side, on the Applies To tab, for "Applies To" choose **Objects**.

8. In the object list, check **Multi-view Part**.

9. In the Classifications list, expand MvPart Type and then select **Air Terminal**.

10. Click the Definition tab.

11. On the far right, click the Add Automatic Property Definition icon (second from the top).

The "Automatic Property Source" dialog appears. Scroll through the list and notice the large collection of available properties. These are all the automatic properties that can possible be queried from Multi-view Parts. They are not all assigned to this Property Set. This is an important distinction. You could, of course, add as many of these as you want to your current Property Set, but some may not be appropriate for an Air Terminal. For example, it is unlikely that the "Hot Water Calculated Flow Rate" would yield a useful result for Air Terminals. Remember that Multi-view Parts are used for several different kinds of elements. This is why we use the additional filter of the Classification when defining them. However, the Classification does not filter the list of available Automatic properties, so it is up to you to select carefully.

12. In the "Automatic Property Source" dialog, check the box for Flow and then click OK.

13. Set the Format of the Flow property to **Flow-CFM** (see the top half of Figure 15.9 below).

The Flow-CFM Property Data Format will append a CFM suffix on the flow value.

14. On the left side of Style Manager, beneath *Property Set Definitions*, select MvPartAir_TerminalStyles (Styles not Objects).

This Property Set Definition was already present and attached to the Air Terminal Styles in the drawing.

15. On the Applies To tab, verify that:

- Applies To is: **Styles and Definitions.**
- The object type selected is: **Multi-view Part Styles.**
- And in the Classifications list, **Air Terminal** is selected.

16. Click the Definition tab.

There is a single property called: SupplierName. This property came from the Multi-view Part Catalog when the part was inserted.

17. Click the Add Manual Property icon (first one).

18. For the Name, input **Type,** and then click OK.

> Note: This "Start With" setting refers to the definition of the property definition. That is, if you have multiple definitions that are similar, you may use this effectively to start with a copy of another definition.

19. Click the Add Automatic Property Definition icon.

20. Check the boxes for the following properties, and then click OK:

- Connection Diameter
- Connection Shape
- Rectangular Connection Height
- Rectangular Connection Width

Notice that each of the checked items will become a separate automatic property. This is a big time saver when you have several properties to add.

21. Uncheck the Visible checkbox for the four properties added in the previous step.

Remember this makes them invisible on the Properties palette. We have added these properties to use in a formula. Therefore, it is not necessary to have them display on the Properties palette.

22. Click the Add Formula Property Definition icon (third one down).

23. For the Name, type **ConnectionSize**.

24. In the Formula box, type the following code:

> **Note:** The text in [brackets] must NOT be typed in directly. These items must be inserted by double-clicking on the property in the Insert Property Definition list below as shown in Figure 15.8. Pay careful attention to the quotes and capitalization. Finally, the %%c is the standard AutoCAD convention to insert a diameter symbol.

```
If UCase("[ConnectionShape]") = "ROUND" Then
                                                                    R
ESULT = "[ConnectionDiameter]%%c"
ElseIf UCase("[ConnectionShape]") = "RECTANGULAR" Then
                                                                    R
ESULT =
"[RectangularConnectionWidth]x[RectangularConnectionHeight
]"
end if
```

This formula simply checks the ConnectionShape property to confirm its value. If the shape is round, it reports the ConnectionDiameter, suffixed with the commonly used diameter symbol (%%c). If the shape is rectangular, it reports width x height. An additional check would be required if we expected to have air terminals with oval connections.

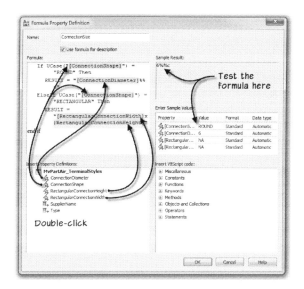

FIGURE 15.8

Insert Properties into a formula by double-clicking on the Property name—typing them will not work

Properties used in the formula will appear in a list in the right middle pane. You can type in sample values there and test if the formula is working. If you get a valid result in the "Sample Result" field, your formula is working. If the formula syntax itself appears in the "Sample Result" field, you have an error. Check your typing, quotes, parenthesis and be certain that you double-clicked the property names from the list at the bottom instead of typing them.

25. Click OK to close the Formula Editor.

26. Change the Format of the ConnectionSize property to **Unformatted** (see the bottom half of Figure 15.9).

> **Note:** The default format is Standard, which is defined to set everything to uppercase. Since we want the x in the rectangular duct size to be lower case, we set this to Unformatted.

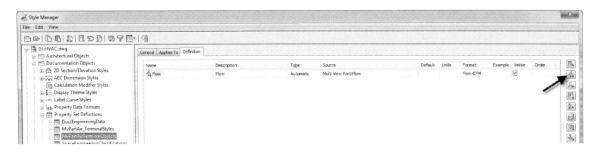

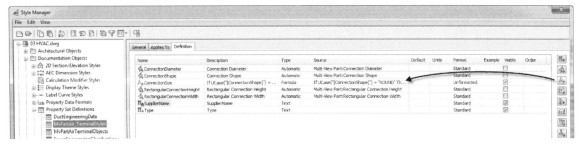

FIGURE 15.9
Configure both an object-based and a style-based Property Set

27. Click OK to close the Style Manager.

28. Select one of the air terminals in the drawing.

29. If the Properties palette isn't open, right-click and choose **Properties**.

30. On the Extended Data tab, Click the Edit Style Property Set Data icon.

31. In the "Edit Property Set Data" dialog, in the Type filed, input **SD-1,** and then click OK (see Figure 15.10).

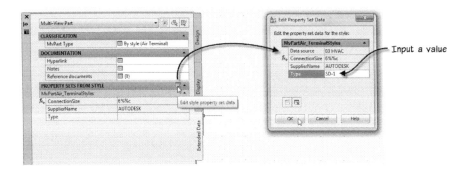

FIGURE 15.10
Edit a Style-based Property via the Properties palette

Since we modified the Type for the air terminal style, all air terminals of this specific style (24 x 24 Inch Square Louver Face Ceiling Diffuser – 6-Inch neck) now have the SD-1 property value associated with them. We will later define a tag for this property, and see the effect of changing the air terminal style.

32. Save the *03 HVAC* file.

In the steps above, we created some properties that will attach to the Air Terminal style, and a Flow property that will attach to Air Terminal instances. In a later exercise, we will create a tag that will display this flow property, along with the Air Terminal's style Type property and the ConnectionSize property.

DEFINE LIGHTING FIXTURE PROPERTY SETS

In this exercise we will introduce Location properties. The Location property will allow us to schedule the physical location (which room) of each lighting fixture, which then can be exported to a spreadsheet to compute the total lighting wattage per square foot of each space.

1. On the QAT, click the Open icon.

2. Browse to the *C:\MasterAME 2012\Chapter15* folder and open the *03 Lighting.dwg* file.

3. On the Manage ribbon, Style & Display tab, click Style Manager.

4. Expand the Documentation Objects folder.

5. Right-click on Property Set Definitions, and choose **New**.

6. For the name, type **DeviceLightingStyles** and then press ENTER.

7. Click the Applies To tab and configure the following:

   ⇨ For Applies To, choose **Styles and Definitions**.

   ⇨ In the object list, check **Device Style**.

   ⇨ Expand the Classifications list, expand Device Type, and then select **Lighting**.

8. Click the Definition tab.

9. Click the Add Manual Property Definition icon.

10. For the Name, input **FixtureType** and then click OK.

11. Click the Add Automatic Property Definition icon.

12. Check the box for **Load** and then click OK.

13. Right-click on *Property Set Definitions*, and choose **New**.

14. For the name, type **DeviceLightingObjects** and then press ENTER.

15. Click the Applies To tab and configure the following:

⇨   For Applies To, choose: **Objects**.

⇨   In the object list, check: **Device**.

⇨   Expand the Classifications list, expand Device Type, and then select **Lighting**.

16. On the Definition tab, click the Add Manual Property Definition icon.

17. For the Name, input **Switchleg** and then click OK.

18. Click the Add Automatic Property Definition icon.

19. Check the box for **Circuit Number** and **Panel Name,** and then click OK.

20. Uncheck the Visible box for both of these properties.

21. Click the Add Location Property Definition icon (fourth one from the top).

22. For the Name, input **SpaceNumber**.

In the "Property Definition" area, beneath Space, you will see the Property Sets attached to Space objects. A Location property is able to read a property from a Space in the proximity of the object (in this case the Device). In other words, the Device can "ask" the nearby Space for the value of one of its properties. For example, the Device can report the number of the Room in which it is located.

23. Beneath Space > SpaceObjects, select **Number** and then click OK.

The Property Set Definition used to define a Location Property should match the Property Set Definition attached to the actual Space objects. In this case, the SpaceObjects Property Set in the *03 Lighting.dwg* file was copied from the *03 Partitions.dwg* file.

24. Click the Add Formula Property Definition icon.

25. For the Name, input **CircuitTag**.

26. In the Formula box, type the following code (see Figure 15.11):

> **Note:** Remember, the text in [brackets] must NOT be typed in directly. These items must be inserted by double-clicking on the property in the Insert Property Definition list at the bottom of the dialog.

```
RESULT = "[PanelName]-[CircuitNumber]"
If Len(Trim("[Switchleg]")) > 0 Then
                                                              R
ESULT = RESULT + "([Switchleg])"
end if
```

FIGURE 15.11
Specify the Property Name and Formula

The first part of the formula shown in Figure 15.11 will show the panel name and circuit number of the circuit; for example, LA-1. The second part of the formula, inside the if/end if clause, checks whether the user has specified a value for the Switchleg property. If they have, it will append, in parentheses, this value to the circuit tag. For example, if they enter "A" (without quotes) as the Switchleg property, the circuit tag would show LA-1(a). The trim function removes any whitespace from the ends of the property value; thus, if the user specified a space for the Switchleg property value, this would be removed before checking the length (len) of the Switchleg property value. Stated another way, if the number of characters in the trimmed Switchleg property value are more than 0, the function will append this value to the circuit tag inside parentheses.

27. Click OK to close the "Formula Property Definition" dialog.

28. In the Order column for the Switchleg property, type **1**.

29. Click OK to close the Style Manager.

In a later exercise, we will use these Property Sets to create tags that will display the lighting fixture type and circuit information for our lighting fixture Devices.

30. Save the *03 Lighting* file.

DEFINE PROPERTY SETS FOR ELECTRICAL PANELS AND TRANSFORMERS

1. On the QAT, click the Open icon.

2. Browse to the *C:\MasterAME 2012\Chapter15* folder and open the *03 Power.dwg* file.

3. Open the Style Manager and expand *Documentation Objects*.

4. Create a New Property Set and name it **PanelObjects**.

5. Click the Applies To tab and configure the following:

⇨ For Applies To, choose **Objects**.

⇨ In the object list, check **Panel**.

⇨ Do not Classify it.

6. Click the Definition tab.

7. Click the Add Automatic Property Definition icon, check **Panel Name**, and then click OK.

8. Create another new Property Set Definition, and name it **MvPartTransformerObjects**.

9. Click the Applies To tab and configure the following:

⇨ For Applies To, choose **Objects**.

⇨ In the object list, check **Multi-view Part**.

⇨ Expand the Classifications list, expand MvPart Type, and then select **Transformer**.

10. Click the Definition tab.

11. Click the Add Automatic Property Definition icon, check **Engineering Part ID**, and then click OK.

12. Click OK again to close Style Manager.

In a later exercise, we will use these Property Sets to build tags to annotate our transformer and panel objects.

13. Save the *03 Power* file.

TAGS

Tags are special Multi-view Blocks that are used to annotate properties of an object. The AutoCAD Blocks used as view blocks in the Multi-view Block use standard AutoCAD attribute definitions to report the values of the object properties. However, unlike standard AutoCAD blocks, we do not edit these attribute values directly. AutoCAD MEP will input the required values automatically.

AutoCAD MEP provides a wizard-like interface for defining a tag. To create a tag, draw the geometry that you want for the tag graphics and use text objects as placeholders for any properties. The wizard will convert the selected items to a Block with the placeholder text becoming attributes. This Block will then be nested within a Multi-view Block definition for the tag. Attach the resultant tag to an object and it will report the Property Set data assigned to the object.

The size of the tag is based on the Annotation Plot Size setting (Application Menu > Utilities > Drawing Setup) as well as the drawing scale. Thus, when defining the tag geometry, it is important to draw the text objects at a height of one (1) unit tall. This will result in the text plotting at the size specified as the "Annotation Plot Size" in the "Drawing Setup" dialog.

In the exercises that follow, we will define tags for the various Property Sets defined above. We will create an Air Terminal Tag, a Lighting Fixture Tag, and an Electrical Panel Tag. We will then store the tags in a content file, and create tool palette tools so the tags may be used in other projects.

In the first set of exercises, we will define an air terminal tag that displays the type, neck size, and air flow, (see Figure 15.12).

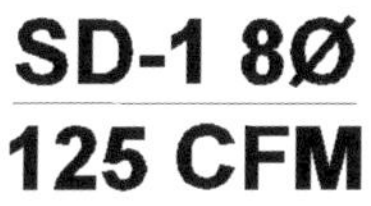

SD-1 8Ø
125 CFM

FIGURE 15.12

A sample air terminal tag

However, first we are going to modify the Property Set to include an extra property that is going to be used to display the information in the tag. As you may be able to discern, if the type and size properties were longer (SD-10 and 12x12), the line would appear too short; additionally, the two separate properties may end up overlapping or appear too close. As such, we will use a formula property to provide the formatting necessary to avoid these issues.

MODIFY THE AIR TERMINAL PROPERTY SET

1. On the QAT, click the Open icon.

2. Browse to the *C:\MasterAME 2012\Chapter15* folder and open the *03 HVAC.dwg* file.

3. Open Style Manager, expand the *Documentation Objects* folder and then the *Property Set Definitions* category.

4. Select the MvPartAir_TerminalStyles Property Set on the left and on the right, click the Definition tab.

5. Click the Add Formula Property Definition icon.

6. For the Name, input: **TagTopLine**.

7. In the Formula box, type the following code:**%%u[Type] [ConnectionSize]** (see Figure 15.13)

Note: Remember, these items must be inserted by double-clicking on the property in the Insert Property Definition list at the bottom of the dialog. First type **%%u**, then double-click the Type property, click next to [Type] in the Formula box, press the SPACEBAR one time and then double-click the ConnectionSize property.

The result will be a concatenation of the Type and ConnectionSize properties separated by a space.

8. Click OK to close the "Formula Property Definition" dialog.

FIGURE 15.13

Create a formula to concatenate two properties in the tag

9. Uncheck the Visible box for the TagTopLine property.

10. Set the Format to **Unformatted**.

11. Click OK to close Style Manager.

We now have a property that will join the two properties together separated by a space. This property will be available to our tags and schedules, but will be invisible to users on the Properties palette.

DEFINE THE SCHEDULE TAG

1. On the Insert tab, on the Block panel, click the Insert Block button.

2. In the "Insert" dialog, click the Browse button.

3. Browse to the *C:\MasterAME 2012\Chapter15* folder, select the *Air Terminal Tag Layout.dwg* file and then click Open.

4. Clear the "Specify On-screen" checkboxes for all options.

    -   Insertion point should be: X=0, Y=0, and Z=0.

    -   Scale should be: X=1, Y=1, and Z=1.

    -   Rotation should be: Angle=0.

5. Check the "Explode" checkbox and then click OK (see Figure 15.14).

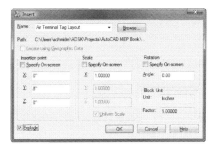

FIGURE 15.14

Options for inserting the Air Terminal Tag Layout block

The inserted and exploded block contains two text items near the origin (0,0,0). You will need to zoom to find the objects. Both objects have a height of 1" and are Justified Middle Center.

We will use these items to create a custom tag. As noted above, the place holder text should be drawn 1" to ensure that it will insert at the correct height when the tag is complete. When the tag is inserted, it is scaled according to the drawing's Annotation Plot Size and the current drawing scale. For example, for an Annotation Plot Size of 3/32" at a scale of 1/8"=1'-0" (1:96), the text in the tag will end up at 9" (1 x 3/32 x 96 = 9).

6. Zoom in on the drawing origin to find the objects.

7. On the Home tab, expand the Annotation panel and click the Create Tag button (see Figure 15.15).

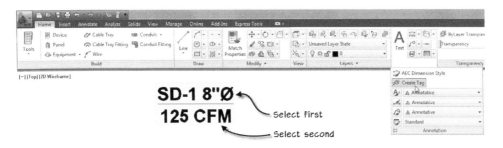

FIGURE 15.15
The Create Tag tool on the expanded Annotation panel

8. At the "Select object(s) to create tag from" prompt, select the two text objects (from the exploded block) and then press ENTER.

   Select the top one first.

9. In the "Define Schedule Tag" dialog, for the Name, type **MyAirTerminalTag**.

The two selected text objects are listed in the table at the bottom of the dialog. You can decide how each should behave in the tag. In the Type column, they default to Text. The other option is Property. If you leave them Text, they will remain static text values and not be associated to the object in any way. To make them report the values of properties from the tagged objects, choose Property.

10. In the Type column, change both rows to **Property**.

11. In the Property Set column make the following changes (see Figure 15.16):

⇨ For the 125 CFM row, choose **MvPartAirTerminalObjects**.

⇨ For the %%uSD-1 8%%C row, choose MvPartAir_TerminalStyles.

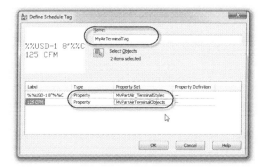

FIGURE 15.16

Specify the Name, and set the Type and Property Set

12. In the Property Definition column, make the following changes (see Figure 15.17):

⇨ For the 125 CFM row, choose **Flow**.

⇨ For the %%uSD-1 8%%C row, choose **TagTopLine**.

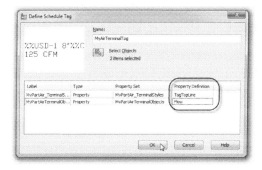

FIGURE 15.17

Specify the Property Definition

13. Click OK.

14. At the "Specify insertion point" prompt, type **0,0,** and then press ENTER.

The two text objects were replaced with an instance of the tag. It will become larger when it is replaced. This is because of the automatic scaling behavior discussed above. The original text was 1" tall and the text in the tag is now 9" tall. Since the tag is not associated with an Air

Terminal object, the tag displays the default value defined by the original text. Use the following steps to inspect the definition of the tag and associated block.

15. On the Manage tab, on the Style & Display panel, click the Style Manager button.

16. Expand the *Multi-Purpose Objects* folder and then select Multi-View Block Definitions.

17. Double-click on MyAirTerminalTag.

18. Click the View Blocks tab.

19. Click on each of the Display Representations (General, General Screened, etc).

Note that only the General Display Representation has a block associated with it. By default, tags created using the Create Tag tool are not visible in Model views (see Figure 15.18).

FIGURE 15.18

The View Blocks tab of the Multi-view Block definition

20. Click OK to close the Style Manager.

21. On the Insert tab, on the Block panel, click the Edit Block button.

22. Select MyAirTerminalTag from the list, and then click OK.

23. On the Properties palette, on the Design tab, look at the Block properties.

Notice that the Annotative property is set to Yes (see Figure 15.19). This is what allows the object to scale when the drawing scale changes.

FIGURE 15.19

The block definition for the tag is set to Annotative

The tag is made up of two Attribute Definition objects. The Attribute Definition Tag property is defined according to the Property Set Definition and Property Name that the tag will display, separated by a colon. In this case, the first Attribute reads: MvPartAir_TerminalStyles:TagTopLine. This means that this attribute automatically reads the TagTopLine property in the MvPartAir_TerminalStyles Property Set Definition.

FIGURE 15.20

The Properties palette showing Attribute Definition Tag and Annotative settings

CAD MANAGER NOTE: If you ever rename a Property Set Definition or Property Name (not recommended), your tags will no longer function. You can fix your tag by editing its view block and then modifying the tag property on the Properties palette. After updating the block in such a case, you will need to update your Multi-view Blocks. You can do this by typing the command MvBlock, and then using the Update option.

24. On the Block Editor tab, on the Close panel, click the Close Block Editor button.

25. If prompted: In the "Block – Changes Not Saved" dialog, click the "Discard the changes and close the Block Editor" option.

ATTACHING OBJECT-BASED PROPERTY SETS

26. Zoom Extents.

27. Select one of the Air Terminals, and then on the Properties palette, click the Extended Data tab.

Note that the MvPartAirTerminalObjects Property Set is not shown.

28. Click the Add Property Sets icon at the bottom of the Properties palette.

29. In the "Add Property Sets" dialog, check only the MvPartAirTerminalObjects Property Set, and then click OK.

You should now see the Flow property displayed in the MvPartAirTerminalObjects Property Set Definition (see Figure 15.21).

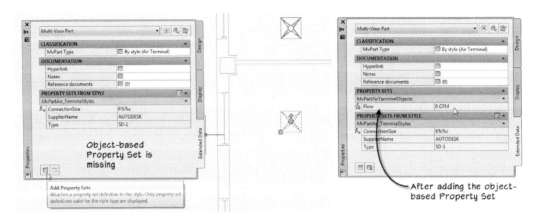

FIGURE 15.21

Object and Style properties for the selected air terminal.

Select another air terminal. Note that this air terminal doesn't have the MvPartAirTerminalObjects Property Set either. Object-based Property Sets must be added individually to each object. This can be time-consuming if you do it manually. Fortunately, there are a few methods to accomplish the task more quickly. In this case, instead of manually attaching the Property Set, we will attach it automatically with our new tag.

30. Deselect the Air Terminal and select the tag.

> **Note:** The tag is still inserted at the origin. Feel free to move it closer to the drawing before continuing.

31. On the Tag ribbon tab, on the General panel, click the Add Selected button.

32. At the "Select object to tag" prompt, select an Air Terminal.

33. At the "Specify location of tag" prompt, click to place the tag.

34. If the "Edit Property Set Data" dialog appears, click OK.

The prompt will repeat; do not exit the command.

> **TIP:** Each time you tag this way, the "Edit Property Set Data" dialog will appear offering you the opportunity to edit manual Property Set values. You can always edit such values after the tagging is complete, therefore you may wish to turn off the display of this dialog. To do so, from the Application Menu choose Options. Click the AEC Content tab and then uncheck the "Display Edit Property Data Dialog During Tag Insertion" checkbox.

35. Right-click and choose **Multiple**.

36. Select the remaining air terminals onscreen and then press ENTER.

If you select one of the objects you previously tagged, you will be asked if you would like to tag the object again. Answer No.

37. If the "Edit Property Set Data" dialog appears, click OK.

38. Press ENTER to end the command.

MODIFY PROPERTIES

39. Deselect the original tag, and select one of the air terminals.

40. On the Equipment tab, on the Modify panel, click the Modify Equipment button.

41. Click the Flow tab.

42. Specify any new flow value, and then click OK.

The MvPartAirTerminalObjects Flow value on the Properties palette Extended Data updates accordingly, as does the value on the tag for the selected air terminal.

43. With the air terminal still selected, on the Equipment tab, on the General panel, click the Edit style button.

44. On the General tab, click the Property Sets button.

Clicking the Property Sets button within the style editor window is the same as clicking the Edit style Property Set data button on the Properties palette.

45. Modify the type to SD-2, and then click OK twice to return to the drawing.

All the tags for the air terminals of this same style update to reflect the type change.

46. With the air terminal still selected, on the Equipment tab, on the Modify panel, click the Modify Equipment button.

47. On the Part tab, select Perforated Face Rectangular Neck Ceiling Diffusers US Imperial, select 10x10 Neck - 24x24 Face from the Part Size Name list, and then click OK (see Figure 15.22).

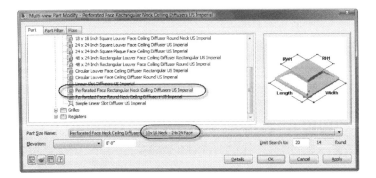

FIGURE 15.22

Choose a different Multi-view Part for the selected object

The tag updates to reflect the new neck size; however, there is no type displayed.

48. With the air terminal still selected, on the Properties palette, click the Edit style Property Set data icon (see Figure 15.23).

FIGURE 15.23

The Device tool on the Home ribbon, Build panel

49. Specify SD-3 in the type box and then click OK.

The tag updates accordingly.

50. Deselect the air terminal and select another air terminal.

51. Using the Modify Equipment tool, select the Perforated Face Neck Ceiling Diffusers - 10x10 Neck - 24x24 Face, and then click OK.

The tag updates, reflecting the type and neck size associated with the style.

Note that each Part Size Name is associated with a different style; thus, if you were to select the 12x12 Neck - 24x24 Face, you would have to assign a Type in the Property Set definition. You can use the same Type assigned to the 10x10, or you can assign a different one.

DEFINE A LIGHTING FIXTURE TAG

In this exercise, we will utilize the Property Set Definition created previously in this chapter, and create a lighting fixture tag that displays the fixture type, circuit, and the switchleg.

1. On the QAT, click the Open icon.

2. Browse to the *C:\MasterAME 2012\Chapter15* folder and open the *03 Lighting.dwg* file.

3. On the Insert tab, on the Block panel, click the Insert Block button.

4. Browse to the *C:\MasterAME 2012\Chapter15* folder, select the *Lighting Tag Template.dwg* file, and then click Open.

5. Check the "Specify On-screen" checkbox for Insertion point.

6. Clear the "Specify On-screen" checkboxes for all Scale and Rotation.

   ⇨ Scale should be: X=1, Y=1 and Z=1.

   ⇨ Rotation should be: Angle=0.

7. Check the "Explode" checkbox and then click OK (see Figure 15.24).

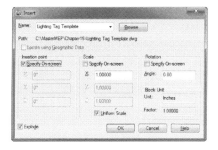

FIGURE 15.24
Insert options for tag template

8. Click OK, and then pick a location to place the block.

9. Zoom in on the inserted text. There are two text items.

10. On the Home tab, expand the Annotation panel and click the Create Tag button (similar to Figure 15.15 above).

11. At the "Select object(s) to create tag from" prompt, select the two text objects (from the exploded block) and then press ENTER.

Select the top one first.

12. In the "Define Schedule Tag" dialog, for the Name, type **MyLightingFixtureTag**.

13. In the FIXTURETYPE row, make the following changes:

⇨   For the Type, choose: **Property**.

⇨   For the Property Set, choose: **DeviceLightingStyles**.

⇨   For the Property Definition, choose: **FixtureType**.

14. In the CIRCUIT row, make the following changes:

⇨   For the Type, choose: **Property**.

⇨   For the Property Set, choose: **DeviceLightingObjects**.

⇨   For the Property Definition, choose: **CircuitTag** (see Figure 15.17).

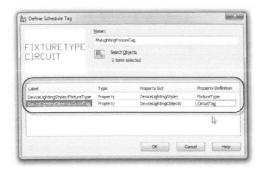

FIGURE 15.25
Configure the settings for the properties in the new tag

15. Click OK to dismiss the "Define Schedule Tag" dialog.

16. At the "Specify Insertion Point" prompt, hold the SHIFT key, right-click, and choose **Insert**.

17. Pick on the top text Insertion point (see Figure 15.26).

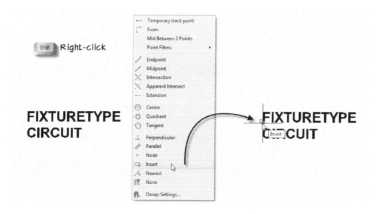

FIGURE 15.26

Specify the insertion point of the tag

The Text object has now been converted into a tag. See if you can find the Multi-view Block Definition for the tag in Style Manager. In the Block Definition associated with the tag, what do you think the tag properties are for the two text objects? If you answered DeviceLightingStyles:FixtureType and DeviceLightingObjects:CircuitTag, you are correct. Inspect the Block Definition if you like.

ASSOCIATE LIGHTING FIXTURE TAGS WITH LIGHTING FIXTURES

18. In this exercise, we will associate the lighting fixture tag we just created with the lighting fixtures in the drawing. We will then specify the Fixture Type property in the Property Set.

19. Select the fixture type tag instance in the drawing created in the previous exercise.

20. On the Tag ribbon tab, on the General panel, click the Add Selected button.

21. Select one of the lighting fixtures

22. Click to place the tag in the approximate location (see Figure 15.27).

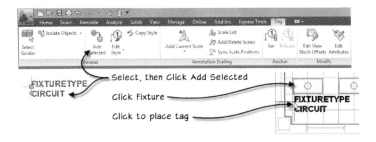

FIGURE 15.27
Use the Add Selected tool to associate a tag with a Device

If the "Edit Property Set Data" dialog appears, click OK.

After you place the tag, there is initially nothing visible since we have not specified a value for the Fixture Type property. To do so, we need to modify the Device Style and specify the value. Let's finish adding the tags first, and then we will address this issue.

23. Right-click, and choose **Multiple**.

24. Select the remaining lighting fixtures, and then press ENTER.

    If a message appears about multiple tags, click No.

    If the "Edit Property Set Data" dialog appears, click OK.

25. Press ENTER to complete the command.

26. Select a lighting fixture, and on the Device tab, on the General panel, click the Edit Style button.

27. In the "Device Style Properties" dialog, on the General tab, click the Property Sets button.

28. In the FixtureType field, type **F3**, and then click OK twice.

The tags should update to reflect the specified FixtureType. If the fixture is not yet circuited, the circuit portion of the tag will report NA-NA (for the panel-circuit name).

29. Erase the original tag—that is, the one not associated with the lighting fixture.

MODIFY THE LIGHTING FIXTURE TYPE

In this exercise, we will modify the lighting fixture style, then specify the fixture type to update the tag.

30. Select one of the lighting fixtures in the room to the right of the corner office at the bottom of the plan.

31. On the Properties palette, click the Design tab.

32. Click the Style preview image.

33. For the Drawing file, choose <**Current Drawing**> and then double-click 2x4 Troffer - 4 Lamp.

Note that the tag still indicates the previous fixture type. This is because the tag does not know how to update itself since the DeviceLightingStyles Property Set is not yet attached to the 4 Lamp Device Style. The following steps will demonstrate how to attach Property Sets manually to the Device Style.

34. With the lighting fixture still selected, on the Device ribbon tab, on the General panel, click the Edit Style button.

35. On the General tab, click Property Sets button.

36. Click the Add Property Sets icon (Figure 15.28).

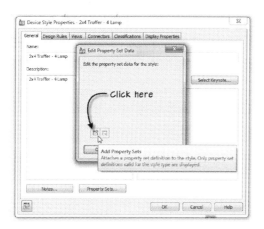

FIGURE 15.28

Click the Add Property Sets button

37. Make sure DeviceLightingStyles is checked, and then click OK.

38. Set the FixtureType to F4, and then click OK.

39. Click OK to close the "Device Style Properties" dialog.

The Tag should update to reflect the specified fixture type. When a tag is placed on an object, the necessary Property Sets are automatically attached. In this case, we changed the Style of the Device to a Style that did not have the Property Set attached, thus, it was necessary to attach the Property Set manually. Another means of attaching Property Sets is by using a Schedule to attach all Property Sets. See the "Scheduling" topic below for more information.

40. Select the remaining fixtures in the room in which you changed an F3 fixture to F4.

41. On the Properties palette, on the design tab, click the Style preview image to select the 4 Lamp fixture.

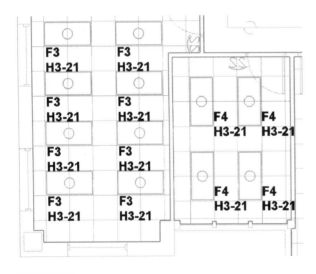

FIGURE 15.29
Change all fixtures in the office to the 4-lamp style

42. Save the *03 Lighting* file.

CREATE A PANEL AND TRANSFORMER TAG

In the *03 Power* drawing, create tags for Panels and Transformers named **MyPanelTag** and **MyTransformerTag**, respectively. For the panel tag, use the Property Set Definition PanelObjects and the PanelName property. For the transformer tag, use the Property Set Definition MvPartTransformerObjects and the EngineeringPartId property. Since you will be creating this from scratch, no blocks exist for you to start with. Instead, start by creating non-annotative text using the RomanS text style, justified middle center, with a height of 1".

SCHEDULING

Schedules are tables listing the Property Set Data associated with scheduled items. The basics of creating a Schedule Style consists of: defining what the table applies to; defining what columns to show in the schedule; and specifying which columns (if any) to sort on. The following exercises will create two separate schedules that, although they may not be used directly on construction documents, may prove useful for energy analysis tasks (see Figure 15.30).

| Fixture Quantity by Type | | | |
Fixture Type	Quantity	Load	Total
F2	2	64	128
F3	3	96	288
F4	9	128	1152
	14		1568

| Fixture Quantity by Space | | | | |
Space	Fixture Type	Quantity	Load	Total
301	F2	2	64	128
301	F4	1	128	128
302	F2	3	96	288
302	F4	6	128	768
303	F4	2	128	256
		14		1568

FIGURE 15.30
Schedules showing the quantity fixtures (left), and the quantity by space (right)

DEFINE A LIGHTING FIXTURE TAKEOFF SCHEDULE

1. Make sure *03 Lighting* is the current drawing.

2. In Style Manager, expand *Documentation Objects,* right-click Schedule Table Styles, and choose **New**.

3. Enter the name **Lighting Device Space Quantity Takeoff Schedule**.

4. Click the Applies To tab.

5. Check the box for Device and, from the Classifications list, beneath Device Type, check Lighting.

6. Click the Columns tab.

7. Click the Add Column button.

8. From the DeviceLightingObjects Property Set, select SpaceNumber and then click OK.

9. Double click on the SpaceNumber column heading (Figure 15.31).

The text becomes editable.

FIGURE 15.31

Double-click on a column heading to edit the text

10. Rename the heading to Space.

11. Repeat the previous steps to add columns for:

⇨ DeviceLightingStyles : FixtureType.

⇨ DeviceLightingStyles : Load.

12. Select the FixtureType column and, at the bottom of the table, click the Modify button.

13. In the Heading box, type **\P** before the "T" in Type; i.e., **Fixture\PType**.

Note: The \P represents a return character; instead of Fixture Type showing up in one row, it
will be split into two rows when the Schedule is placed in a drawing as shown in Figure
15.30 above.

---

14. Check the "Include Quantity Column" box.

A Quantity column will appear first in the list.

15. Pick the Quantity column heading and drag-and-drop it on the Fixture Type column.

This causes the Quantity column to be placed between the Fixture Type and the Load
columns (see Figure 15.32).

FIGURE 15.32

Drag and drop column headings to reorder them

16. Select the Quantity column and then click the Modify button.

17. Check the Total box and then click OK.

18. Click the Add Formula Column button, and then specify the following formula (see Figure 15.33):

⇨   In the Heading box, type: **Total**.

⇨   Check the Total box.

⇨   At the bottom, in the "Insert Property Definition" box, expand DeviceLightingStyles and then double-click Load.

⇨   In the Formula box, to the right of [DeviceLightingStyles:Load], type a multiply symbol (*).

⇨   At the bottom corner of the dialog, click the Insert Quantity button.

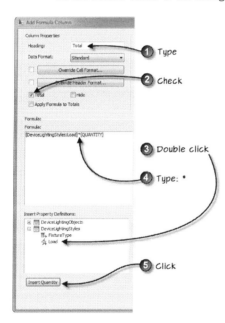

FIGURE 15.33
Specify the Total formula

19. Click OK to close the "Add Formula Column" dialog.

The resulting columns are shown in Figure 15.34.

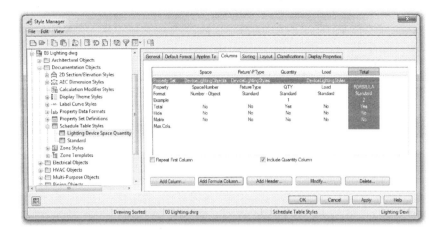

FIGURE 15.34

Columns for the Schedule style

20. Click the Sorting tab.

21. Click the Add button.

22. Select DeviceLightingObjects:SpaceNumber, and then click OK.

23. Click the Add button again.

24. Select DeviceLightingStyles:FixtureType, and then click OK.

Note: Both properties should sort ascending by default.

25. On the left side of Style Manager, select Lighting Device Space Quantity Takeoff Schedule, and then click the Copy icon at the top.

26. Click the Paste icon (see Figure 15.35).

This will create a new Schedule style named: Lighting Device Space Quantity Takeoff Schedule (2).

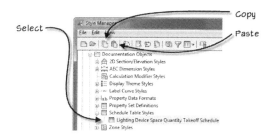

FIGURE 15.35

Copy a Schedule Table Style in Style Manager

27. Select the new copy and on the General tab, rename it to **Lighting Device Quantity Takeoff Schedule**.

28. On the Columns tab, select the Space column and then click the Delete button. In the "Remove Columns/Headers" dialog, click OK (see Figure 15.36).

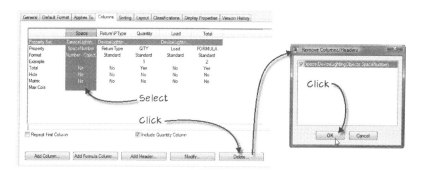

FIGURE 15.36

Delete the Space column

29. Click OK to close the Style Manager.

PLACE SCHEDULES

1. On the Annotate tab, on the Scheduling panel, click the Schedules drop-down button and select any schedule.

Note: In this case, we're not really going to use the selected schedule; we will change the style on the Properties palette.

---

2. On the Properties palette, click the Design tab and then specify the following:

⇨ For the Style, choose: Lighting Device Quantity Takeoff Schedule.

⇨ For the Title, type: **Lighting Device Quantity Takeoff Schedule**.

⇨ From the Update Automatically option, choose: Yes.

⇨ From the Add new objects automatically option, choose: Yes.

3. At the "Select objects" prompt, type **ALL** and then press ENTER.

4. Press ENTER again to end the selection process.

5. Pick a point to place the schedule.

6. Press ENTER again to accept the default scale for the schedule.

We are given a complete list of all lighting fixtures totaled by style. The quantity of each style appears in the Quantity column with a grand total at the bottom.

7. Select one of the lighting fixtures, and, on the Properties palette, change the style to 2x4 Troffer - 2 Lamp.

Lighting Device Quantity Takeoff Schedule			
Fixture Type	Quantity	Load	Total
?	1	?	?x1
F3	32	96	3072
F4	4	128	512
	37		3584

FIGURE 15.37

The Question Marks (?) in the schedule indicate objects are missing Property Sets

> Note: The 2 Lamp fixture shows in the schedule as a Question Mark (?) (Figure 15.37). This is because we have not yet attached the DeviceLightingStyles Property Set to the Style.

8. Right-click, and choose **Deselect All**.

9. Select the Schedule.

10. On the Schedule Table tab, on the Modify panel, click the Add All Property Sets button.

> Note: The 2 Lamp fixture shows no Fixture Type in the tag or in the schedule. This is because we have not yet specified a value. But the question marks have disappeared indicating that the Property Set is now attached.

11. Open the Style Manager.

12. Expand *Electrical Objects* and select Device Styles.

13. Select the 2x4 Troffer - 2 Lamp style.

14. On the General tab click the Property Sets button.

15. For the FixtureType, input **F2**, and then click OK (see Figure 15.38).

16. Click OK to close Style Manager.

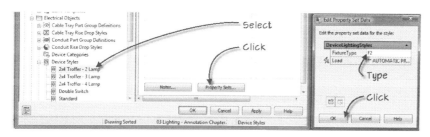

FIGURE 15.38
Modify the Style Property Set in Style Manager

Note that the schedule updates to reflect the Fixture Type F2, as do any tags associated with the 2 Lamp fixtures.

17. Select the Schedule and, on the Properties palette, Design tab, change the style to *Lighting Device Space Quantity Takeoff Schedule*.

Note that the schedule updates to provide a breakdown of the quantity and type of fixtures per space.

18. Save *03 Lighting*.

LABELS

Labels are similar to tag objects in that they report properties of an object. Labels are unique in that their intent is to be used only for duct, pipe, plumbing line, schematic line, conduit, and cable tray segments. The most common use of Labels is to show the size and/or the system designation of these components. The next most commonly used option is to show flow arrows. However, several other options exist as well, and are described below:

LABEL STYLE ANNOTATION TAB—OPTIONS

The Label Style Annotation tab contains the following Options (see Figure 15.39):

Block graphics—define a block to be shown along the object.

Objects Properties—display the size of the object.

Object Props. And Abbr. System Name—shows the size and system abbreviation.

Abbreviated System Name—shows the system abbreviation.

Component Description—displays the description from the segment style.

Component Name—displays the name of the object (22.0 x 10.0 in. Rectangular Duct US Imperial).

Flow/Run Arrows—display a flow arrow along the segment.

Property Set—displays a property from a Property Set. Note that when using this option, the Property Set does not automatically attach to the object, you need to do it manually. If the Property Set is not attached to the object, the label just displays N/A.

String—display user defined text. Note that the text is the same for all instances of this particular option, regardless of the size/system of the labeled object.

FIGURE 15.39

Label Style Annotation options

LABEL STYLE ANNOTATION TAB—TEXT OPTIONS

The text options section allows you to specify the text style and layering of the Labels.

Text style—defines what AutoCAD text style to apply to the object.

> **Note:** The text style should be non-annotative, and have a height of 0 as shown in Figure 15.40 to allow the label style to control the height.

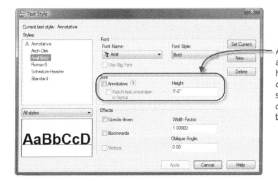

A text style that is not annotative and has 0 height allows the drawing annotation and scale settings to control the height of the Label.

FIGURE 15.40

Text Style definition for Label annotations

Mask Display of Curve—causes the label to hide part of the labeled object, as can be seen in samples a, b, and e.

Use layer key from system definition—causes the label to be placed on the same layer as the labeled object. You can override the Label display component of the Label Style to select a different layer and/or color for the label if the color of the system is not suitable for text. Note that, regardless of the layer of the label, if the labeled object is not visible because its layer is turned off, the label will be hidden as well— you may need to regenerate the drawing for the effect.

Use layer override from system definition—can be used in conjunction with or independent of the use layer key from system definition option. If used in conjunction with the Use layer key option, the label will end up on the same layer as the object, as before. If used independently, the label can end up on a system specific annotation layers, such as M-Anno-Std-Supp or M-Anno-Std-Rtrn.

LABEL STYLE ANNOTATION TAB—DIMENSION STRING OPTIONS

These settings specify how object sizes are displayed; that is, the x separator results in text such as 24x12, whereas a / would result in text such as 24/12. The Diameter option set to "%%c results in text such as 4"θ. Prefix and Suffix are not commonly used.

Gap paper size affects how much masking occurs with the label. Sample "a" below has a gap of 1/32", whereas sample "b" has a gap of 1/16", and sample "c" has a gap of 1/64".

LABEL STYLE OFFSET TAB

On the Label Style Offset tab several offset options are available (see Figure 15.41). Example "e" is displayed using the Force To Horizontal Justification. Without this option, the label follows the orientation of the object. Examples a, b, c, and e all use the None offset option. Example "d" is using the Force Outside of Bounding Box option. The Auto-Adjust for 1-Line option places the label above the segment (as in sample d), but close to the center line of the object. The Force Outside and Auto-Adjust for 1-Line options make use of the Clear Distance Paper Offset distance setting. In sample d, there is a 1/16" gap between the edge of the duct, and the label.

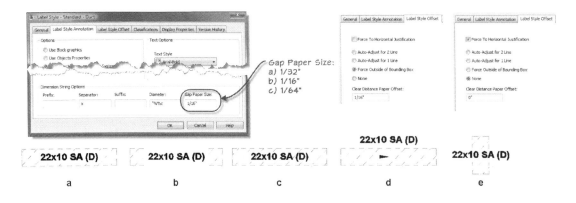

FIGURE 15.41
Label Style Offset settings and examples

The Classifications tab allows you to classify your label styles to be able to control their visibility using Display Set Display Options. In order to classify the labels, a classification definition (Style Manager > Multi-Purpose Objects > Classification Definitions) must first be defined that Applies To: Label Curves.

Finally, the Display Properties tab has the same functionality of all style-based objects. Refer to Chapter 2 for information on display control hierarchy. Labels have a single display component called Label.

MODIFY A LABEL STYLE

1. On the QAT, click the Open icon.

2. Browse to the C:\MasterAME 2012\Chapter15 folder and open the *03 Telecom.dwg* file.

3. Set your Workspace to Electrical.

4. Select one of the cable tray Labels and on the Label ribbon tab, click the Edit Style button.

5. On the Label Style Annotation tab, click the Use Object Props. And Abbr. System Name option, and then click OK.

Notice that the size of the Cable Tray is now included in the Label.

6. Deselect the Label.

7. Select the DEMO cable tray segment and on the Cable Tray ribbon tab, click the Edit System Style button.

8. On the Design Rules tab, set the abbreviation to **D**, and then click OK.

The Label updates to read "D" in place of "Demo."

9. Deselect the Cable Tray.

10. Select the EXISTING cable tray segment, and on the Cable Tray ribbon tab, click the Edit System Style button.

11. On the Design Rules tab, set the abbreviation to E, and then click OK.

12. Deselect the Cable Tray.

13. Select one of the segments labeled Standard, and on the Cable Tray ribbon tab, click the Edit System Style button.

The Standard system has no abbreviation set. When no abbreviation is set, the Label will use the system name instead. You can circumvent this by entering a space in the abbreviation field.

14. On the Design Rules tab, click in the abbreviation field, press the SPACEBAR, and then click OK.

Notice that the word Standard has now been removed from all of the Labels.

15. Deselect All, and then select one of the Labels.

16. On the Label ribbon tab, on the General panel, click the Edit Style button.

17. On the Label Style Annotation tab, set the Gap Paper Size to 1/64", and then click OK.

The Label now displays without breaking the edge of the Cable Tray.

18. Save and close *03 Telecom*.

ANNOTATION CONTENT

Annotation content (excluding Labels) is typically inserted from a content file and, as such, it is not necessary to have the styles defined in your drawing templates. By placing the annotation content in a centralized content file, you can use the content from any file. Many Property Sets, Tags, and Schedules are defined in the following files (and others), stored in the default content folder:

C:\ProgramData\Autodesk\MEP 2012\enu\Styles\Imperial

- *Mechanical Equipment Tags & Schedules (US Imperial).dwg*
- *Mechanical Tags (US Imperial).dwg*
- *Electrical Equipment Tags & Schedules (US Imperial).dwg*
- *Electrical Tags (US Imperial).dwg*
- *Plumbing Equipment Tags & Schedules (US Imperial).dwg*
- *Plumbing Tags (US Imperial).dwg*

There are numerous tool palette tools that are predefined utilizing the styles in these files. When the tool palette tools are defined, their paths are saved relative to the Tool Catalog Content Root Path defined in the "Options" dialog on the AEC Content tab (see Figure 15.42). This allows you to move the content to an alternate location, and not have to redefine your tool palette tools.

FIGURE 15.42

The root location of Content is specified in Options

For example, the Bottom of Duct Label (no slope) style on the HVAC Tool Palette Group on the Annotation palette is defined to retrieve its style from the *Mechanical Tags (US Imperial).dwg* in the *\Styles\Imperial* subfolder of the Tool Catalog Content Root Path (see Figure 15.43). To see this for yourself on this or any tool, right-click the tool on the palette and choose Properties. The locations of the style and any required Property Sets will appear as shown in the figure.

> **Note:** The annotation object Bottom of Duct Label (no slope) is actually a Multi-view Block tag, not a Label.

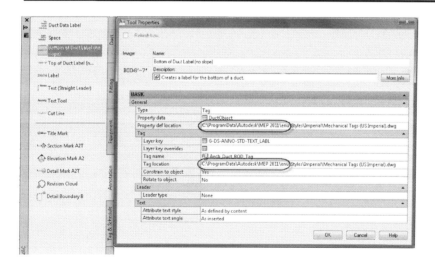

FIGURE 15.43

Tag tool properties

This is important to keep in mind as you centralize your own Property Sets, Tags, and Schedules. The files you save these in, especially once you define the tool palette tools to insert these components, should reside within your Tool Catalog Content Root Path to ensure flexibility if and when you decide to relocate your content (such as from a local machine to a central server).

DEFINE AN ANNOTATION CONTENT FILE

1. Create a new drawing based on the *Aecb Model (US Imperial Ctb).dwt* template file.

2. From the Application menu, click Save As.

3. On the left side, click the Content icon.

4. At the top of the dialog, click the New Folder icon.

5. Select the New Folder created, press F2, type **My Content** for the new name and then press ENTER (see Figure 15.44).

Note: Verify the Tool Catalog Content Root Path in Options, on the AEC Content tab.

FIGURE 15.44
Create a new folder in the Save Drawing As window

6. Double-click the new *My Content* folder to open it.

7. Save the file as: My Property Sets Tags and Schedules.dwg.

COPY STYLES TO ANNOTATION CONTENT FILE

In this exercise, we will copy the annotation styles created earlier in this chapter to the centralized content file. Continue with the *My Property Sets Tags and Schedules.dwg* file open.

8. On the QAT, click the Open icon.

9. Browse to the *C:\MasterAME 2012\Chapter15* folder and open the *03 Lighting.dwg* file.

10. Click the Open icon again, browse to the *C:\MasterAME 2012\Chapter15* folder and open the *03 Power.dwg* file.

11. Open the Style Manager.

12. Expand *03 Lighting.dwg*, expand and the *Documentation Objects* folder and then Schedule Table Styles (see Figure 15.45).

13. On the right side of Style Manager, select the **Lighting Device Space Quantity Takeoff Schedule**, hold down the CTRL key, and then select the **Lighting Device Quantity Takeoff Schedule**.

14. Click the Copy icon.

15. On the left side, select My Property Sets Tags and Schedules.dwg.

16. Click the Paste icon.

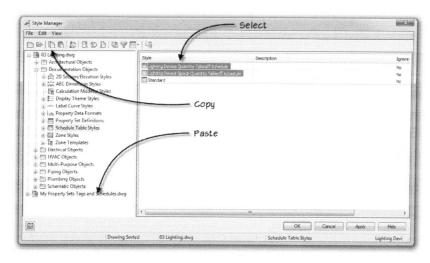

FIGURE 15.45
Copy and paste the schedule styles to the content file

Note: It is not necessary to select the proper node in the destination drawing; the Styles will be organized automatically according to their type.

---

17. Expand My Property Sets Tags and Schedules.dwg; expand Documentation Objects and then Schedule Table Styles.

You should see the copied schedules.

18. Expand Property Set Definitions.

Among others, you should see DeviceLightingObjects and DeviceLightingStyles.

Since the schedules are dependent on the Property Set Definitions DeviceLightingObjects and DeviceLightingStyles, these are copied to the destination drawing automatically. The same does *not* hold true for tags.

19. Expand 03 Lighting then Multi-Purpose Objects, and then select Multi-view Block Definitions.

20. Select MyLightingTypeTag and then click the Copy icon.

21. Select My Property Sets Tags and Schedules.dwg and click the Paste icon.

22. Expand *03 Power* drawing and repeat the process to copy the **PanelObjects** and **MvPartTransformerObjects** Property Set Definitions to *My Property Sets Tags and Schedules.dwg*.

23. Also copy the **MyPanelTag** and **MyTransformerTag** Multi-view Block Definitions (under Multi-Purpose Objects) from *03 Power* to *My Property Sets Tags and Schedules.dwg*.

COPY AIR TERMINAL TAG AND PROPERTY SET DEFINITIONS

In this exercise, you are on your own. Copy the MyAirTerminalTag from *03 HVAC* to *My Property Sets Tags and Schedules.dwg*. This tag is dependent on the Property Set Definitions MvPartAir_TerminalStyles and MvPartAirTerminalObjects, so make sure you copy them as well.

24. Save the *My Property Sets Tags and Schedules.dwg* file when you are finished copying styles to it.

CREATE TOOL PALETTE TOOLS FOR TAGS AND SCHEDULES

In this exercise, we will create Tool palette Tools for the Tags and Schedules created previously in this chapter. To do so, first we will place an instance of each tag in the drawing. We will then create a Tool Palette, and drag the tags and schedules to it.

When a project is active in Project Browser, any Tool palette modifications will be specific to that project. This can cause confusion because you may create palettes with one project active, change projects, and not be able to find the tools you created. To avoid this, first close the active project (if any).

25. On the Project Navigator, click the Project tab.

26. Click the Close Current Project icon. If a "Project Browser – Close Project Files" dialog appears, choose the "Close all project files" option (see Figure 15.46).

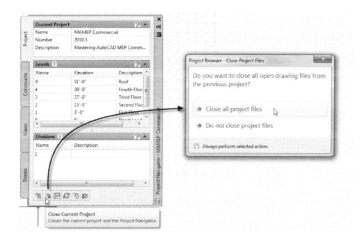

FIGURE 15.46

Close the current project

Make sure My Property Sets Tags and Schedules.dwg is the current drawing.

27. On the Insert tab, on the Block panel, click the Multi-view Block button.

28. On the Properties palette (on the Design tab) from the Definition list, choose **MyLightingFixtureTag**.

29. Pick any point in the drawing to place the Tag.

30. On the Properties palette, from the Definition list, choose **MyAirTerminalTag**.

31. Pick any point in the drawing to place the Tag.

32. Create instances of the Panel and Transformer tags as well.

33. Save the drawing.

When creating Tool palette tools, it is best to save just before creating them. If you don't, AutoCAD MEP may warn you that the tools you are attempting to create aren't yet saved in the drawing.

34. With the Electrical Tool palette group current, right-click on the Tool palettes titlebar, and choose: **New Palette**.

Where you right-click depends on the state of the Tool palettes (see Figure 15.47).

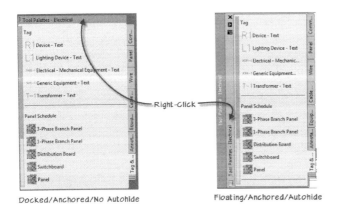

Docked/Anchored/No Autohide   Floating/Anchored/Autohide

FIGURE 15.47

Right click on the palette group title to create a new palette

35. Enter the name **My Annotations** (see Figure 15.48).

FIGURE 15.48

Type the palette name

You can right-click on the new palette to rename it if you don't get it right the first time. Note the information tip, "Learn more about customizing tool palettes." If you had a project current, it would indicate Learn more about customizing *project* tool palettes (emphasis added).

36. In the drawing, select the MyLightingFixtureTag instance.

37. Select any part of the tag (not on a grip); drag the Block onto the Palette (see Figure 15.49).

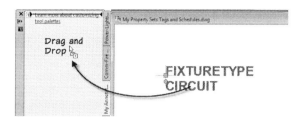

FIGURE 15.49

Drag-and-drop a tag instance onto the palette

38. Drag the Panel tag and the Transformer tag to the palette as well.

39. Open the Style Manager.

40. Expand Documentation Objects and then Schedule Table Styles.

41. Select and drag the Lighting Device Quantity Takeoff Schedule onto the My Annotations palette (see Figure 15.50).

42. Repeat for the Lighting Device Space Quantity Takeoff Schedule.

FIGURE 15.50

Drag a schedule style from Style Manager onto the palette

43. Click OK to close the Style Manager.

44. Right-click on one of the Schedule tools, and choose Properties (see Figure 15.51).

45. Set each of the following properties to **Yes**, and then click OK.

⇨   Update automatically.

⇨   Add new objects automatically.

⇨   Scan XREFs.

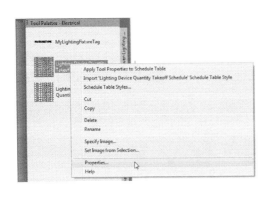
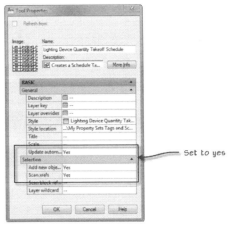

Set to yes

FIGURE 15.51

Modify the Schedule tool properties

46. Optionally, you may specify a default Title for the schedule.

47. Repeat the above steps for the other Schedule tool.

Schedules will update automatically when the drawing is opened. However, by default, as you make changes to objects in a drawing, the schedule doesn't update automatically, for performance reasons. Unless you have extremely large datasets with many objects in a schedule, you can probably safely set update automatically to Yes.

When placing a schedule, you are prompted for which objects to schedule. Generally, enabling "add new objects" automatically avoids the extra steps of manually reselecting objects to schedule later.

The scan XREFs option is very useful in scenarios where you want to schedule objects from multiple drawings. To do so, you can XREF in each drawing to be scheduled into a single drawing. Note, however, that for performance reasons, you may want to unload any dependent XREFs (referenced architectural backgrounds, other trades, etc.). However, any drawings linked for space location properties should be loaded). For a multi-story building, offsetting the drawings from one another (instead of stacking vertically) will help minimize the computation of hidden lines—which, for scheduling purposes, aren't necessary.

When creating and modifying Tool palettes and tools, it is not necessary (or even possible) to manually save any changes to the tools or palettes themselves. The tools and palettes are automatically saved every time you close AutoCAD MEP.

TEST THE TOOLS

48. Create a new drawing.

49. Place several instances of several different lighting fixture styles.

50. Tag the lighting fixtures with the My Lighting Fixture Tag.

51. Edit the lighting fixture style's Property Sets to specify a FixtureType.

52. Schedule the fixtures using the Lighting Device Quantity Takeoff Schedule.

CREATE HVAC ANNOTATIONS PALETTE

This exercise is intended to reinforce what you have learned providing minimal steps.

53. With the HVAC Tool palette group current, create a new palette, and name it appropriately.

54. Add the My Air Terminal Tag to the palette.

55. Create a new drawing, and test the tool.

COPY THE CABLE TRAY LABEL TO THE CONTENT FILE

It is good practice to centralize all your customized annotations. With this in mind, there is one last bit of content we created in this chapter that we could copy to our content file. Copy the Standard - Cable Tray Label Curve Style from the *03 Telecom* drawing to the *My Property Sets Tags and Schedules.dwg*. You can even create a tool palette tool using a procedure similar to that used to create a schedule tool on the tool palette. This way, when you are establishing your company standard template or creating tool palette tools, you have all your standard objects in one place.

56. Close and save the *My Property Sets Tags and Schedules.dwg* file.

CONTENT BROWSER

When we created the tool palette tools, we essentially defined a shortcut to the specific bit of content in the *My Property Sets Tags and Schedules.dwg* file. A tool is a shortcut with a series of Property Settings specific to the type of content. A tool palette contains a collection of such tools.

Centralizing the content file so others can use it is a straightforward task. A bit less obvious is how you would share tool palette tools so others may have easy access to the styles in the content file.

To share tool palettes and tools with other AutoCAD MEP users, use the Content Browser. The Content Browser doesn't actually contain the content styles themselves; it is an organized collection of the palettes and tools that provide access to the content styles. Thus, sharing tools requires both the content file and the tool palette tools themselves. In this topic, we will demonstrate how you can create a Catalog with our custom tools.

The Content Browser is accessed from the Insert ribbon tab. When the Content Browser is opened, a series of catalogs are displayed.

CREATE A CONTENT LIBRARY

You can create your own Content Library and your own Catalogs. In this exercise, you will create your own Content Library, Catalog, and Tool palettes.

1. On the Home tab, on the Build panel, click the Tools drop-down button and choose Content Browser (see Figure 15.52).

FIGURE 15.52

Content Browser on the Home tab

2. At the bottom-left corner, click the "add or create" catalog icon.

3. In the "Add Catalog" dialog, choose the "Create a new catalog" option.

4. For the name, type **MAMEP Tools**.

5. Click the Browse button and choose a folder on your network server that all users can access (see Figure 15.53).

   In the figure, a local folder called *C:\MasterAME 2012\Catalog* is used. If you are not ready to save tools to your server, feel free to experiment in the location shown here for now.

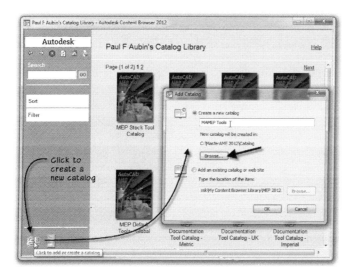

FIGURE 15.53

Create a new Catalog

6. Click OK to create the catalog.

   The new MAMEP Tools catalog will appear among the others in your library. It will have a blank blue image. You can change this if you wish.

7. Right-click MAMEP Tools and choose **Properties**.

8. Right-click the blue image and choose **Specify Image**.

9. Browse to the *C:\MasterAME 2012\Chapter15* folder, select *MAMEP-Catalog.png,* and then click Open (see Figure 15.54).

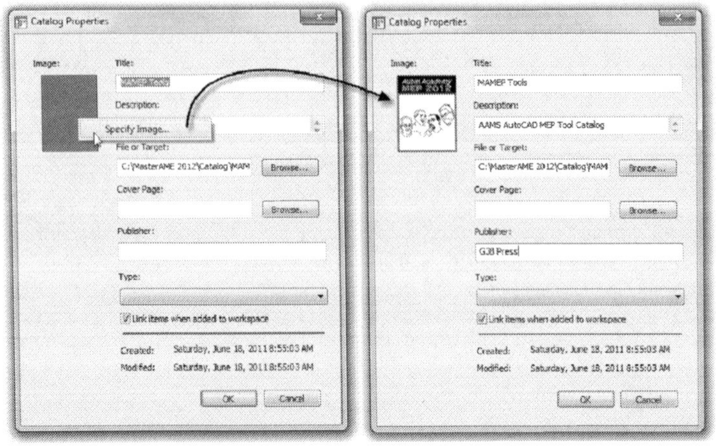

FIGURE 15.54

Specify a new image for your catalog

10. Click OK to close the "Catalog Properties" dialog.

COPY TOOL PALETTES TO THE TOOL CATALOG

This exercise demonstrates how to copy tool palettes to the new catalog.

11. Click on the MAMEP Tools catalog.

    The catalog is currently empty.

12. Right-click on the Content Browser titlebar and choose Always on Top if it is not already
    selected (see Figure 15.55).

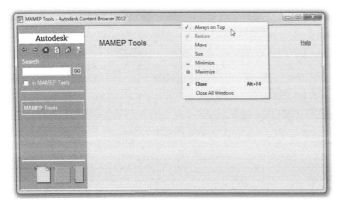

FIGURE 15.55

Set the Content Browser to Always on Top

13. Return to AutoCAD MEP.

14. Make sure the Electrical Tool Palette Group is active.

15. Select the My Annotations palette to activate it.

16. Click and hold the left mouse button on the My Annotations palette tab. Drag it over the Content Browser window where the MAMEP Tools Catalog should still be open and on top, and then release the mouse button (see Figure 15.56).

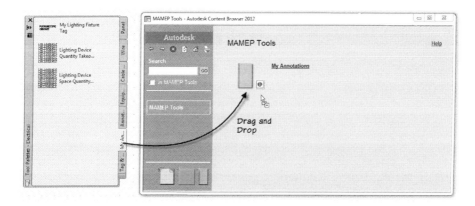

FIGURE 15.56
Drag-and-drop a palette into the catalog

The tools are now copied to the location where you specified the Catalog. If created on a network drive, other users can access it there as well. For another user to access it, they must open their Content Browser, click the "add or create" catalog icon, and choose the "Add an existing catalog or web site" option. Browse to the network location and open the catalog. Next, using the eyedropper icon on the palette, they would drag-and-drop the palette into their workspace. This will add the palette to their tool palettes. Assuming that the *My Property Sets Tags and Schedules.dwg* drawing file is also saved to the server, and accessible by your coworker, the tools should function flawlessly.

If you would like, you can test it on your own system. Simply iDrop the palette back into AutoCAD MEP (see Figure 15.57).

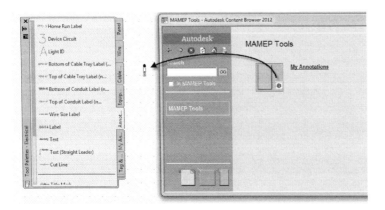

FIGURE 15.57

iDrop the palette into your Workspace

At the bottom of the palette there is now a refresh icon as shown in Figure 15.58. This indicates that the palette came from a catalog. Any tools added to the Catalog will appear in each user's local copies when the palette is refreshed with this icon.

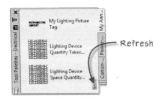

Refresh

FIGURE 15.58

Refresh icon on palette indicates it originated from a catalog

This refresh mechanism provides users a way to update their palettes with master versions stored in Content Browser. This helps distribute tools that help annotate drawings with the Property Sets, Tags, and Schedules you learned about in this chapter.

SUMMARY

- ✓ Property Sets are used to define the type of information that may be used in schedules and tags.

- ✓ Tags are Multi-view Blocks that contain attributes that reference properties within a Property Set.

- ✓ Schedules are used to tabulate properties that are associated with objects and/or their styles.

- ✓ Labels are used to annotate segments such as ducts, pipes, conduit, cable tray, plumbing line, and schematic lines.

- ✓ Annotation content should be centralized for ease of management.

- ✓ Content Browser may be used to share tool palettes with other users.

11229127R0

Made in the USA
Lexington, KY
18 September 2011